Identification of Dynamical Systems with Small Noise

Mathematics and Its Applications

Volume 300

Identification of Dynamical Systems with Small Noise

by

Yu. Kutoyants
Département de Mathématiques,
Faculté des Sciences,
Université du Maine,
Le Mans, France

SPRINGER-SCIENCE+BUSINESS MEDIA, B.V.

A C.I.P. Catalogue record for this book is available from the Library of Congress.

ISBN 978-94-010-4444-8 ISBN 978-94-011-1020-4 (eBook)
DOI 10.1007/978-94-011-1020-4

Printed on acid-free paper

Originally published by Kluwer Academic Publishers in 1994
Softcover reprint of the hardcover 1st edition 1994

To Nadejda,

Maria and Ekaterina

Contents

Introduction

Small noise is
a good noise.

In this work, we are interested in the problems of estimation theory concerned with observations of the diffusion-type process

$$dX_t = S_t(X)\,dt + \varepsilon\,dW_t, \qquad X_0 = x_0,\ 0 \le t \le T, \tag{0.1}$$

where W_t is a standard Wiener process and $S_t(\cdot)$ is some nonanticipative smooth function. By the observations $X = \{X_t, 0 \le t \le T\}$ of this process, we will solve some of the problems of identification, both parametric and nonparametric. If the trend $S(\cdot)$ is known up to the value of some finite-dimensional parameter $S_t(X) = S_t(\theta, X)$, where $\theta \in \Theta \subset R^d$, then we have a parametric case. The nonparametric problems arise if we know only the degree of smoothness of the function $S_t(X), 0 \le t \le T$ with respect to time t. It is supposed that the diffusion coefficient ε is always known. In the parametric case, we describe the asymptotical properties of maximum likelihood (MLE), Bayes (BE) and minimum distance (MDE) estimators as $\varepsilon \to 0$ and in the nonparametric situation, we investigate some kernel-type estimators of unknown functions (say, $S_t(\cdot), 0 \le t \le T$).

The asymptotic in such problems of estimation for this scheme of observations was usually considered as $T \to \infty$, because this limit is a direct analog to the traditional limit ($n \to \infty$) in the classical mathematical statistics of i.i.d. observations.

The limit $\varepsilon \to 0$ in (0.1) is interesting for the following reasons. Let

$$\frac{dx_t}{dt} = S_t(x), \ \ x_0,\ 0 \le t \le T \tag{0.2}$$

be a dynamic system on the right side of which acts the white Gaussian noise $\varepsilon \dot{W}_t$. Then the state X_t of the system is described by the equation

$$\frac{dX_t}{dt} = S_t(X) + \varepsilon\,\dot{W}_t, \ \ X_0 = x_0,\ 0 \le t \le T,$$

which is adopted to write as a stochastic differential

$$dX_t = S_t(X)\,dt + \varepsilon\,dW_t, \ \ X_0 = x_0,\ 0 \le t \le T.$$

This differential is understood as a shortened form of the integral equation

$$X_t = x_0 + \int_0^t S_v(X)\,dv + \varepsilon\,W_t, \ \ 0 \le t \le T,$$

which also explains the notion of *small noise* because, for the white noise $\varepsilon \dot{W}_t$, we have $\mathbf{E}(\varepsilon \dot{W}_t)^2 = \infty$.

This scheme of observation is interesting for some applied problems. For example, if the behavior of a real dynamic system is described by Equation (0.2) and a small, say thermal, noise perturbs its unknown right side, then the problem of the identification of this system naturally arises.

From the mathematical point of view, this scheme is attractive for two reasons. First of all, it is simple because, under the Lipschitz conditions, the (nonasymptotic!) inequality

$$\sup_{0 \leq t \leq T} |X_t - x_t| \leq C\varepsilon \sup_{0 \leq t \leq T} |W_t|$$

holds with probability 1. This allows us to know *how far from the limit* we are at any given moment. Secondly, this scheme is sufficiently complicated (nonlinear and nongaussian) and the statistical procedures are nontrivial. All these allow us to illustrate the opportunities of some well-developed methods of modern mathematical statistics (introduce the minimax bounds on risks of estimators due to L. Le Cam, J. Hajek and P. Millar and use the powerful methods of asymptotical statistics due to I. Ibragimov, R. Khas'minskii etc).

Besides the usual conventions of notation and terminology, Chapter 1 contains some facts from Itô calculus and asymptotic theory of estimation. It is assumed that the reader is familiar with the traditional course of Ito's stochastic calculus (e.g. [46], [97]).

Chapter 2 is devoted to the parametric problems concerning observations of the diffusion-type process

$$dX_t = S_t(\theta, X)\, dt + \varepsilon\, dW_t, \quad X_0 = x_0,\, 0 \leq t \leq T, \tag{0.3}$$

where $\theta \in \Theta$, Θ is an open, bounded subset of R^d and $\varepsilon \to 0$. It begins with the minimax theorem of Hajek-Le Cam on risks of all estimators of the parameter θ and then presents the ***standard*** theorem: if the problem is regular, i.e. the function $S_t(\cdot)$ is smooth, and the equation

$$G(y, \theta) \equiv \int_0^T [S_t(y, x) - S_t(\theta, x)]^2\, dt = 0, \qquad y \in \Theta, \tag{0.4}$$

has a unique solution $y = \theta$, then the MLE $\hat{\theta}_\varepsilon$ and the Bayes estimators are uniformly consistent and asymptotically normal

$$\mathcal{L}_\theta\{\varepsilon^{-1}(\hat{\theta}_\varepsilon - \theta)\} \Rightarrow \mathcal{N}(0, I(\theta)^{-1}),$$

where

$$I(\theta) = \int_0^T \dot{S}_t(\theta, x)\dot{S}_t(\theta, x)^T\, dt.$$

Here and in the sequel, a point above a letter means differentiation with respect to θ, the superscript T means transposition, and x_t is a solution of the equation

$$\frac{dx_t}{dt} = S_t(\theta, x), \quad x_0,\, 0 \leq t \leq T. \tag{0.5}$$

These estimators are also local asymptotic minimax (LAM), i.e. Hajek-Le Cam's bound is reached by these estimators.

Then, for the special linear scheme of a *large signal*

$$dX_t = \frac{\theta}{\varepsilon} X_t \, dt + dW_t, \quad X_0 = x_0, \quad 0 \le t \le T,$$

which, of course, differs from (0.5), the asymptotic behavior of the MLE $\hat{\theta}_\varepsilon$ is investigated and four different limits are presented. Particularly, if $\theta > 0$ and $x_0 \neq 0$, then as $\varepsilon \to 0$

$$\mathcal{L}_\theta\{(\hat{\theta}_\varepsilon - \theta)\, \varepsilon^{-\frac{1}{2}}\, e^{\frac{\theta T}{\varepsilon}}\} \Rightarrow \mathcal{N}\left(0, \frac{2\theta}{x_0^2}\right)$$

and if $\theta > 0$ and $x_0 = 0$, then

$$\mathcal{L}_\theta\{(\hat{\theta}_\varepsilon - \theta)\, 2\, \theta\, e^{\frac{\theta T}{\varepsilon}} \Rightarrow Cauchy.$$

In the rest of Chapter 2, the properties of the MLE and BE are investigated under *nonstandard* conditions. The regular case could be described as a situation with a smooth trend $S_t(\cdot)$, a deterministic initial value X_0, a known function $S_t(\theta, X), 0 \le t \le T$, a unique solution of the equation $G(y, \theta) = 0$, and where θ is not on the border of the set Θ. Rejecting, step by step, these conditions, we obtain the following results.

The existence of even a first derivative of $S(\cdot)$ with respect of θ is not a necessary condition for the asymptotic normality of the MLE and BE. To illustrate this, we consider the linear process

$$dX_t = X_{\theta t}\, dt + \varepsilon\, dW_t, \quad X_0 = x_0 \neq 0,\ 0 \le t \le T$$

with the unknown parameter $\theta \in (0,1)$. The process $X_{\theta t}$ is as smooth as a Wiener process, so the derivative $\dot{S}(\theta, X)$ does not exist in any sense. Nevertheless, the MLE and BE are asymptotically normal:

$$\mathcal{L}_\theta\{\varepsilon^{-1}(\hat{\theta}_\varepsilon - \theta)\} \Rightarrow \mathcal{N}\left(0, \left(\int_0^T t^2\, x_{\theta^2 t}^2 dt\right)^{-1}\right)$$

and LAM. We obtain a similar result we have and for the more simple *delay*-estimation problem, when $S_t(\theta, X) = X_{t-\theta}$ and for some other generalizations.

Another extreme case of smoothness is the existence of k derivatives of the function $S_t(\theta, X)$ on θ at the point θ_0, when the first $k-1$ of these derivatives are equal to zero and the k-th derivative is not. In this situation, we obtain limits like (k even)

$$\mathcal{L}\{\varepsilon^{-\frac{1}{k}}(\hat{\theta}_\varepsilon - \theta_0)\} \Longrightarrow \mathcal{L}\{(\xi_1 \wedge \xi_2 \wedge 0)^{\frac{1}{k}}\},$$

where ξ_1 and ξ_2 are some Gaussian variables and $a \wedge b = \min(a, b)$.

If the initial value X_0 is a random variable independent of the Wiener process $W_t, 0 \le t \le T$, then the MLE is asymptotically mixing normal and LAM in an appropriate sense.

The next problem concerns the misspecification of the underlying model. Let the observed process be

$$dX_t = b_t(X)\,dt + \varepsilon\,dW_t,\quad X_0 = x_0,\ 0 \le t \le T,$$

but the statistician does not know this true equation and constructs the MLE on the basis of the parametric model

$$dX_t = S_t(\theta, X)\,dt + \varepsilon\,dW_t, \qquad X_0 = x_0, \qquad 0 \le t \le T$$

with $\theta \in \Theta$. Denote by x_t the solution of the equation

$$\frac{dx_t}{dt} = b_t(x), \qquad x_0$$

and define the *true* value

$$\theta^* = \arg\min_{\theta\in\Theta} \int_0^T [S_t(\theta, x) - b_t(x)]^2 dt.$$

Then under natural conditions, the MLE is *consistent*:

$$\mathbf{P}_b - \lim_{\varepsilon\to 0} \hat{\theta}_\varepsilon = \theta^*$$

and the difference $\varepsilon^{-1}(\hat{\theta}_\varepsilon - \theta^*)$ is asymptotically normal. The behavior of a Bayes estimator is similar.

The condition that (0.3) has a unique solution is obviously a necessary condition for the existence of a consistent estimator. Suppose now that this equation has k different solutions $y = \theta = \theta_1, \theta_2, \ldots, \theta_k$, so the functions $S_t(\theta_1, x), \ldots, S_t(\theta_k, x), 0 \le t \le T$ coincide. Denote

$$\begin{aligned} \xi_j &= I_j^{-1/2} \int_0^T \dot{S}_t(\theta_j, x) dW_t, \quad I_j = \int_0^T \dot{S}_t(\theta_j, x)^2 dt, \\ H_j &= \{\omega : |\xi_j| > \max_{i\ne j} |\xi_i|\}, \qquad p_j = \mathbf{P}(H_j) \end{aligned}$$

and introduce the discrete random variable

$$\hat{\theta} = \sum_{j=1}^k \theta_j \chi_{\{H_j\}}.$$

It is established that

$$\mathbf{P}_{\theta_1} - \lim_{\varepsilon\to 0} \hat{\theta}_\varepsilon = \hat{\theta}.$$

So the MLE converges to the value θ_j with probability p_j. The Bayes estimator converges to the weighted sum $\tilde{\theta} = \sum_{j=1}^k \theta_j q_j$, where q_j are some random variables and q_j are also described. For example, if the prior density of θ is constant and $I_1 = I_j, j = 2, \ldots, k$, then $\tilde{\theta} = 1/k \sum_{j=1}^k \theta_j$.

In all of the above-mentioned problems, the true value θ was an inner point of the set Θ. If $\Theta = [\alpha, \beta)$ and $\theta = \alpha$, then the MLE $\hat{\theta}_\varepsilon$ is also consistent but asymptotically *half-normal*:

$$\lim_{\varepsilon\to 0} \mathbf{P}_\alpha^{(\varepsilon)} \{I(\alpha+)^{1/2} \varepsilon^{-1} (\hat{\theta}_\varepsilon - \alpha) < y\} = M(y),$$

where the distribution function $M(y) = \Phi(y)$ is the Gaussian law for $y \geq 0$ and $M(y) = 0$ for $y < 0$.

Chapter 3 is devoted to the expansions of the MLE and some of its characteristics by the powers of the diffusion coefficient. It begins with the expansion of the MLE $\hat{\theta}_\varepsilon$ itself in the case of the diffusion observations

$$dX_t = S(\theta, X_t)\, dt \;+\; \varepsilon\, dW_t, \qquad X_0 = x_0,\ 0 \leq t \leq T.$$

The MLE has the following representation

$$\hat{\theta}_\varepsilon = \theta + \left\{ \sum_{j=1}^{k} \psi_j \varepsilon^j + \eta \varepsilon^{k+1/2} \right\} \chi_{\{\mathcal{A}\}} + \zeta \chi_{\{\mathcal{A}^c\}},$$

where the random variables $\psi_j, j = 1, \ldots, k$ do not depend on ε, the variable ζ is small in a certain sense, $|\eta| < 1$, and the probability $\mathbf{P}_\theta^{(\varepsilon)}(\mathcal{A}^c)$ is exponentially small.

The first term is a Gaussian random variable:

$$\psi_1 = I(\theta)^{-1} \int_0^T \dot{S}(\theta, x)\, dW_t$$

and the second is a bilinear functional of the Wiener process:

$$\begin{aligned}
\psi_2 \;&=\; I(\theta)^{-3} \left\{ I(\theta) \int_0^T \dot{S}(\theta, x_t) dW_t \int_0^T \ddot{S}(\theta, x_t) dW_t \right. \\
&-\; 3/2 \left(\int_0^T \dot{S}(\theta, x_t) dW_t \right)^2 \int_0^T \dot{S}(\theta, x_t) \ddot{S}(\theta, x_t) dt \\
&-\; 2I(\theta) \int_0^T \dot{S}'(\theta, x_t) \dot{S}(\theta, x_t) x_t^{(1)} dt \int_0^T \dot{S}(\theta, x) dW_t \\
&+\; \left. I(\theta)^2 \int_0^T \dot{S}'(\theta, x_t) x_t^{(1)} dW_t \right\}.
\end{aligned}$$

Here the prime denotes the derivative of $S(\theta, x)$ with respect to x and $x_t^{(1)}$ is the Gaussian process from the equation

$$dx_t^{(1)} = S'(\theta, x_t)\, x_t^{(1)} dt + dW_t, \quad x_0^{(1)} = 0,\ 0 \leq t \leq T.$$

Then we discuss the possible generalization of this expansion and obtain the expansion of the distribution function

$$\begin{aligned}
&\mathbf{P}_\theta^{(\varepsilon)} \{ \varepsilon^{-1}\, I(\theta)^{-1/2}\, (\hat{\theta}_\varepsilon - \theta) < x \} = \\
&\quad = \Phi(x) + \varepsilon\, R\, n(x) \;+\; \varepsilon\, (R - Q)(1 - x^2) n(x) + O(\varepsilon^{3/2})
\end{aligned}$$

and the moments of the MLE. Here $\Phi(x)$ and $n(x)$ are the distribution function and the density of the $\mathcal{N}(0, 1)$ law.

The problems of nonparametric estimation are considered in Chapter 4. Suppose that the trend coefficient $S_t(X), 0 \leq t \leq T$ of the diffusion-type process

$$dX_t = S_t(X)\, dt \;+\; \varepsilon dW_t, \qquad X_0 = x_0, 0 \leq t \leq T \tag{0.6}$$

is unknown and we have to estimate the function $S_t(x), 0 \le t \le T$ by these observations. Here x_t is a solution of Equation (0.2). The kernel-type estimator

$$\hat{S}_t = \frac{1}{\varphi_\varepsilon} \int_0^T G\left(\frac{\tau - t}{\varphi_\varepsilon}\right) dX_\tau$$

is introduced and, under certain regularity conditions, it is proved that this estimator is consistent and its rate of convergence is optimal in the appropriate sense.

We also consider the problem of the estimation of the function $S(x)$, $a \le x \le b$ by the observations

$$dX_t = S(X_t)\, dt \; + \; \varepsilon\, dW_t, \qquad X_0 = x_0,\; 0 \le t \le T.$$

The solution of the limit system

$$\frac{dx_t}{dt} = S(x_t), \qquad x_0$$

is, of course, unknown to the observer, so the problem of the choice of the constants a and b arises. We suppose that $x_0 > 0$ and $S(x_0) > 0$ and, under some additional conditions, prove that the kernel-type estimator

$$\hat{S}_\varepsilon(x) = \frac{1}{\varphi_\varepsilon} \int_0^T G\left(\frac{t - \tau_\varepsilon(x)}{\varphi_\varepsilon}\right) dX_t$$

is consistent on some nondegenerated set of x. Here

$$\tau_\varepsilon(x) = \inf\{t : X_t \ge x,\; 0 \le t \le T\}.$$

Sometimes, the linear scheme is used in control problems

$$dX_t = \theta(t)\, X_t\, dt \; + \; \varepsilon\, dW_t, \qquad X_0 = x_0,\; 0 \le t \le T,$$

where the function $\{\theta(t), 0 \le t \le T\}$ is unknown and should be estimated. We propose a similar kernel-type estimator which is consistent and has an optimal rate of convergence.

The last problem is concerned with the estimation of the nonperturbed state $\{x_t, 0 \le t \le T\}$ by the observations of the system (0.6). It is natural to consider the observations $\{X_t, 0 \le t \le T\}$ as an estimator, so X_t estimates x_t and the following question arises : is it possible to construct some other estimators $\hat{X}_t$ of x_t (using, say, other observations $X_s, s \ne t$ too), which are better than X_t? First we introduce a low bound on the risk of any estimators and then prove that this bound is achieved for the estimator X_t . Let us denote $l(\cdot)$ as the loss function and introduce the function

$$F(t) = \int_0^t S(x_s)^{-2}\, ds.$$

Then a minimax bound of the type

$$\lim_{\varepsilon \to 0} \inf_{\hat{X}} \sup \mathbf{E} l(\varepsilon^{-1}(\hat{X} - x)) \ge \mathbf{E} l(S(x) W(F))$$

is introduced and we have equality for the estimator $\{X_t, 0 \le t \le T\}$, so this estimator is LAM.

Chapter 5 is devoted to the disorder problems for the diffusion-type observations. Suppose that the observed process $X_t, 0 \le t \le T$ has two different differentials on $[0,\tau]$ and $[\tau, T]$, so for $t \in [0,\tau]$

$$dX_t = h_t(\theta, X)dt + \varepsilon dW_t, \quad X_0 = x_0,$$

and for $t \in (\tau, T]$

$$dX_t = g_t(\theta, X)dt + \varepsilon dW_t, \qquad X_\tau,$$

where θ is an unknown (but *smooth*) parameter. The problem of simultaneous estimation of (τ, θ) has the following answer. The MLE $(\hat{\tau}_\varepsilon, \hat{\theta}_\varepsilon)$ is consistent, the vector $(\varepsilon^{-2}(\hat{\tau}_\varepsilon - \tau), \varepsilon^{-1}(\hat{\theta}_\varepsilon - \theta))$ converges weakly to the random vector (ζ^*, ξ^*), where ζ^* is defined by the maximum of a Wiener process with a linear trend, and the random variable ξ^* is Gaussian and independent on ζ^*.

Then a k-dimensional diffusion-type process $\{X_t^{(1)}, \ldots, X_t^{(k)}, 0 \le t \le T\}$ was considered, when every component $X_t^{(j)}$ has an unknown moment $\tau^{(j)}$ of switching and the vector $(\tau^{(1)}, \ldots, \tau^{(k)})$ depends upon the unknown vector-parameter θ. The problem is to estimate θ by observation. Such a model could be reasonable if, for example, we have k detectors of seismic waves and the signal from an earthquake achieves them at different moments, but these moments are defined by the position of the center and the time of the earthquake. The asymptotics of the MLE and the BE for this model are described. The next problem is also concerned with disorder, but for a misspecified model. It is supposed that the trend of the observed process is not known, either before or after the switching. On the basis of false model, the MLE is constructed and its consistence is established.

Chapter 6 describes the properties of the estimators for partially observed systems. It begins at the linear system of Kalman filtering when the equation of observation

$$dX_t = b_t(\theta)\, Y_t\, dt + \varepsilon\, dW_t, \quad X_0 = 0, 0 \le t \le T$$

and the state equation

$$dY_t = a_t(\theta)\, Y_t\, dt + \varepsilon\, \sigma_t(\theta) dV_t, \qquad Y_0 = y_0 \ne 0, 0 \le t \le T$$

are linear. Here the functions $a(\cdot)$, $b(\cdot)$ and $\sigma(\cdot)$ are known, the Wiener processes W_t and V_t are independent, and the finite-dimensional parameter θ is require to estimate. Under regularity conditions, it is proved that the MLE and the BE are consistent, asymptotically normal, and LAM.

The nonlinear analog of this scheme is

$$\begin{aligned} dX_t &= b_t(\theta, Y_t)\, dt + \varepsilon\, dW_t, \quad X_0 = 0, \\ dY_t &= a_t(\theta, Y_t)\, dt + \varepsilon\, dV_t, \quad Y_0 = y_0,\ 0 \le t \le T. \end{aligned}$$

The likelihood ratio is difficult to find in an explicit form, so we introduce another estimators of θ based on the linearization of the system in the vicinity of the nonperturbed solution of the equation

$$\frac{dy_t}{dt} = a(\theta, y_t), \quad y_0, \quad 0 \le t \le T,$$

which provides the system

$$dX_t = [b_t(\theta, y_t) + b_t'(\theta, y_t)(Y_t - y_t)]\, dt + \varepsilon dW_t,$$

$$dY_t = [a_t(\theta, y_t) + a_t'(\theta, y_t)(Y_t - y_t)]\, dt + \varepsilon dV_t$$

and then we behave as if this scheme was really true. So we obtain the misspecified model mentioned in the second chapter and, using the Kalman filter, now construct *pseudo*-MLE. This estimator is consistent and asymptotically normal.

The disorder in partially observed systems was considered separately in two cases. The first corresponds to switching in the equation of observation:

$$dX_t = h_t\, Y_t dt + \varepsilon\, dW_t, \quad X_0 = 0, 0 \leq t \leq \tau$$

and

$$dX_t = g_t\, Y_t\, dt + \varepsilon\, dW_t, \quad X_\tau,\ \tau < t \leq T,$$

where Y_t is the process satisfying the state equation

$$dY_t = b_t\, Y_t\, dt + \varepsilon\, \sigma_t\, dV_t, \qquad Y_0 = y_0, 0 \leq t \leq T.$$

It is established that the MLE $\hat{\tau}_\varepsilon$ is consistent and the limit distribution of $\varepsilon^{-2}(\hat{\tau}_\varepsilon - \tau)$ is described. The limit distribution of the Bayes estimators $\varepsilon^{-2}(\tilde{\tau}_\varepsilon - \tau)$ differs from the limit distribution of the MLE.

In the second problem, the observed process is

$$dX_t = f_t\, Y_t dt + \varepsilon\, dW_t, \quad X_0 = 0,\ 0 \leq t \leq T$$

and the state equation switches at the time τ, so

$$dY_t = h_t\, Y_t dt + \varepsilon\, dV_t, \quad Y_0 = y_0, 0 \leq t \leq \tau$$

and

$$dY_t = g_t\, Y_t dt + \varepsilon\, dV_t, \quad Y_\tau, \tau < t \leq T.$$

It is interesting to note that in this *disorder-type* problem, the MLE and the BE are asymptotically normal with the *normal rate*, so $\varepsilon^{-1}(\hat{\tau}_\varepsilon - \tau) \Rightarrow \mathcal{N}$.

The last Chapter 7 is devoted to minimum distance estimators. Let us denote θ_0 as the true value of θ, $\{x_t(\theta), 0 \leq t \leq T\}$, $\theta \in \Theta$ the set of solutions of the deterministic system (0.5), and $\| * \|$ the norm in a certain Banach space $\mathcal{B}$. The first minimum distance estimator (MDE) is defined by the equality

$$\theta_\varepsilon^* = \arg \inf_{\theta \in \Theta} \| X - x(\theta) \| . \tag{0.7}$$

The second MDE is introduced as

$$\theta_\varepsilon^{**} = \arg \inf_{\theta \in \Theta} \| X - \hat{X}(\theta) \|,$$

where

$$\hat{X}_t = x_0 + \int_0^t S_v(\theta, X)\, dv.$$

The third MDE θ_ε^{***} is introduced with the help of the nonparametric estimator $\hat{S}_t$.

The fourth $\breve{\theta}_\varepsilon$ and fifth θ_ε^o are defined by (0.7) with *sup* and L_1 norms correspondingly.

First we establish their exponential consistency and then we study their limit distributions. If $\mathcal{B}$ is a Hilbert space with the inner product $\langle f, g\rangle$, then the MDE θ_ε^* is, for example, asymptotically normal :

$$\mathbf{P}_{\theta_0} - \lim_{\varepsilon\to 0} \varepsilon^{-1}(\theta_\varepsilon^* - \theta_0) = \xi,$$

where ξ is a Gaussian vector

$$\xi = J(\theta_0)^{-1}\langle x^{(1)}(\theta_0), \dot{x}(\theta_0)\rangle,$$

($x_t^{(1)}$ is a Gaussian process) and the matrix

$$J(\theta) = \langle \dot{x}(\theta), \dot{x}(\theta)^T\rangle.$$

The estimators θ_ε^{**} and θ_ε^{***} are also asymptotically normal, but the estimators $\breve{\theta}_\varepsilon$ and θ_ε^o have non-Gaussian limits. For example,

$$\mathbf{P}_{\theta_0} - \lim_{\varepsilon\to 0} \varepsilon^{-1}(\breve{\theta}_\varepsilon - \theta_0) = \zeta,$$

$$\zeta = \arg \inf_{u\in R^d} \sup_{0\le t\le T} |x_t^{(1)}(\theta_0) - (u, \dot{x}(\theta_0))|.$$

It is interesting to note that for the Ornstein-Uhlenbeck process

$$dX_t = \theta_0\, X_t\, dt \;+\; \varepsilon\, dW_t, \quad X_0 = x_0 \neq 0, \theta_0 > 0, 0 \le t \le T,$$

the limit random variables $\zeta = \zeta_T$ for two different metrics L_1 and *sup* are asymptotically normal (as $T \to \infty$) with the *same parameters*:

$$\mathcal{L}\{T\zeta_T\} \Rightarrow \mathcal{N}\left(0, \frac{1}{2\, x_0^2\, \theta_0}\right)$$

The properties of MDEs in nonstandard (similar to §2.4 - §2.8) situations are also described.

The first terms of the asymptotic expansion of the θ_ε^* are

$$\begin{aligned}\varepsilon^{-1}(\theta_\varepsilon^* - \theta) &= J^{-1}\langle x^{(1)}, \dot{x}\rangle + \varepsilon\, J^{-3}\{J\, \langle x^{(1)}, \dot{x}\rangle\, \langle x^{(1)}, \ddot{x}\rangle\; - \\ &- \frac{3}{2}\, \langle \dot{x}, \ddot{x}\rangle\, \langle x^{(1)}, \dot{x}\rangle^2 + \frac{1}{2}\, J^2\, \langle x^{(2)}, \dot{x}\rangle\} + o(\varepsilon).\end{aligned}$$

Here $J = J(\theta)$ and $x_t^{(2)}$ is a second derivative of X_t on ε at the point $\varepsilon = 0$. The first terms of the asymptotic expansion of the distribution function are also presented there.

The problem of the robustness of the estimators is considered within the following framework. Let the observed process be from some nonparametric vicinity of a suggested parametric model, then it is possible to define an asymptotic minimax bound at the risk of any estimators where the max is taken over this nonparametric vicinity. It is shown that this bound could be chosen in such a way that the MDEs asymptotically achieve it and so are local asymptotic minimax in this problem. Thus, the MDEs are sometimes better than the MLEs.

My initial interests in the statistics of stochastic processes were founded at two Moscow seminars : on *Statistics and Control of Stochastic Processes* (Steklov Institute of Mathematics) held by N.V. Krylov, R.Sh. Liptser and A.N. Shiryayev and on *Statistics* (Institute of Problems of Information Transmission) held by R.Z. Khas'minskii. The author is indebted to all the participants of these seminars.

This work was began at Yerevan State University then I had the opportunity to continue it at the Laboratory of Probability of the University Paris VI, INRIA (Sophia Antipolis) and at the Laboratory "Statistique et Modélisation" of the University Paris VI. The work was finished in the University of Maine (Le Mans).

The author express his gratitude to J. Jacod, for reading some parts of the book and for his useful comments. His thanks are also due to V. Arman for the careful typing of this text.

Chapter 1

Auxiliary Results

In this chapter we recall some notions of probability theory and mathematical statistics which we shall use in the sequel. Especial attention is paid to the stochastic integral and differential equation. For the convenience of references, we also cite here several results of asymptotic theory of estimation.

1.1 Some notions of probability theory

Let us denote by $\Omega = \{\omega\}$ the space of elementary events, $\mathcal{F}$ as a σ-algebra of the subsets of Ω, and $\mathbf{P}$ as a probability measure on $\mathcal{F}$, i.e. a nonnegative, σ-additive function of sets of $\mathcal{F}$ with norming $\mathbf{P}\{\Omega\} = 1$. The triple $\{\Omega, \mathcal{F}, \mathbf{P}\}$ is a probability space. We say that the set $A \subset \Omega$ has probability zero if there exists a set $B \in \mathcal{F}$ such that $A \subset B$ and $\mathbf{P}\{B\} = 0$. We complete the σ-algebra $\mathcal{F}$ by all sets of measure zero and shall consider only completed probability spaces. Let $(\mathcal{X}, \mathcal{B})$ be a measurable space and $X = X(\omega)$ – a measurable mapping

$$X : \Omega \longrightarrow \mathcal{X}, \tag{1.1}$$

then X is called a random element. If $(\mathcal{X}, \mathcal{B}) = (\mathbf{R}, \mathcal{B}(\mathbf{R}))$, where $\mathbf{R}$ is the real line and $\mathcal{B}(\mathbf{R})$ is a σ-algebra of Borelian subsets of $\mathbf{R}$, then X is called a random variable and if $(\mathcal{X}, \mathcal{B}) = (\mathbf{R}^d, \mathcal{B}(\mathbf{R}^d))$, where $\mathbf{R}^d$ is a d-dimensional Euclidean space with a Borelian σ-algebra $\mathcal{B}(\mathbf{R}^d)$, then X is a d-dimensional random vector. Denote $\mathbf{T} = \{t\}$ as a nonempty set and suppose that the random variable $X_t = X_t(\omega)$ be defined for any $t \in \mathbf{T}$, then the family $\{X_t, t \in \mathbf{T}\}$ is called a random process. The image of the measure $\mathbf{P}$ under the mapping (1.1) will be denoted by $\mathbf{P}_X$ and $\mathbf{E}$ will be the symbol of mathematical expectation with respect to this measure:

$$\mathbf{E} f(X) = \int_{\mathcal{X}} f(x)\, \mathbf{P}_X(dx).$$

Let $\mathcal{X}$ be a metric space with the metric $d(\cdot,\cdot)$ and for any $\varepsilon \in (0,1]$ on the probability space $(\Omega, \mathcal{F}, \mathbf{P})$ will be defined the random elements ξ, ξ_ε, $\varepsilon \in (0,1]$ with values in $\mathcal{X}$. We say that ξ_ε converges to ξ in probability as $\varepsilon \to 0$ if for any $\delta > 0$

$$\lim_{\varepsilon \to 0} \mathbf{P}\{d(\xi_\varepsilon, \xi) > \delta\} = 0$$

and denote it as

$$\mathbf{P}-\lim_{\varepsilon\to 0}\xi_\varepsilon = \xi.$$

Denote also by $\mathbf{P}_\xi^{(\varepsilon)}$ and $\mathbf{P}_\xi$ the distributions of ξ_ε and ξ, respectively. We say that ξ_ε converges to ξ (or $\mathbf{P}_\xi^{(\varepsilon)}$ converges to $\mathbf{P}_\xi$) weakly (or in distribution) if for any continuous, bounded function $f(\cdot)$ on $\mathcal{X}$, we have

$$\mathbf{E}f(\xi_\varepsilon) = \int_{\mathcal{X}} f(x)\mathbf{P}_\xi^{(\varepsilon)}(dx) \to \int_{\mathcal{X}} f(x)\mathbf{P}_\xi(dx) = \mathbf{E}f(\xi)$$

and shall denote this type of convergence as

$$\mathcal{L}\{\xi_\varepsilon\} \Longrightarrow \mathcal{L}\{\xi\}.$$

In particular, if ξ is a Gaussian vector with the mean $\mathbf{a}$ and a covariance matrix $\mathbf{V}$, then we shall write

$$\mathcal{L}\{\xi_\varepsilon\} \Longrightarrow \mathcal{N}(\mathbf{a},\mathbf{V}).$$

Suppose now that a set $\Theta = \{\theta\}$ of indices is given and the random elements $\xi(\theta)$ and $\xi_\varepsilon(\theta), \varepsilon \in (0,1]$ are defined for every $\theta \in \Theta$. Denote $\mathbf{P}_\theta$ and $\mathbf{P}_\theta^{(\varepsilon)}, \varepsilon \in (0,1]$ as the corresponding measures.

We shall say that $\xi_\varepsilon(\theta)$ converges uniformly on Θ to $\xi(\theta)$ in probability, as $\varepsilon \to 0$ if for any $\delta > 0$

$$\lim_{\varepsilon\to 0}\sup_{\theta\in\Theta}\mathbf{P}\{d(\xi_\varepsilon(\theta),\xi(\theta)) > \delta\} = 0$$

and we say that $\xi_\varepsilon(\theta)$ converges uniformly on Θ to $\xi(\theta)$ in distribution as $\varepsilon \to 0$ if for any continuous bounded function $f(\cdot)$ on $\mathcal{X}$

$$\lim_{\varepsilon\to 0}\sup_{\theta\in\Theta}\left|\int_{\mathcal{X}} f(x)\mathbf{P}_\theta^{(\varepsilon)}(dx) - \int_{\mathcal{X}} f(x)\mathbf{P}_\theta(dx)\right| = 0.$$

Sometimes to emphasize the dependence of the distribution ξ_ε on θ, we shall write

$$\mathcal{L}_\theta\{\xi_\varepsilon\} \Longrightarrow \mathcal{L}_\theta\{\xi\}.$$

Suppose now that Θ is an open subset of $\mathbf{R}^d$ and $\xi(\theta)$, $\theta \in \Theta$ is a random element with values in $L_2[0,T]$ the space of square integrable functions with the norm

$$\|f\| = \left\{\int_0^T f(t)^2\,dt\right\}^{1/2}.$$

We say that $\xi(\theta)$ is continuous in probability at point $\theta_0 \in \Theta$, in $L_2[0,T]$ if

$$\mathbf{P}-\lim_{\theta\to\theta} d(\xi(\theta),\xi(\theta_0)) = 0,$$

where we set $d(\xi(\theta),\xi(\theta_0)) = \|\xi(\theta) - \xi(\theta_0)\|$. In the same way, we define the differentiability of $\xi(\theta)$ with respect to θ. We say that $\xi(\theta)$ is differentiable at the point

θ_0 in probability in $L_2[0,T]$ if there exists a vector-process $\dot{\xi}(\theta) = \{\dot{\xi}_t(\theta), 0 \le t \le T\}$ where

$$\dot{\xi}_t(\theta_0) = \{\dot{\xi}_t^{(1)}(\theta_0), \ldots, \dot{\xi}_t^{(k)}(\theta_0)\}, \quad \dot{\xi}_t^{(j)}(\theta_0) \in L_2[0,T], j = 1, \ldots, k$$

such that

$$\|\xi(\theta_0 + h) - \xi(\theta_0) - (h, \dot{\xi}(\theta_0))\| = o_p(|h|),$$

where $(\cdot,\cdot)$ and $|\cdot|$ are the inner product and the norm in $\mathbf{R}^d$ and $o_p(|h|)$ is small in probability, i.e., $\mathbf{P} - \lim_{|h|\to 0} o_p(|h|)|h|^{-1} = 0$. We call the stochastic vector-process $\dot{\xi}(\theta)$ a derivative of the process $\xi(\theta) \in L_2[0,T]$ in probability at the point θ_0. Similarly, we define the derivatives of higher orders.

We say that the random elements ξ_ε converge to ξ with probability 1 if

$$\mathbf{P}\{\omega : \lim_{\varepsilon\to 0} \xi_\varepsilon = \xi\} = 1$$

with the appropriate notations for the uniform convergence and uniform derivability.

1.2 Stochastic integral

Assume that a complete probability space $\{\Omega, \mathcal{F}, \mathbf{P}\}$ is given and a family $\hat{\mathcal{F}} = \{\mathcal{F}_t, 0 \le t \le T\}$ of σ-algebras $\mathcal{F}_t$ satisfies the "usual conditions" (filtration): $\mathcal{F}_s \subset \mathcal{F}_t \subset \mathcal{F}, s \le t, \mathcal{F}_t = \mathcal{F}_{t+} = \bigcap_{s>t} \mathcal{F}_s$, every $\mathcal{F}_t$ is augmented by sets of zero-probability from $\mathcal{F}$. The random process $X_t(\omega), 0 \le t \le T$ is called *measurable* if for any Borelian set $B \in \mathcal{B}(\mathbf{R})$, the inclusion

$$\{(\omega, t) : X_t(\omega) \in B\} \in \mathcal{F} \otimes \mathcal{B}_{[0,T]}$$

is valid.

The measurable stochastic process $\{X_t(\omega), 0 \le t \le T\}$ is called *adapted* (or $\mathcal{F}_t$ - adapted) if for every $t \in [0,T]$ the random variable $X_t(\omega)$ is $\mathcal{F}_t$-measurable. The stochastic process $\{X_t(\omega), 0 \le t \le T\}$ is called *progressively measurable* if for any $t \in [0,T]$ the inclusion

$$\{(\omega, s \le t) : X_s(\omega) \in B\} \in \mathcal{F}_t \otimes \mathcal{B}_{[0,t]}$$

is valid. Here $\mathcal{B}_{[0,t]}$ is a σ-algebra of Borelian subsets of $[0,t]$. Let us introduce $\mathcal{P}_0$ as the *class of progressively mesurable processes* $\{f_t(\omega), 0 \le t \le T\}$, such that

$$\mathbf{P}\left\{\int_0^T f_t(\omega)^2 dt < \infty\right\} = 1.$$

The adapted stochastic process $W_t, 0 \le t \le T$ is a (standard) *Wiener process* if with probability 1:

- $W_0 = 0$,
- $W_t, 0 \le t \le T$ *is continuous,*

- *the conditional mathematical expectation* $\mathbf{E}(W_t|\mathcal{F}_s) = W_s, 0 \le s \le t \le T$,
- $\mathbf{E}[(W_t - W_s)^2|\mathcal{F}_s] = t - s, 0 \le s \le t \le T$.

The Wiener process is a Gaussian process with zero mean and covariance function $\mathbf{E}W_tW_s = t \wedge s$. Here $t \wedge s = \min(t, s)$.

We say that the function $e_t(\omega) \in \mathcal{P}_0$ is simple if there exists a subdivision of $0 = t_0 < t_1 < \ldots < t_m = T$ of $[0, T]$ such that $e_t(\omega) = \alpha_i$ if $t \in [t_{i+1}, t_i)$ and α_i is $\mathcal{F}_{t_i}$ measurable. For the simple function $e_t(\omega)$ the stochastic integral is defined as

$$I_T(e) = \int_0^T e_t(\omega)\, dW_t = \sum_{i=0}^{m-1} \alpha_i \left[W_{t_{i+1}} - W_{t_i}\right].$$

The class of simple functions is dense in $\mathcal{P}_0$, so for any $f \in \mathcal{P}_0$ it is possible to find a sequence of simple functions $\{e_t^{(n)}(\omega), 0 \le t \le T\}$ such that

$$\mathbf{P} - \lim_{n\to\infty} \|f - e^{(n)}\| = 0.$$

Here $\|\cdot\|$ is a $L_2[0, T]$ norm.

The *stochastic (Itô) integral* of the function $f \in \mathcal{P}_0$ we define as a limit in probability of the sequence of corresponding integrals of simple functions:

$$I_T(f) = \int_0^T f_t(\omega)dW_t = \mathbf{P} - \lim_{n\to\infty} I_T(e^{(n)}). \tag{1.2}$$

Denote also by $\mathbf{M}$ the class of stochastic processes $\{f_t(\omega), 0 \le t \le T\} \in \mathcal{P}_0$ satisfying the condition

$$\mathbf{E}\int_0^T f_t(\omega)^2 dt < \infty. \tag{1.3}$$

The stochastic integral $I_T(f)$ for $f \in \mathbf{M}$ has the following properties: with probability 1

$$\mathbf{E}\left(I_t(f)|\mathcal{F}_s\right) = I_s(f), \quad 0 \le s \le t \le T. \tag{1.4}$$

Particularly

$$\mathbf{E}I_t(f) = 0. \tag{1.5}$$

Moreover,

$$\mathbf{E}I_t(f)I_s(g) = \mathbf{E}\int_0^{t\wedge s} f_v(\omega)g_v(\omega)dv, \tag{1.6}$$

where $f, g \in \mathbf{M}$ and

$$\mathbf{E}[I_T(f)^2|\mathcal{F}_0] = \mathbf{E}\left[\int_0^T f(t)^2\, dt|\mathcal{F}_0\right]$$

For any $\delta > 0$ and $\gamma > 0$

$$\mathbf{P}\left\{\sup_{0\le t\le T}\left|\int_0^t f_s(\omega)dW_s\right| > \delta\right\} \le \frac{\gamma}{\delta^2} + \mathbf{P}\{\|f\|^2 > \gamma\}. \tag{1.7}$$

Let $f \in \mathbf{M}$ and

$$\int_0^T \mathbf{E}|f_t(\omega)|^{2m} dt < \infty$$

for some integer $m \geq 1$, then

$$\mathbf{E}(I_T(f))^{2m} \leq [m(2m-1)]^m T^{m-1} \int_0^T \mathbf{E}(f_t(\omega))^{2m} dt. \tag{1.8}$$

For any $f \in \mathcal{P}_0$

$$\mathbf{E} \exp\left\{\int_0^T f_t dW_t - \frac{1}{2}\int_0^T f_t^2 dt\right\} \leq 1. \tag{1.9}$$

If $f \in \mathcal{P}_0$ and with probability 1

$$\int_0^T f_t^2 dt \geq \sigma^2,$$

then the stopping time

$$\tau = \inf\left\{t : \int_0^t f_s^2 ds \geq \sigma^2\right\}$$

is finite with probability 1 and

$$\int_0^\tau f_t dW_t = \tilde{W}(\upsilon), \tag{1.10}$$

where $\tilde{W}(\cdot)$ is some other Wiener process.

Assume now that the set $\Theta \in \mathbf{R}^d$ and the stochastic process $f_t(\theta,\omega) \in \mathcal{P}_0$ is measurable in the following sense : for any $B \in \mathcal{B}(\mathbf{R})$

$$\{(\omega, s \leq t, \theta) : \ f_s(\theta,\omega) \in B\} \in \mathcal{F}_t \otimes \mathcal{B}_{[0,t]} \otimes \mathcal{B}(\Theta),$$

where $\mathcal{B}(\Theta)$ is Borel σ-algebra of the subsets of Θ. Denote $\mathcal{P}$ as class of such functions. Let $\Theta = (\alpha, \beta)$ and $f_t(\theta,\omega) \in \mathcal{P}$, then the random function

$$\eta(\theta) = \int_0^T f_t(\theta,\omega) dW_t, \quad \theta \in (\alpha,\beta)$$

is defined.

Lemma 1.1. *Let the stochastic process $f_t(\theta,\omega), 0 \leq t \leq T$ have two continuous with probability 1 derivatives $\dot{f}_t(\theta,\omega)$ and $\ddot{f}_t(\theta,\omega)$ with respect to θ for almost all t and*

$$\mathbf{E}\int_0^T \dot{f}_t(\theta,\omega)^2 dt < \infty, \quad \sup_{\theta\in\Theta} \mathbf{E}\int_0^T \ddot{f}_t(\theta,\omega)^2 dt < \infty.$$

Then the stochastic process $\eta(\theta), \alpha < \theta < \beta$ is continuously differentiable w.r.t. θ with probability 1 and the process

$$\dot{\eta}(\theta) = \int_0^T \dot{f}_t(\theta,\omega) dW_t$$

is its derivative.

Proof. Below we shall omit the variable ω. The stochastic process $\dot\eta(\theta)$ satisfies the inequality

$$\begin{aligned}\mathbf{E}[\dot\eta(\theta+h)-\dot\eta(\theta)]^2 &= \mathbf{E}\int_0^T [\dot f_t(\theta+h)-\dot f_t(\theta)]^2 dt = \\ &= \mathbf{E}\int_0^T \left(\int_0^h \ddot f_t(\theta+v)dv\right)^2 dt \le h^2 \sup_\theta \int_0^T \mathbf{E}\ddot f(\theta,t)^2 dt \le Ch^2.\end{aligned}$$

Therefore it has a continuous with probability 1 modification ([47], §4.5). By the Fubini theorem ([97], theorem 5.15), we can write

$$\begin{aligned}\eta(\theta+h)-\eta(\theta) &= \int_0^T [f_t(\theta+h)-f_t(\theta)]dW_t = \\ &= \int_0^T \left(\int_0^h \dot f_t(\theta+v)dv\right) dW_t = \int_0^h \dot\eta(\theta+v)dv\end{aligned}$$

and using the continuity of $\dot\eta(\theta)$, we have, with probability 1, the convergence

$$\frac{\eta(\theta+h)-\eta(\theta)}{h} = \frac{1}{h}\int_0^h \dot\eta(\theta+v)dv \to \dot\eta(\theta).$$

In a similar way and using induction, we can prove the following lemma.

Lemma 1.2. *Let the function $f_t(\theta,\omega)\in\mathcal{P}$ have $k+1$ continuous with probability 1 derivatives $f_t^{(j)}(\theta)=\partial^j/\partial\theta^j\ f_t(\theta)\in\mathbf{M}, j=1,\dots,k+1$ on θ and*

$$\sup_{\theta\in\Theta}\int_0^T \mathbf{E}\left(f_t^{(k+1)}(\theta)\right)^2 dt < \infty,$$

then the stochastic process $\eta(\theta)$ has k continuous derivatives on θ and with probability 1

$$\frac{d^j}{d\theta^j}\eta(\theta) = \int_0^T f_t^{(j)}(\theta)dW_t, \qquad j=1,\dots,k.$$

Of course, $f_t(\theta)=f_t(\theta,\omega)$. We also need the following lemma.

Lemma 1.3. *Let $\eta(\theta),\theta\in[0,1]$ be a continuous random process satisfying the condition: there exist nonnegative functions $g(h)$ and $q(c,h), h\ge 0$ such that*

$$\mathbf{P}\{\eta(\theta+h)-\eta(\theta)>cg(h)\}\le q(c,h) \tag{1.11}$$

and

$$\begin{aligned}G &= \sum_{n=1}^{\infty} g(2^{-n}) < \infty, \\ H(c) &= \sum_{n=1}^{\infty} 2^{n-1}q(c,2^{-n}) < \infty,\end{aligned}$$

then

$$\mathbf{P}\left\{\sup_{0\le\theta\le1}[\eta(\theta)-\eta(0)]>N\right\}\le H\left(\frac{N}{G}\right). \tag{1.12}$$

Proof. Introduce the following collections $A_1, A_2, \dots$ of intervals from $[0,1]$:

$$\begin{aligned}
A_1 &= \left\{\left[0,\frac{1}{2}\right]\right\};\\
A_2 &= \left\{\left[0,\frac{1}{4}\right];\left[\frac{2}{4},\frac{3}{4}\right]\right\};\\
A_3 &= \left\{\left[0,\frac{1}{8}\right];\left[\frac{2}{8},\frac{3}{8}\right];\left[\frac{4}{8},\frac{5}{8}\right];\left[\frac{6}{8},\frac{7}{8}\right]\right\};\\
&\dots\dots
\end{aligned}$$

To obtain the collection A_n we have to partition the interval $[0,1]$ into subintervals of length 2^{-n} and from these 2^n intervals take all odd counting from the left to the right. Any interval $[0,x)$ of $(0,1]$ could be present as a countable union of subintervals from $A_1, A_2, \dots$, where we take no more than one interval from each A_n.

Suppose that on each $A_n, n = 1,2,\dots$, the condition

$$\eta\left(\frac{2k+1}{2^n}\right)-\eta\left(\frac{2k}{2^n}\right)\le cg(2^{-n}),\quad k=0,1,\dots,2^{n-1}-1 \tag{1.13}$$

holds, i.e. the increment of $\eta(\theta)$ on every A_n is not more than $cg(2^{-n})$. Then from the continuity of $\eta(\theta)$, it follows that

$$\sup_{0\le\theta\le1}[\eta(\theta)-\eta(0)]\le c\sum_{n=1}^{\infty}g(2^{-n})=cG.$$

If this inequality is broken, then there exists at least one interval A_n, where (1.13) was violated. So taking in account that every A_n consists of 2^{n-1} intervals, we obtain

$$\mathbf{P}\left\{\sup_{0<\theta\le1}[\eta(\theta)-\eta(0)]\ge cG\right\}\le\sum_{n=1}^{\infty}2^{n-1}q(c,2^{-n})=H(c). \tag{1.14}$$

So if we put $N = Gc$ then from (1.14) follows (1.11).

If $\theta\in[\alpha,\beta]$, then (1.11) takes the form

$$\mathbf{P}\left\{\sup_{\alpha<\theta\le\beta}[\eta(\theta)-\eta(\alpha)]>N\right\}\le H\left(\frac{N}{G(\beta-\alpha)}\right). \tag{1.15}$$

Lemma 1.4. *Let $f\in\mathcal{P}_0$ and for some $\gamma>0$ and $p\ge1$*

$$M_1\triangleq\mathbf{E}\exp\left\{\gamma\left(\int_0^T f_t^2dt\right)^{\frac{1}{p}}\right\}<\infty,$$

then for $N \geq N_0 = (2\gamma)^{\frac{p}{p-1}}$

$$\mathbf{P}\left\{\int_0^T f_t dW_t > N\right\} \leq (1 + M_1)\exp\left\{-\gamma N^{-\frac{1}{p}}\right\}.$$

Proof. Below, we use Tchebyshev's inequality and (1.9).

$$\mathbf{P}\left\{\int_0^T f_t dW_t > N\right\} \leq \mathbf{P}\left\{\int_0^T f_t dW_t - \frac{1}{2}\int_0^T f_t^2 dt > \frac{N}{2}\right\} +$$

$$+\mathbf{P}\left\{\int_0^T f_t^2 dt > N\right\} \leq$$

$$\leq \exp\left(-\frac{N}{2}\right)\mathbf{E}\exp\left\{\int_0^T f_t dW_t - \frac{1}{2}\int_0^T f_t^2 dt\right\} +$$

$$+\exp\left\{-\gamma N^{\frac{1}{p}}\right\}\mathbf{E}\exp\left\{\gamma\left(\int_0^T f_t^2 dt\right)^{\frac{1}{p}}\right\} \leq$$

$$\leq (M_1 + 1)\exp\left\{-\gamma N^{\frac{1}{p}}\right\}$$

because $M_1 \geq 1$ and $N \geq N_0$.

Put

$$\eta(\theta) = \int_0^T f_t(\theta)\, dW_t.$$

Lemma 1.5. *Let the measurable stochastic process* $\{f_t(\theta), 0 \leq t \leq T\} \in \mathcal{P}$ *for all* $\theta \in [\alpha, \beta]$ *be differentiable with probability 1 and for some* $\gamma_0 > 0$ *and* $p > 1$*, the derivative satisfies the condition*

$$M \triangleq \mathbf{E}\exp\left\{\gamma_0 \sup_{\alpha \leq \theta \leq \beta}\left(\int_0^T \dot{f}_t(\theta)^2 dt\right)^{\frac{1}{p}}\right\} < \infty. \tag{1.16}$$

Then for $N \geq N_0 = \frac{4p-2}{\gamma}$

$$\mathbf{P}\left\{\sup_{\alpha < \theta \leq \beta}[\eta(\theta) - \eta(\alpha)] > N\right\} \leq C\left(\frac{N}{\beta - \alpha}\right)^{-\frac{1}{p}}\exp\left\{-\kappa\left(\frac{N}{\beta - \alpha}\right)^{\frac{1}{p}}\right\}, \tag{1.17}$$

where $\kappa = \gamma_0(2 - \sqrt{2})^{\frac{1}{p}}$ *and* $C > 0$ *is constant.*

Proof. First, we estimate the probability of increments of $\eta(\theta)$. As in Lemma 1.4, we have ($\Delta f = f_t(\theta + h) - f_t(\theta)$)

$$\mathbf{P}\{\eta(\theta + h) - \eta(\theta) > cg(h)\} \leq$$

$$\leq \mathbf{P}\left\{\int_0^T \frac{\Delta f}{h} dW_t - \frac{1}{2}\int_0^T\left(\frac{\Delta f}{h}\right)^2 dt > \frac{cg(h)}{2h}\right\} +$$

$$+\mathbf{P}\left\{\int_0^T\left(\frac{\Delta f}{h}\right)^2 dt > \frac{cg(h)}{2h}\right\} \leq$$

$$\leq \exp\left\{-\frac{cg(h)}{2h}\right\} + \mathbf{P}\left\{\sup_\theta \int_0^T \dot{f}_t(\theta)^2 dt > \frac{cg(h)}{h}\right\} =$$

$$= \exp\left\{-\frac{cg(h)}{2h}\right\} + \mathbf{P}\left\{\gamma_0 \sup_\theta\left(\int_0^T \dot{f}_t(\theta)^2 dt\right)^{\frac{1}{p}} > \gamma_0\left(\frac{cg(h)}{h}\right)^{\frac{1}{p}}\right\} \leq$$

$$\leq \exp\left\{-\frac{cg(h)}{2h}\right\} + M\exp\left\{-\gamma_0\left(\frac{cg(h)}{h}\right)^{\frac{1}{p}}\right\}.$$

Put $g(h)=\sqrt{h}$, then, if we choose $\gamma_0 < 2c^{\frac{1-p}{p}}$, we have

$$q(c,h) \leq (M+1)\exp\left\{-\gamma_0 c^{\frac{1}{p}} h^{-\frac{1}{2p}}\right\},$$

and the value of G is

$$G = \sum_{n=1}^\infty g((\beta-\alpha)2^{-n}) = \sqrt{\beta-\alpha}\sum_{n=1}^\infty 2^{-\frac{n}{2}} = \frac{\sqrt{\beta-\alpha}}{\sqrt{2}-1}.$$

To evaluate the sum

$$H(c) = \sum_{n=1}^\infty 2^{n-1}(M+1)\exp\left\{-\gamma_0(\beta-\alpha)^{-\frac{1}{2p}} c^{\frac{1}{p}} 2^{\frac{n}{2p}}\right\}$$

we consider the integral

$$Q(c) = \int_1^\infty 2^x \exp\left\{-\gamma c^{\frac{1}{p}} 2^{\frac{x}{2p}}\right\} dx.$$

Changing the variable $z=(c^2 2^x)^{\frac{1}{2p}}$ we obtain

$$Q(c) = \frac{2p}{c^2\ln 2}\int_{N_0}^\infty z^{2p-1} e^{-\gamma z} dz$$

with $N_0 = (2c^2)^{\frac{1}{2p}}$. The last integral, which is the tail of gamma function, we integrate by parts:

$$\begin{aligned}\int_{N_0}^\infty z^{2p-1}e^{-\gamma z}dz &= \gamma^{-1}N_0^{2p-1}e^{-\gamma N_0} + \frac{2p-1}{\gamma}\int_N^\infty e^{-\gamma z}z^{2p-2}dz < \\ &< \gamma^{-1}N_0^{2p-1}e^{-\gamma N_0} + \frac{2p-1}{\gamma N_0}\int_{N_0}^\infty z^{2p-1}e^{-\gamma z}dz.\end{aligned}$$

This yields

$$\int_{N_0}^\infty z^{2p-1}e^{-\gamma z}dz < \frac{N_0^{2p-1}}{\gamma\left(1-\frac{2p-1}{\gamma N_0}\right)}e^{-\gamma N_0} < \frac{2}{\gamma}N_0^{2p-1}e^{-\gamma N_0}$$

for $N_0 > \gamma^{-1}(4p-2) = N_1$.

Hence

$$Q(c) < \frac{4p}{\gamma c^2 \ln 2}(2c^2)^{\frac{2p-1}{2p}} \exp\left\{-\gamma(2c^2)^{\frac{1}{2p}}\right\}$$

and finally

$$\begin{aligned} H(c) &< (M+1)\frac{p2^{3-\frac{1}{2p}}}{\gamma \ln 2} c^{-\frac{1}{p}} \exp\left\{-\gamma(2c^2)^{\frac{1}{2p}}\right\} < \\ &< \frac{8(M+1)p}{\gamma_0 \ln 2}(\beta-\alpha)^{\frac{1}{2p}} c^{-\frac{1}{p}} \exp\left\{-\gamma_0(\beta-\alpha)^{-\frac{1}{2p}} 2^{\frac{1}{2p}} c^{\frac{1}{p}}\right\}. \end{aligned}$$

Now putting $c = \frac{N}{G}$, by Lemma 1.3 we obtain

$$\begin{aligned} &\mathbf{P}\left\{\sup_{\alpha<\theta\le\beta}[\eta(\theta)-\eta(\alpha)] > N\right\} \le \\ &\le \frac{8p(M+1)}{\gamma_0 \ln 2}\left(\frac{N}{\beta-\alpha}\right)^{-\frac{1}{p}} (\sqrt{2}-1)^{\frac{1}{p}} \exp\left\{-\gamma_0(2-\sqrt{2})^{\frac{1}{p}}\left(\frac{N}{\beta-\alpha}\right)^{\frac{1}{p}}\right\} \le \\ &\le C\left(\frac{N}{\beta-\alpha}\right)^{-\frac{1}{p}} \exp\left\{-\kappa\left(\frac{N}{\beta-\alpha}\right)^{-\frac{1}{p}}\right\}. \end{aligned}$$

Remark 1.1. We shall use Lemmas 1.4 and 1.5 in the estimation of probabilities like

$$\mathbf{P}\left\{\sup_{\alpha<\theta<\beta}\left|\int_0^T f_t(\theta)dW_t\right| > N\right\}.$$

Without mentioning this every time, we shall do it in the following way. Denote $\nu(\theta) = -\eta(\theta)$, then

$$\begin{aligned} &\mathbf{P}\left\{\sup_{\alpha<\theta<\beta}|\eta(\theta)| > N\right\} \le \mathbf{P}\left\{\sup_{\alpha<\theta<\beta}\eta(\theta) > N\right\} + \mathbf{P}\left\{\inf_{\alpha<\theta<\beta}\eta(\theta) < -N\right\} \le \\ &\le \mathbf{P}\left\{\eta(\alpha) > \frac{N}{2}\right\} + \mathbf{P}\left\{\sup_{\alpha<\theta<\beta}[\eta(\theta)-\eta(\alpha)] > \frac{N}{2}\right\} + \\ &\qquad + \mathbf{P}\left\{\nu(\alpha) > \frac{N}{2}\right\} + \mathbf{P}\left\{\sup_{\alpha<\theta<\beta}[\nu(\theta)-\nu(\alpha)] > \frac{N}{2}\right\} \le \\ &\le 2(M_1+1)\exp\left\{-\gamma\left(\frac{N}{2}\right)^{\frac{1}{p}}\right\} + 4C\left(\frac{N}{2(\beta-\alpha)}\right)^{-\frac{1}{p}} \exp\left\{-\kappa\left(\frac{N}{2(\beta-\alpha)}\right)^{\frac{1}{p}}\right\}. \end{aligned}$$

Choosing N_1, we can always reduce this sum to

$$\mathbf{P}\left\{\sup_{\alpha<\theta<\beta}|\eta(\theta)| > N\right\} < C\exp\left\{-\tilde{\kappa}N^{\frac{1}{p}}\right\}$$

for $N > N_1$.

If the measurable function $f_t(\theta), \theta \in [\alpha, \beta]$ is nonrandom and differentiable in $L_2[0,T]$ with

$$K^2 = \sup_{\alpha \leq \theta \leq \beta} \int_0^T \dot{f}_t(\theta)^2 dt < \infty,$$

then for the Gaussian process

$$\eta(\theta) = \int_0^T f_t(\theta) dW_t, \quad \alpha \leq \theta \leq \beta$$

M. Burnashev [13] has proved the following estimate:

Lemma 1.6.

$$\mathbf{P}\left\{\sup_{\alpha<\theta<\beta} [\eta(\theta) - \eta(\alpha)] > N\right\} \leq 8\Phi\left(\frac{N}{K(\beta-\alpha)}\right), \tag{1.18}$$

where

$$\Phi(x) = \frac{1}{\sqrt{2\pi}} \int_x^\infty e^{-\frac{y^2}{2}} dy.$$

Lemma 1.7. *Let $f \in \mathcal{P}_0$ and*

$$M \triangleq \mathbf{E} \exp\left\{\gamma \left(\int_0^T f_t^2 dt\right)^{\frac{1}{p}}\right\} < \infty$$

for some $\gamma > 0$ and $p > 1$, then

$$\mathbf{P}\left\{\sup_{0 \leq t \leq T} \int_0^t f_s dW_s > N\right\} \leq (1+M) \exp\left\{-\gamma_0 N^{\frac{2}{p+1}}\right\}, \tag{1.19}$$

where $\gamma_0 = \gamma \wedge \frac{1}{2}$.

Proof. Below we use Doob's inequality [98].

$$\mathbf{P}\left\{\sup_{0 \leq t \leq T} \left(\exp\left\{\int_0^t g_s dW_s - \frac{1}{2}\int_0^t g_s^2 ds\right\}\right) \geq A\right\} \leq$$
$$\leq A^{-1} \mathbf{E} \exp\left\{\int_0^T g_s dW_s - \frac{1}{2}\int_0^T g_s^2 ds\right\} \leq A^{-1}$$

for $g \in \mathcal{P}_0$ and $A > 0$ and Tchebyshev's inequality:

$$\mathbf{P}\left\{\sup_{0 \leq t \leq T} \int_0^t f_s dW_s > N\right\} \leq$$
$$\leq \mathbf{P}\left\{\sup_{0 \leq t \leq T} \left(\int_0^t \lambda f_s dW_s - \int_0^t \frac{(\lambda f_s)^2}{2} ds\right) + \int_0^T \frac{(\lambda f_s)^2}{2} ds > \lambda N, \int_0^T f_s^2 ds < N^a\right\} +$$
$$+ \mathbf{P}\left\{\int_0^T f_s^2 ds \geq N^a\right\} \leq \mathbf{P}\left\{\sup_{0 \leq t \leq T} \exp\left\{\lambda \int_0^t f_s dW_s - \frac{\lambda^2}{2}\int_0^t f_s^2 ds\right\} >\right.$$

$$> \exp\left(\lambda N - \frac{\lambda^2}{2}N^a\right)\Big\} + \mathbf{P}\left\{\gamma\left(\int_0^T f_s^2 ds\right)^{\frac{1}{p}} > \gamma N^{\frac{a}{p}}\right\} \leq$$

$$\leq \exp\left\{-\left(\lambda N - \frac{\lambda^2}{2}N^a\right)\right\} + \exp\left\{-\gamma N^{\frac{a}{p}}\right\}\mathbf{E}\exp\left\{\gamma\left(\int_0^T f_s^2 ds\right)^{\frac{1}{p}}\right\}.$$

Minimizing $\lambda N - \frac{1}{2}\lambda^2 N^a$ on λ we obtain the value $\lambda^* = N^{1-a}$, hence

$$\mathbf{P}\left\{\sup_{0\leq t\leq T}\int_0^t f_s dW_s > N\right\} \leq \exp\left\{-\frac{1}{2}N^{2-a}\right\} + M\exp\left\{-\gamma N^{\frac{a}{p}}\right\}.$$

Now choosing the same rate of convergence of these two exponentials, we get the value $a = \frac{2p}{p+1}$ and (1.19) is proved.

For the bounded random functions f_t, i.e. $\sup_{0\leq t\leq T}|f_t| \leq K$ with probability 1, we have the well-known estimate

$$\mathbf{P}\left\{\sup_{0\leq t\leq T}\int_0^t f_s dW_s > N\right\} \leq \exp\left\{-\frac{N^2}{2K^2T}\right\}. \tag{1.20}$$

We present the *central limit theorem* for the stochastic integral in the scheme of series. Suppose that for every $\varepsilon \in (0,1]$ we are given a vector stochastic process $\{f_t^{(\varepsilon)}, 0 \leq t \leq T\} \in \mathcal{P}_0$. Below, $\|\sigma\|^*$ denotes the matrix norm of σ.

Lemma 1.8. *If*

$$\mathbf{P} - \lim_{\varepsilon\to 0}\int_0^T f_t^{(\varepsilon)} f_t^{(\varepsilon)T} dt = \sigma, \tag{1.21}$$

then

$$\mathcal{L}\left\{\int_0^T f_t^{(\varepsilon)} dW_t\right\} \Longrightarrow \mathcal{N}(0,\sigma).$$

Proof can be found in [69], theorem 3.3.6.

A uniform variant of this theorem is the following. We are given a nonempty set Θ and for every $\theta \in \Theta$ and $\varepsilon \in (0,1]$ the vector stochastic process $\{f_t^{(\varepsilon)}(\theta), 0 \leq t \leq T\} \in \mathcal{P}$.

Lemma 1.9. *If the convergence*

$$\mathbf{P} - \lim_{\varepsilon\to 0}\int_0^T f_t^{(\varepsilon)}(\theta) f_t^{(\varepsilon)}(\theta)^T dt = \sigma(\theta), \tag{1.22}$$

is uniform on $\theta \in \Theta$ *and*

$$\sup_{\theta\in\Theta}\|\sigma(\theta)\|^* < \infty,$$

then the stochastic integral

$$\eta_\varepsilon(\theta) = \int_0^T f_t^{(\varepsilon)}(\theta) dW_t$$

is asymptotically normal uniformly on $\theta \in \Theta$ *with the parameters* $(0, \sigma(\theta))$.

Proof see [69], theorem 3.3.7.

The continuous stochastic process $X = \{X_t, \mathcal{F}_t, 0 \le t \le T\}$ is called *Itô process* with respect to the Wiener process $\{W_t, \mathcal{F}_t, 0 \le t \le T\}$ if there exist two adapted processes $\{a_t(\omega), \mathcal{F}_t, 0 \le t \le T\}$ and $\{b_t(\omega), \mathcal{F}_t, 0 \le t \le T\}$ such that

$$\begin{aligned} \mathbf{P}\left\{\int_0^T |a_t(\omega)|dt < \infty\right\} &= 1, \\ \mathbf{P}\left\{\int_0^T |b_t(\omega)|^2 dt < \infty\right\} &= 1 \end{aligned}$$

and with probability 1

$$X_t = X_0 + \int_0^t a_s(\omega)ds + \int_0^t b_s(\omega)dW_s. \tag{1.23}$$

For brevity, this process is often written in the differential form

$$dX_t = a_t(\omega)dt + b_t(\omega)dW_t, \quad X_0, \quad 0 \le t \le T, \tag{1.24}$$

which is understood as an abbreviated form of (1.23) and is called the *stochastic differential* of the process X.

Let $\{f_t(\omega), \mathcal{F}_t, 0 \le t \le T\}$ belong to $\mathcal{P}_0$ process. The stochastic integral

$$I_T(f) = \int_0^T f_t(\omega)dX_t \tag{1.25}$$

of this function with respect to the Itô process X with stochastic differential (1.24) could be understood as the sum

$$\int_0^T f_t(\omega)dX_t = \int_0^T f_t(\omega)a_t(\omega)dt + \int_0^T f_t(\omega)b_t(\omega)dW_t. \tag{1.26}$$

So it is defined for the functions f satisfying the condition

$$\mathbf{P}\left\{\int_0^T \left[|f_t(\omega)a_t(\omega)| + f_t(\omega)^2 b_t(\omega)^2\right] dt < \infty\right\} = 1. \tag{1.27}$$

The definition (1.26) is not convenient for our statistical problems because X is an observed process and W is not observed, so we need another construction of (1.25) directly based on the process X . We define the integral (1.25) as a limit of integral sums

$$I_T(f_n) = \sum_{i=0}^{m} f^{(n)}_{t_i^{(n)}}(\omega)[X_{t_{i+1}^{(n)}} - X_{t_i^{(n)}}] + f^{(n)}_{t_{m+1}}(\omega)[X_T - X_{t_{m+1}^{(n)}}],$$

where $t_{m+1}^{(n)} < T$ and $f_t^{(n)}, 0 \le t \le T$ is a sequence of simple functions, approximating $f_t(\omega)$ in the following sense:

$$\begin{aligned} \mathbf{P} - \lim_{n\to\infty} \int_0^T \Big\{|a_t(\omega)||f_t^{(n)}(\omega) - f_t(\omega)| + \\ + b_t^2(\omega)|f_t^{(n)}(\omega) - f_t(\omega)|^2\Big\} dt = 0. \end{aligned}$$

Proof see [97], lemma 4.13.

Another inequality-type lemma is

Lemma 1.12. *Let $u(t)$ be a differentiable function satisfying the inequality*

$$\frac{du(t)}{dt} \leq v(t)u(t) + r(t), \quad 0 \leq t \leq T, \quad u(0),$$

where $v(t), r(t)$ are continuous functions, then

$$u(t) \leq u(0) \exp\left\{\int_0^t v(s)ds\right\} + \int_0^t \exp\left\{\int_s^t v(z)dz\right\} r(s)ds. \tag{1.39}$$

The proof could be found in [49], lemma 1.2.1.

In this work, we deal with the diffusion-type process

$$dX_t = S_t(X)dt + \varepsilon dW_t, \quad X_0 = x_0, \quad 0 \leq t \leq T, \tag{1.40}$$

where $\varepsilon \in (0,1]$ and $\varepsilon \to 0$.

Denote by $x = \{x_t, 0 \leq t \leq T\}$ the solution of the nonperturbed (1.40) process:

$$\frac{dx_t}{dt} = S_t(x), \quad x_0, \quad 0 \leq t \leq T. \tag{1.41}$$

The following lemma plays an important role in our investigation.

Lemma 1.13. *Suppose that the functional $S_t(\cdot), t \in [0,T]$ satisfies the condition $\mathcal{L}$, then with probability 1*

$$|X_t - x_t| \leq C\varepsilon \sup_{0 \leq s \leq t} |W_s| \tag{1.42}$$

and

$$\sup_{0 \leq t \leq T} |X_t - x_t| \leq C\varepsilon \sup_{0 \leq t \leq T} |W_t| \tag{1.43}$$

with some constant $C = C(L_1, L_2, K_0, T)$.

Proof. Denote $u(t) = |X_t - x_t|$, then by condition (1.34)

$$\begin{aligned} u_t &\leq \int_0^t |S_v(X) - S_v(x)|dv + \varepsilon|W_t| \leq \\ &\leq L_1 \int_0^t \left[\int_0^v u_r dK(r)\right] dv + L_2 \int_0^t u_v dv + \varepsilon \sup_{0 \leq s \leq t} |W_s|. \end{aligned}$$

Therefore (1.42) follows from (1.38) with $C_0 = \varepsilon \sup_{0 \leq s \leq t} |W_s|, C_1 = L_2, C_2 = L_1$ if we put

$$C = \exp\{(L_1 K_0 + L_2)T\}. \tag{1.44}$$

Note also that by using Lemma 1.11, we can estimate

$$\begin{aligned}
\mathbf{E}|X_t - x_t|^2 &\le 2\mathbf{E}\left(\int_0^t [S_v(X) - S_v(x)]dv\right)^2 + 2\varepsilon^2 \mathbf{E}W_t^2 \le \\
&\le 2t\int_0^t \mathbf{E}\left(L_1\int_0^v |X_r - x_r|dK(v) + L_2|X_v - x_v|\right)^2 dv + 2\varepsilon^2 t \le \\
&\le 4TL_1^2\int_0^t \mathbf{E}\left(\int_0^v |X_r - x_r|dK(r)\right)^2 dv + \\
&\quad + 4TL_2^2\int_0^t \mathbf{E}|X_v - x_v|^2 dv + 2\varepsilon^2 T \le \\
&\le 4T\, L_1^2 K_0 \int_0^t\int_0^v \mathbf{E}|X_r - x_r|^2 dK(r)dv + \\
&\quad + 4TL_2^2\int_0^t \mathbf{E}|X_v - x_v|^2 dv + 2\varepsilon^2 T. \qquad (1.45)
\end{aligned}$$

So, denoting $u(t) = \mathbf{E}|X_t - x_t|^2$, from (1.45) and (1.38) we obtain the estimate

$$\sup_{0\le t\le T} \mathbf{E}|X_t - x_t|^2 \le C\varepsilon^2 \qquad (1.46)$$

with the constant

$$C = 2T\exp\{(4L_2^2 + 4L_1^2K_0^2)T^2.$$

The estimate (1.46) can also be obtained directly from (1.42).

The Wiener process has the property [97]

$$\mathbf{P}\left\{\sup_{0\le t\le T} W_t > N\right\} = 2\mathbf{P}\{W_T > N\}. \qquad (1.47)$$

So we can write

$$\begin{aligned}
\mathbf{P}\left\{\sup_{0\le t\le T} |W_t| > N\right\} &< \mathbf{P}\left\{\sup_{0\le t\le T} W_t > N\right\} + \mathbf{P}\left\{\inf_{0\le t\le T} W_t < -N\right\} = \\
&= \mathbf{P}\left\{\sup_{0\le t\le T} W_t > N\right\} + \mathbf{P}\left\{\sup_{0\le t\le T} (-W_t) > N\right\} = \\
&= 4\mathbf{P}\{W_T > N\} \qquad (1.48)
\end{aligned}$$

because $-W_t$ is also a Wiener process. For the Gaussian random variable ξ with the parameters $(0,1)$, we have

$$\begin{aligned}
\mathbf{P}(\xi > N) &= \frac{1}{\sqrt{2\pi}}\int_N^\infty e^{-\frac{x^2}{2}}dx = \frac{1}{\sqrt{2\pi}}\int_0^\infty e^{-\frac{(y+N)^2}{2}}dy = \\
&= e^{-\frac{N^2}{2}}\frac{1}{\sqrt{2\pi}}\int_0^\infty e^{-\frac{y^2}{2}-yN}dy < \\
&< e^{-\frac{N^2}{2}}\frac{1}{\sqrt{2\pi}}\int_0^\infty e^{-\frac{y^2}{2}}dy = \frac{1}{2}e^{-\frac{N^2}{2}}.
\end{aligned}$$

At the same time ([55], p. 17)

$$\mathbf{P}(\xi > N) \le \frac{2}{(\sqrt{N^2+2}+N)\sqrt{2\pi}}e^{-\frac{N^2}{2}} < \frac{1}{N\sqrt{2\pi}}e^{-\frac{N^2}{2}}.$$

So

$$\mathbf{P}(\xi > N) \le \min\left(\frac{1}{2}, \frac{1}{N\sqrt{2\pi}}\right) e^{-\frac{N^2}{2}}.$$

Hence

$$\mathbf{P}\left(\sup_{0\le t\le T} |W_t| > N\right) \le \min\left(2, \frac{4}{N}\sqrt{\frac{T}{2\pi}}\right) e^{-\frac{N^2}{2T}}. \tag{1.49}$$

Using the inequalities (1.48), (1.49), we prove

Lemma 1.14. *For any $\lambda > 0$*

$$\mathbf{E}\exp\left\{\lambda \sup_{0\le t\le T} |W_t|\right\} \le 1 + \lambda\sqrt{8\pi T} e^{\frac{T\lambda^2}{2}} \tag{1.50}$$

and for $\lambda < \frac{1}{2T}$

$$\mathbf{E}\exp\left\{\lambda \sup_{0\le t\le T} |W_t|^2\right\} \le 1 + 2\lambda + \frac{8\lambda T}{\sqrt{1-2\lambda T}}. \tag{1.51}$$

Proof. Denote $F(x)$ as the probability distribution function of $\sup_{0\le t\le T} |W_t|$. Then integrating by parts for $\lambda = 1$ we have.

$$\begin{aligned}
\mathbf{E}\exp\left(\sup_{0\le t\le T} |W_t|\right\} &= \int_0^\infty e^x dF(x) = -\int_0^\infty e^x d[1-F(x)] = \\
&= 1 + \int_0^\infty e^x[1-F(x)]dx \le 1 + 4\int_0^\infty e^x \mathbf{P}\{W_T > x\}dx \le \\
&\le 1 + 2\int_0^\infty \exp\left(x - \frac{x^2}{2T}\right) dx = \\
&= 1 + 2e^{\frac{T}{2}} \int_0^\infty \exp\left\{-\frac{1}{2T}(X-T)^2\right\} dx < \\
&< 1 + 2e^{\frac{T}{2}} \int_\infty^\infty \exp\left\{-\frac{1}{2T}(X-T)^2\right\} dx = \\
&= 1 + \sqrt{8\pi T} e^{\frac{T}{2}}.
\end{aligned}$$

Hence, after multiplication by λ, we obtain (1.50).

To derive (1.51) we first integrate by parts and then use both estimates in (1.49).

$$\begin{aligned}
&\mathbf{E}\exp\left\{\lambda \sup_{0\le t\le T} |W_t|^2\right\} = \int_0^\infty e^{\lambda x^2} dF(x) = \\
&= 1 + 2\lambda \int_0^\infty [1-F(x)] x e^{\lambda x^2} dx \le \\
&\le 1 + 4\lambda \int_0^1 x \exp\left\{-\left(\frac{1}{2T} - \lambda\right) x^2\right\} dx + \\
&\qquad + 8\lambda\sqrt{\frac{T}{2\pi}} \int_1^\infty \exp\left\{-\left(\frac{1}{T} - 2\lambda\right)\frac{x^2}{2}\right\} dx \le
\end{aligned}$$

$$\leq 1 + 2\lambda + 8\lambda\sqrt{\frac{T}{2\pi}} \int_{-\infty}^{\infty} \exp\left\{-\frac{x^2}{2\sigma^{*2}}\right\} dx =$$
$$= 1 + 2\lambda + 8\lambda\sqrt{\frac{T}{2\pi}} \frac{\sqrt{2\pi T}}{\sqrt{1-2\lambda T}},$$

where we denote $\sigma^{*2} = (T^{-1} - 2\lambda)^{-1}$.

Let us denote the $\mathbf{P}^{(\varepsilon)}$ and $\mathbf{P}_0^{(\varepsilon)}$ measures induced in $(\mathcal{C}_T, \mathcal{B}_T)$ by the diffusion-type process $\{X_t, 0 \leq t \leq T\}$ satisfying (1.40) and by the process $\{Y_t = x_0 + \varepsilon W_t, 0 \leq t \leq T\}$. We shall take the likelihood ratio formulas from the following theorem.

Theorem 1.2. *The conditions*

$$\mathbf{P}\left\{\int_0^T S_t(X)^2 dt < \infty\right\} = 1, \quad \mathbf{P}\left\{\int_0^T S_t(Y)^2 dt < \infty\right\} = 1 \qquad (1.52)$$

are necessary and sufficient for the equivalence of the measures $\mathbf{P}^{(\varepsilon)}$ and $\mathbf{P}_0^{(\varepsilon)}$ and if they are fulfilled, then the Radon-Nikodym derivative

$$\frac{d\mathbf{P}^{(\varepsilon)}}{d\mathbf{P}_0^{(\varepsilon)}}(Y) = \exp\left\{\frac{1}{\varepsilon^2}\int_0^T S_t(Y)dY_t - \frac{1}{2\varepsilon^2}\int_0^T S_t(Y)^2 dt\right\}. \qquad (1.53)$$

Proof see [97], theorem 7.7.

Suppose that for every $\theta \in \Theta$(Θ is a nonempty set), we are given a diffusion-type process

$$dX_t = S_t(\theta, X)dt + \varepsilon dW_t, \quad X_0 = x_0, \quad 0 \leq t \leq T \qquad (1.54)$$

and the trend $S_t(\theta, X)$ satisfies the conditions $\mathcal{L}$, then conditions (1.52) are also fulfilled and for any two measures $\mathbf{P}_{\theta_1}^{(\varepsilon)}, \mathbf{P}_{\theta_2}^{(\varepsilon)}$ corresponding to solutions of (1.54) with $\theta = \theta_1$ and $\theta = \theta_2$ by the chain rule and (1.53), we have

$$\begin{aligned}\frac{d\mathbf{P}_{\theta_1}^{(\varepsilon)}}{d\mathbf{P}_{\theta_2}^{(\varepsilon)}}(X) &= \exp\left\{\frac{1}{\varepsilon^2}\int_0^T [S_t(\theta_1, X) - S_t(\theta_2, X)]dX_t - \right.\\ &\left. - \frac{1}{2\varepsilon^2}\int_0^T [S_t(\theta_1, X)^2 - S_t(\theta_2, X)^2]dt\right\} \end{aligned} \qquad (1.55)$$

and with $\mathbf{P}_{\theta_2}^{(\varepsilon)}$ probability 1 (on realizations of (1.54) with $\theta = \theta_2$) we can write

$$\begin{aligned}\frac{d\mathbf{P}_{\theta_1}^{(\varepsilon)}}{d\mathbf{P}_{\theta_2}^{(\varepsilon)}}(X) &= \exp\left\{\frac{1}{\varepsilon}\int_0^T [S_t(\theta_1, X) - S_t(\theta_2, X)]dW_t - \right.\\ &\left. - \frac{1}{2\varepsilon^2}\int_0^T [S_t(\theta_1, X) - S_t(\theta_2, X)]^2 dt\right\}. \end{aligned} \qquad (1.56)$$

This last formula will be used in our work as a *likelihood ratio*.

1.3 On asymptotic estimation theory

Below, we would like to illustrate certain notions and methods of estimation theory on the model of observation of the diffusion-type process. Assume that on the basic probability space $\{\Omega, \mathcal{F}, \mathbf{P}\}$ with the filtration $\{\mathcal{F}_t, 0 \le t \le T\}$ we are given a Wiener process $\{W_t, \mathcal{F}_t, 0 \le t \le T\}$ and for every $\theta \in \Theta$, Θ is some subset of $\mathbf{R}^d$, the diffusion-type process

$$dX_t = S_t(\theta, X)dt + \varepsilon dW_t, \quad X_0 = x_0, \quad 0 \le t \le T, \tag{1.57}$$

is defined. In this work we always assume, that the functions $S_t(\theta, \cdot), \theta \in \Theta$ satisfy the conditions $\mathcal{L}$, so Equation (1.57) has a unique strong solution and all the measures $\mathbf{P}_\theta^{(\varepsilon)}, \theta \in \Theta$ induced by the processes (1.57) in the measurable space $(C_T, \mathcal{B}_T)$ are equivalent. Here ε is an "index of series", $\varepsilon \in (0,1]$, asymptotic corresponds to $\varepsilon \to 0$, so we have a sequence of processes (1.57) indexed by ε. As usual in asymptotic problems, if one has an observation (1.57) with *small* ε, then one can embed this problem in a sequence of problems with $\varepsilon \to 0$ and then apply the obtained result to the original one.

An *estimator* θ_ε of parameter θ is defined as a measurable mapping

$$\theta_\varepsilon : \quad C_T \longrightarrow \bar{\Theta}, \tag{1.58}$$

where $\bar{\Theta}$ is a closure of Θ and we assume that Θ is supplied with σ-algebra of Borelian subsets in $\mathbf{R}^d$.

We say that the estimator θ_ε is *consistent*, if for any $\theta \in \Theta$ and $\delta > 0$,

$$\lim_{\varepsilon \to 0} \mathbf{P}_\theta^{(\varepsilon)}\{|\theta_\varepsilon - \theta| > \delta\} = 0$$

and denote this convergence as

$$\mathbf{P}_\theta - \lim_{\varepsilon \to 0} \theta_\varepsilon = \theta.$$

We say that the estimator θ_ε is uniformly consistent on compacts $\mathbf{K} \subset \Theta$ (or *uniformly consistent*) if for any compact $\mathbf{K} \subset \Theta$ and $\delta > 0$

$$\lim_{\varepsilon \to 0} \sup_{\theta \in \mathbf{K}} \mathbf{P}_\theta^{(\varepsilon)}\{|\theta_\varepsilon - \theta| > \delta\} = 0.$$

We shall study the asymptotic properties of three types of estimators: maximum likelihood, Bayesian, and minimum distance.

The *maximum likelihood estimator* (MLE) $\hat{\theta}_\varepsilon$ is defined as a solution of the equation

$$L(\hat{\theta}_\varepsilon, \theta_1; X) = \sup_{\theta \in \bar{\Theta}} L(y, \theta_1, X), \tag{1.59}$$

where the likelihood ratio

$$L(y, \theta_1; X) = \frac{d\mathbf{P}_y^{(\varepsilon)}}{d\mathbf{P}_{\theta_1}^{(\varepsilon)}}(X), \quad y \in \Theta \tag{1.60}$$

and θ_1 is some fixed value, $\theta_1 \in \Theta$.

If Equation (1.59) has more than one solution, then we take any one of them as $\hat{\theta}_\varepsilon$. The functions $L(y, \theta_1, X), y \in \Theta$ in our problems will be continuous on y, so the solution of (1.59) will always exist.

Particularly if the process (1.57) is linear on $\theta \in \mathbf{R}^d$:

$$dX_t = (\theta_0, f_t(X))dt + \varepsilon dW_t, \quad X_0 = x_0, \quad 0 \le t \le T,$$

and the matrix

$$I_T = \int_0^T f_t(X) f_t^T(X) dt$$

is positive definite with $\mathbf{P}_{\theta_0}^{(\varepsilon)}$ probability 1, then the MLE

$$\hat{\theta}_\varepsilon = I_T^{-1} \int_0^T f_t(X) dX_t.$$

To introduce a *Bayesian estimator* we need a function $l(y), y \in \mathbf{R}^d$ which is

1. *nonnegative, continuous at point* 0 *and* $l(0) = 0$, *but is not identically* 0;

2. *symmetric*: $l(y) = l(-y)$;

3. *the set* $\{y : l(y) < c\}$ *are convex for all* $c > 0$.

The functions $l(\cdot)$ will be called *loss functions.* Denote by $\mathbf{W}$ the class of loss functions satisfying 1–3. The notations $\mathbf{W}_p, \mathbf{W}_{e,2}$ will be used to denote the functions $l(\cdot) \in \mathbf{W}$ having a polynomial ($\mathbf{W}_p$) and exponential ($\mathbf{W}_{e,2}$) majorants, i.e., $l(\cdot) \in \mathbf{W}_p$ if there exist $C > 0$ and $p > 0$ such that $l(y) \le C|y|^p$ and $l(\cdot) \in \mathbf{W}_{e,2}$ if $l(y) \le \exp(\gamma |y|^2)$ for any $\gamma > 0$ as $|y| \to \infty$.

In *Bayessian approach* supposed, that the unknown parameter is a random variable with a known "prior" distribution having a density $\pi(y), y \in \Theta$, with respect to Lebesgue measure, then using a loss function $l(\cdot) \in \mathbf{W}$, we can define a *Bayesian risk* of the estimator $\theta_\varepsilon = \theta_\varepsilon(X)$ as

$$R_l(\theta_\varepsilon) = \int_\Theta \pi(y) \left(\int_{\mathcal{C}_T} l(\theta_\varepsilon(X) - y) d\mathbf{P}_\theta^{(\varepsilon)} \right) dy.$$

The Bayesian estimator (BE) $\tilde{\theta}_\varepsilon$ is defined as a solution of the equation

$$R_l(\tilde{\theta}_\varepsilon) = \min_{\theta_\varepsilon} R_l(\theta_\varepsilon). \tag{1.61}$$

If this equation has several solutions, then any of them can be chosen as $\tilde{\theta}_\varepsilon$.

In the case of the quadratic loss function $l(x) = |x|^2$, the BE $\tilde{\theta}_\varepsilon$ is a conditional mathematical expectation (*posterior mean*)

$$\tilde{\theta}_\varepsilon = \mathbf{E}(\theta | X_t, 0 \le t \le T) = \int_\Theta y \, p(y | X_t, 0 \le t \le T) \, dy, \tag{1.62}$$

where the "posterior" density

$$p(y|X_t, 0 \le t \le T) = \left(\int_\Theta \pi(v) L(v, \theta_1, X) dv\right)^{-1} \pi(y) L(y, \theta_1; X).$$

The Bayesian approach could be also used in problems with nonrandom θ. In this case, $\pi(\cdot)$ is a chosen function and (1.61) is a rule of the calculation of the estimator. This is the so-called *generalized Bayesian approach*.

Denote by $\{x_t(\theta), 0 \le t \le T\}, \theta \in \Theta$ the family of solutions of limit ($\varepsilon = 0$) equations: $x_t = x_t(\theta)$

$$\frac{dx_t}{dt} = S_t(\theta, x), \quad x_0, \quad 0 \le t \le T \tag{1.63}$$

and introduce a Hilbert space

$$L_2(\mu) = \left\{ f(t), 0 \le t \le T : \|f\|^2 = \int_0^T f(t)^2 \mu(dt) < \infty \right\},$$

where $\mu(\cdot)$ is a finite on $[0, T]$ measure.

The *minimum distance estimator* (MDE) θ_ε^* could be introduced as a solution of the equation

$$\|X - x(\theta_\varepsilon^*)\| = \min_{\theta \in \bar{\Theta}} \|X - x(\theta)\| \tag{1.64}$$

and if there are more than one solution, then we call any of them a MDE. The functions $x_t(\theta), 0 \le t \le T$ in our problems are rather regular, so our norm in (1.64) will be a continuous function of θ.

The other types of MDE are introduced in §7.2.

To formulate *Cramér-Rao inequality* (in the one-dimensional case, $d = 1$) we denote by $b_\varepsilon(\theta)$ the bias of the estimator θ_ε, i.e. $b_\varepsilon(\theta) = \mathbf{E}_\theta \theta_\varepsilon - \theta$ and suppose that the differentiation below, are valid:

$$\begin{aligned}
\frac{\partial}{\partial \theta} \mathbf{E}_\theta \theta_\varepsilon &= \frac{\partial}{\partial \theta} \mathbf{E}_{\theta_1} \left(L(\theta, \theta_1, X) \theta_\varepsilon \right) = \mathbf{E}_{\theta_1} \left(\theta_\varepsilon \frac{\partial}{\partial \theta} L(\theta, \theta_1, X) \right) = \\
&= \mathbf{E}_{\theta_1} \left(\theta_\varepsilon L(\theta, \theta_1, X) \left\{ \frac{1}{\varepsilon^2} \int_0^T \dot{S}_t(\theta, X) dX_t - \frac{1}{\varepsilon^2} \int_0^T \dot{S}_t(\theta, X) S_t(\theta, X) dt \right\} \right) = \\
&= \mathbf{E}_\theta \left[\theta_\varepsilon \left(\frac{1}{\varepsilon} \int_0^T \dot{S}_t(\theta, X) dW_t \right) \right] = \mathbf{E}_\theta \left((\theta_\varepsilon - \mathbf{E}_\theta \theta_\varepsilon) \left(\frac{1}{\varepsilon} \int_0^T \dot{S}_t(\theta, X) dW_t \right) \right) \le \\
&\le \left\{ \mathbf{E}_\theta (\theta_\varepsilon - \mathbf{E}_\theta \theta_\varepsilon)^2 \mathbf{E}_\theta \left(\frac{1}{\varepsilon^2} \int_0^T \dot{S}_t(\theta, X)^2 dt \right) \right\}^{\frac{1}{2}}.
\end{aligned}$$

So, denoting

$$I_\varepsilon(\theta) = \mathbf{E}_\theta \int_0^T \dot{S}_t(\theta, X)^2 dt$$

and using the equality $\mathbf{E}_\theta(\theta_\varepsilon - \mathbf{E}_\theta \theta_\varepsilon)^2 = \mathbf{E}_\theta(\theta_\varepsilon - \theta)^2 - b_\varepsilon(\theta)^2$, we can write

$$\mathbf{E}_\theta (\theta_\varepsilon - \theta)^2 \ge \frac{[1 + \dot{b}_\varepsilon(\theta)]^2}{I_\varepsilon(\theta)} \varepsilon^2 + b_\varepsilon(\theta)^2. \tag{1.65}$$

Here and in the sequel, the dot denotes differentiating with respect to θ. The strict derivation of (1.65) could be found in [98] or [67].

The Cramér-Rao inequality (1.65) for unbiased estimators ($b_\varepsilon(\theta) = 0$) is

$$\mathbf{E}_\theta(\theta_\varepsilon - \theta)^2 \geq \frac{\varepsilon^2}{I_\varepsilon(\theta)}.$$

This last inequality suggest defining the asymptotically efficient estimators θ_ε as satisfying the equality:

$$\lim_{\varepsilon\to 0} \mathbf{E}_\theta(\varepsilon^{-1}(\theta_\varepsilon - \theta))^2 = \frac{1}{I(\theta)}, \tag{1.66}$$

where $I(\theta) = \int_0^T \dot{S}_t(\theta, x)^2 dt$ for all $\theta \in \Theta$. Unfortunately, this definition cannot be accepted because of the existence of the phenomena of superefficiency. As was shown by L. Hodges (see [91], or [53] §2.13), some estimators have asymptotic variance below $I(\theta)^{-1}$.

We define asymptotically efficient estimators with the help of *Hajek-Le Cam's inequality (bound)* (see [53] §2.11).

A family $\{\mathbf{P}_\theta^{(\varepsilon)}, \theta \in \Theta\}$ is called *locally asymptotically normal* (LAN) at point $\theta_0 \in \Theta$ as $\varepsilon \to 0$ if, for some nondegenerate $d \times d$ matrix $\varphi_\varepsilon = \varphi_\varepsilon(\theta_0)$ and any $u \in \mathbf{R}^d$, the representation

$$Z_\varepsilon(u) \triangleq L(\theta_0 + \varphi_\varepsilon(\theta_0)u, \theta_0; X) = \exp\left\{(u, \Delta_\varepsilon) - \frac{1}{2}|u|^2 + \psi_\varepsilon(u, \theta_0)\right\} \tag{1.67}$$

is valid, where

$$\mathcal{L}_{\theta_0}\{\Delta_\varepsilon\} \Longrightarrow \mathcal{N}(0, I) \quad as \quad \varepsilon \to 0, \tag{1.68}$$

I is the unit $d \times d$ matrix and for any $u \in \mathbf{R}^d$

$$\mathbf{P}_{\theta_0} - \lim_{\varepsilon\to 0} \psi_\varepsilon(u, \theta_0) = 0. \tag{1.69}$$

If the family $\{\mathbf{P}_\theta^{(\varepsilon)}, \theta \in \Theta\}$ is LAN in all points $\theta_0 \in \Theta$, then we call it *LAN in* Θ.

A family $\{\mathbf{P}_\theta^{(\varepsilon)}, \theta \in \Theta\}$ is called *uniformly asymptotically normal* if for any compact $\mathbf{K} \subset \Theta$, some nonnegative matrix $\varphi_\varepsilon(\theta)$ and arbitrary sequences $\theta_n \in \mathbf{K}, \varepsilon_n \to 0$ and $u_n \to u \in \mathbf{R}^d$ such that $\theta_n + \varphi_{\varepsilon_n}(\theta_n)u_n \in \Theta$, the representation

$$\begin{aligned} Z_{\varepsilon_n,\theta_n}(u) &\triangleq L(\theta_n + \varphi_{\varepsilon_n}(\theta_n)u_n, \theta_n; X) = \\ &= \exp\left\{(\Delta_n, u) - \frac{1}{2}|u|^2 + \psi_{\varepsilon_n}(u_n, u, \theta_n)\right\} \end{aligned}$$

is valid, where

$$\mathcal{L}_{\theta_n}\{\Delta_n\} \Longrightarrow \mathcal{N}(0, I)$$

and

$$\lim_{n\to\infty} \mathbf{P}_{\theta_n}\{|\psi_{\varepsilon_n}(u_n, u, \theta_n)| > \delta\} = 0.$$

J.Hajek [48] proved the following theorem.

Theorem 1.3. *Let the family $\{\mathbf{P}_\theta^{(\varepsilon)}, \theta \in \Theta\}$ satisfy the LAN condition at the point $\theta = \theta_0$ with the normalizing matrix φ_ε and let tr $\varphi_\varepsilon\varphi_\varepsilon^T \to 0$ as $\varepsilon \to 0$. Then for any estimator θ_ε, any loss function $l(\cdot) \in \mathbf{W}_{e,2}$, and any $\delta > 0$, the inequality*

$$\liminf_{\varepsilon\to 0} \sup_{|\theta-\theta_0|<\delta} \mathbf{E}_\theta l(\varphi_\varepsilon^{-1}(\theta_0)(\theta_\varepsilon - \theta_0)) \geq$$
$$\geq \frac{1}{(2\pi)^{\frac{d}{2}}} \int_{\mathbf{R}^d} l(x) \exp\left(-\frac{|x|^2}{2}\right) dx. \tag{1.70}$$

Proof. See also L. Le Cam [92] and I. Ibragimov, R. Khas'minskii [53], §2.12. Note, that a one-dimensional ($d = 1$) version of this theorem was proved in [48] and the multidimensional case was considered in [53].

The inequality (1.70) is valid for all estimators θ_ε and $\delta > 0$, and could be rewritten as

$$\lim_{\delta\to 0} \liminf_{\varepsilon\to 0} \inf_{\theta_\varepsilon} \sup_{|\theta-\theta_0|<\delta} \mathbf{E}_\theta l(\varphi_\varepsilon(\theta_0)^{-1}(\theta_\varepsilon - \theta)) \geq \mathbf{E} l(\Delta), \tag{1.71}$$

where $\mathcal{L}\{\Delta\} = \mathcal{N}(0, I)$.

Using this inequality, we define the *asymptotically efficient estimator* for the loss function $l(\varphi_\varepsilon^{-1}x)$ at point θ_0 as an estimator $\bar{\theta}_\varepsilon$ satisfying the equality

$$\lim_{\delta\to 0} \liminf_{\varepsilon\to 0} \sup_{|\theta-\theta_0|<\delta} \mathbf{E}_\theta l(\varphi_\varepsilon(\theta_0)^{-1}(\bar{\theta}_\varepsilon - \theta)) = \mathbf{E} l(\Delta). \tag{1.72}$$

We shall also call it *a locally asymptotically minimax* (LAM) estimator. Detailed discussion of the different definitions of *asymptotic efficiency* of estimators and their links with this can be found in ([53], Chapter 2).

We also need several general results of I. Ibragimov and R. Khas'minskii concerning the asymptotic behavior of MLEs and BEs, which we formulate for the model (1.57), i.e., assuming that the observed process is (1.57), the sample space $\mathcal{C}_T$ and $\{\mathbf{P}_\theta^{(\varepsilon)}, \theta \in \Theta\}$ the family of measures induced in $(\mathcal{C}_T, \mathcal{B}_T)$ by the processes (1.57). Define the normalizing matrices family $\{\varphi_\varepsilon(\theta), \theta \in \Theta\}$, assume that $\varphi_\varepsilon(\theta) > 0$ (positive definite), and tr $\varphi_\varepsilon(\theta)\varphi_\varepsilon(\theta)^T \to 0$ as $\varepsilon \to 0$. Introduce a normalized *likelihood ratio process*

$$Z_\varepsilon(u) = L(\theta + \varphi_\varepsilon(\theta)u, \theta; X), \quad u \in U_\varepsilon = \varphi_\varepsilon(\theta)^{-1}(\Theta - \theta)$$

and denote $C_0(\mathbf{R}^d)$ as normalized space of functions continuous on $\mathbf{R}^d$ and vanishing at infinity with the norm *sup*.

Denote by $\mathbf{G}$ the set of functions $\{g(y), y \geq 0\}$ with the properties

$$\lim_{y\to\infty} y^N e^{-g(y)} = 0 \tag{1.73}$$

for any $N > 0$. We begin with the MLE $\hat{\theta}_\varepsilon$.

Theorem 1.4. *Let the parameter set Θ be an open convex subset of $\mathbf{R}^d$, functions $Z_\varepsilon(u)$ be continuous with probability 1 possessing the following properties:*

1. *For any compact $\mathbf{K} \subset \Theta$, there correspond the numbers $a(\mathbf{K}) = a$ and $B(\mathbf{K}) = B$ and functions $g^{\mathbf{K}}(y) = g(y) \in \mathbf{G}$ such that*

   (a) there exist the numbers $\alpha > d$ and $m \geq \alpha$ such that for $\theta \in \mathbf{K}$

$$\sup_{|u_1|\leq R, |u_2|\leq R, u_1, u_2 \in U_\varepsilon} |u_2 - u_1|^{-\alpha} \mathbf{E}_\theta \left| Z_\varepsilon^{\frac{1}{m}}(u_2) - Z_\varepsilon^{\frac{1}{m}}(u_1) \right|^m \leq$$
$$\leq B(1 + \mathbf{R}^a); \tag{1.74}$$

   (b) for all $u \in U_\varepsilon$ and $\theta \in \mathbf{K}$

$$\mathbf{E}_\theta Z_\varepsilon^{\frac{1}{2}}(u) \leq e^{-g(|u|)}; \tag{1.75}$$

2. *Uniformly in $\theta \in \mathbf{K}$ for $\varepsilon \to 0$ the finite-dimensional distributions of the random functions $Z_\varepsilon(u)$ converge to the finite-dimensional distributions of random functions $Z(u)$, where $Z(\cdot) \in \mathbf{C}_0(\mathbf{R}^d)$;*

3. *The limit functions $Z(u)$ with probability 1 attain the maximum at unique point $\hat{u}(\theta)$.*

Then uniformly in $\theta \in \mathbf{K}$, the distribution of random variables $\varphi_\varepsilon(\theta)^{-1}(\hat{\theta}_\varepsilon - \theta)$ converge to the distribution of $\hat{u}(\theta)$ and for any continuous loss function $l(\cdot) \in \mathbf{W}_p$ we have uniformly in $\theta \in \mathbf{K}$

$$\lim_{\varepsilon \to 0} \mathbf{E}_\theta l(\varphi_\varepsilon(\theta)^{-1}(\hat{\theta}_\varepsilon - \theta)) = \mathbf{E} l(\hat{u}(\theta)). \tag{1.76}$$

Proof in [53], theorem 1.10.1 is based on the following consideration. The random functions $\{Z_\varepsilon(u), u \in U_{\varepsilon,\theta} = \{u \ : \theta + \varphi_\varepsilon(\theta)u \in \Theta\}\}$ can be prolonged over all space $\mathbf{C}_0(\mathbf{R}^d)$ of continuous functions as decreasing to zero outside the set $U_{\varepsilon,\theta}$. Denote by $\Xi_{\varepsilon,\theta}$ the measure induced in $\mathbf{C}_0(\mathbf{R}^d)$ by these functions. Condition 2 of the Theorem 1.4 provides the convergence of marginal distributions of $Z_\varepsilon(\cdot)$ to the marginal distributions of $Z(\cdot)$ uniformly in $\theta \in \mathbf{K}$ and Condition 1 allows to verify the (uniform in θ) tightness of the family $\Xi_{\varepsilon,\theta}$. So the measures $\Xi_{\varepsilon,\theta}$ converge weakly to the measure Ξ of the process $Z(\cdot)$ and for any functional F continuous in $\mathbf{C}_0(\mathbf{R}^d)$ we have uniformly in θ

$$\mathbf{P}_\theta^{(\varepsilon)}\{F(Z_\varepsilon) < y\} \to \mathbf{P}\{F(Z) < y\}$$

provided $\mathbf{P}\{F(Z) = y\} = 0$. For each open bounded parallelepiped $D \in \mathbf{R}^d$ we define two functionals

$$L_D(\psi) = \sup_{u \in D} \psi(u), \qquad L_{D^c}(\psi) = \sup_{u \in D^c} \psi(u)$$

where $D^c = \mathbf{R}^d \backslash D$. These are clearly continuous functionals.

Denote by u_ε the set in $\mathbf{R}^d$ consisting of all points of absolute maximum of the function $Z_\varepsilon(\cdot)$, then from the uniform in $\theta \in \mathbf{K}$ weak convergence of the measures

$$\Xi_\varepsilon \Longrightarrow \Xi$$

follows

$$\begin{aligned} \mathbf{P}_\theta^{(\varepsilon)}\{u_\varepsilon \subset D\} &= \mathbf{P}_\theta^{(\varepsilon)}\{L_D(Z_\varepsilon) - L_{D^c}(Z_\varepsilon) > 0\} \\ &\longrightarrow \mathbf{P}\{L_D(Z) - L_{D^c}(Z) > 0\} = \mathbf{P}\{\hat{u}(\theta) \subset D\}. \end{aligned}$$

Particularly, if the falimy $\{\mathbf{P}_\theta^{(\varepsilon)}, \theta \in \Theta\}$ is LAN, then the limiting process is

$$Z(u) = \exp\left\{(u,\Delta) - \frac{1}{2}|u|^2\right\}, \quad u \in \mathbf{R}^d, \quad \mathcal{L}\{\Delta\} = \mathcal{N}(0,I),$$

i.e., the MLE $\hat{u}_\varepsilon = \varphi_\varepsilon(\theta)^{-1}(\hat{\theta}_\varepsilon - \theta)$ is asymptotically normal (see Theorem 1.6 below). A similar theorem was proved for the Bayesian estimator $\tilde{\theta}_\varepsilon$ with respect to the loss function $l(\varphi_\varepsilon(\theta)^{-1}(x-\theta))$ and prior density $\pi(\cdot)$. We assume that $l(\cdot) \in \mathbf{W}_p$ and, moreover, for all H sufficiently large and γ sufficiently small

$$\inf_{|u|>H} l(u) - \sup_{|u|\le H^\gamma} l(u) \ge 0 \tag{1.77}$$

and $\pi(\cdot)$ is a continuous positive bounded function. Denote this class of loss functions by $\tilde{\mathbf{W}}$.

Theorem 1.5. *Assume that the normalized likelihood ratio $Z_\varepsilon(u)$ has the following properties:*

1. *For any compact $\mathbf{K} \subset \Theta$, there correspond the numbers $a(\mathbf{K}) = a, B(\mathbf{K}) = B$ and functions $g^{\mathbf{K}}(y) \in \mathbf{G}$ such that*

   (a) *for some $\alpha > 0$ and all $\theta \in \mathbf{K}$*

   $$\sup_{|u_1|\le R, |u_2|\le R, u_1,u_2 \in U_\varepsilon} |u_2 - u_1|^{-\alpha} \mathbf{E}_\theta \left| Z_\varepsilon^{\frac{1}{2}}(u_2) - Z_\varepsilon^{\frac{1}{2}}(u_1)\right|^2 \le \\ \le B(1 + \mathbf{R}^a);$$

   (b) *for all $u \in U_\varepsilon$ and $\theta \in \mathbf{K}$*

   $$\mathbf{E}_\theta Z_\varepsilon^{\frac{1}{2}}(u) \le e^{-g(|u|)};$$

2. *The finite-dimensional distributions of $Z_\varepsilon(u)$, uniformly in $\theta \in \mathbf{K}$ as $\varepsilon \to 0$, converge to the marginal distributions of the random functions $Z(u)$;*

3. *The loss function $l(\cdot) \in \tilde{W}$ and the random function*

   $$\psi(y) = \int_{\mathbf{R}^d} l(y-u) \frac{Z(u)}{\int_{\mathbf{R}^d} Z(v)dv} du \tag{1.78}$$

   attains its (absolute) minimum value at the unique point $\tilde{u}(\theta)$.

Then the distribution of random variables $\varphi_\varepsilon(\theta)^{-1}(\tilde{\theta}_\varepsilon - \theta)$ converges uniformly in $\theta \in \mathbf{K}$ to the distribution of $\tilde{u}(\theta)$ and for any continuous loss function $l(\cdot) \in \mathbf{W}_p$ we have, uniformly in $\theta \in \mathbf{K}$,

$$\lim_{\varepsilon\to 0} \mathbf{E}_\theta l(\varphi_\varepsilon(\theta)^{-1}(\tilde{\theta}_\varepsilon - \theta)) = \mathbf{E}l(\tilde{u}(\theta)). \tag{1.79}$$

Proof. see [53], theorem 1.10.2.

In the regular cases when the family $\{\mathbf{P}_\theta^{(\varepsilon)}, \theta \in \Theta\}$ is LAN we shall use the following versions of these two theorems.

Theorem 1.6. *Let the parameter set Θ be an open convex subset of $\mathbf{R}^d$ and the functions $Z_\varepsilon(u)$ be continuous with probability 1 possessing the following properties:*

1. *For any compact $\mathbf{K} \subset \Theta$*

   (a) $\lim_{\varepsilon\to 0}\sup_{\theta\in\mathbf{K}}\ tr\, \varphi_\varepsilon(\theta)\varphi_\varepsilon(\theta)^T = 0$;

   (b) *for any $N > 0$*

$$\sup_{\varepsilon\in(0,1]}\sup_{\theta\in\mathbf{K}}\sup_{u\in U_\varepsilon} |u|^N \mathbf{E}_\theta Z_\varepsilon^{\frac{1}{2}}(u) < \infty;$$

   (c) *for some $\alpha > d, m > 0, B = B(\mathbf{K}), a = a(\mathbf{K})$*

$$\sup_{\theta\in\mathbf{K}}\ \sup_{|u_1|\le R, |u_2|\le R, u_1,u_2\in U_\varepsilon} |u_2 - u_1|^{-\alpha}\mathbf{E}_\theta\left|Z_\varepsilon^{\frac{1}{m}}(u_2) - Z_\varepsilon^{\frac{1}{m}}(u_1)\right|^m \le$$
$$\le B(1 + \mathbf{R}^a); \tag{1.80}$$

2. *The family $\{\mathbf{P}_\theta^{(\varepsilon)}, \theta \in \Theta\}$ is uniformly asymptotically normal.*

Then uniformly in $\theta \in \mathbf{K}$ the maximum likelihood estimator

(a) is consistent:

$$\mathbf{P}_\theta - \lim_{\varepsilon\to 0} \hat{\theta}_\varepsilon = \theta;$$

(b) is asymptotically normal:

$$\mathcal{L}_\theta\{\varphi_\varepsilon(\theta)^{-1}(\hat{\theta}_\varepsilon - \theta)\} \Longrightarrow \mathcal{N}(0, I);$$

(c) all the moments of the random variable $\varphi_\varepsilon(\theta)^{-1}(\hat{\theta}_\varepsilon - \theta)$ converge as $\varepsilon \to 0$ to the corresponding moments of Δ.

The proof is reduced to verification of the conditions of Theorem 1.4 (see [53], theorem 3.1.1).

REMARK 1.2. Denote by $\mathbf{\Phi}(\mathbf{K})$ the class of matrices $\{\varphi_\varepsilon(\theta), \theta \in \Theta\}$ such that for any compact $\mathbf{K} \subset \Theta$ the convergence

$$\lim_{\varepsilon\to 0} \varphi_\varepsilon(\theta_2)^{-1}\varphi_\varepsilon(\theta_1) = B(\theta_1, \theta_2) \tag{1.81}$$

is uniform in $\theta_i \in \mathbf{K}$ and the matrix $B(\theta_1, \theta_2)$ is continuous in θ_1.

If $\varphi_\varepsilon(\theta) \in \Phi(\mathbf{K})$ and the conditions of Theorem 1.4 are fulfilled, then the MLE is asymptotically efficient for any loss function $l(\varphi_\varepsilon(\theta_0)^{-1}x), l(\cdot) \in \mathbf{W}_p$ ([53], theorem 3.1.3).

Properties of Bayesian estimators in regular cases we describe for loss functions $l(\cdot) \in \tilde{\mathbf{W}}$ and continuous, positive, bounded prior density $\pi(\cdot)$. Moreover, we assume that the sets $\{y : l(y) < c\}$ are bounded for sufficiently small $c > 0$.

Theorem 1.7. *Let the conditions 1-2 of Theorem 1.6 be fulfilled but with some positive α in (1.80) and $\varphi_\varepsilon(\theta) \in \Phi(\mathbf{K})$. Then, uniformly in $\theta \in \mathbf{K}$, the Bayesian estimator*
(a) is consistent:

$$\mathbf{P}_\theta - \lim_{\varepsilon \to 0} \tilde{\theta}_\varepsilon = \theta;$$

(b) is asymptotically normal:

$$\mathcal{L}_\theta\{\varphi_\varepsilon(\theta)^{-1}(\tilde{\theta}_\varepsilon - \theta)\} \Longrightarrow \mathcal{N}(0, I);$$

(c) has converging moments $\mathbf{E}|\varphi_\varepsilon(\theta)^{-1}(\tilde{\theta}_\varepsilon - \theta)|^p$ to the moments $\mathbf{E}|\Delta|^p$;
(d) LAM for any loss function $l(\varphi_\varepsilon(\theta_0)^{-1}x)$ provided $l(\cdot) \in \mathbf{W}_p$;

The proof is a verification of the conditions of Theorem 1.5 (see [53], theorem 3.2.1).

Chapter 2

Asymptotic Properties of Estimators in Standard and Nonstandard Situations

Below we describe the properties of the MLEs $\hat{\theta}_\varepsilon$ and the BEs $\tilde{\theta}_\varepsilon$ of parameter θ, constructed by observations of the diffusion-type process

$$dX_t = S_t(\theta, X)dt + \varepsilon dW_t,\ X_0 = x_0,\ 0 \le t \le T$$

in a usual (standard) situation of a regular statistical experiment (the trend $S(\cdot)$ is a smooth function of its arguments, etc.) when it is possibly a consistent and asymptotically normal estimation of the unknown finite–dimensional parameter $\theta \in \Theta$ and, in nonstandard but "close to regular" situations when trend is not differentiable with respect to θ, the initial value X_0 is a random variable, the observations correspond to another equation, trend is the same for the different values of an unknown parameter, and the true value lies on the boundary of Θ. The problems of parameter estimation for the linear schemes are considered for a slightly different model which is not included in the general scheme of observations and corresponds to the "large signals" observed in white Gaussian noise.

2.1 LAM bound on the risks of estimators

We are given a probability space $\{\Omega, \mathcal{F}, \mathbf{P}\}$, a family $\hat{\mathcal{F}}$, an open, convex set $\Theta \subset \mathbf{R}^d$ and for every $\theta \in \Theta$ the diffusion-type process $\{X_t, \mathcal{F}_t, 0 \le t \le T\}$ with the stochastic differential

$$dX_t = S_t(\theta, X)dt + \varepsilon dW_t,\ X_0 = x_0,\ 0 \le t \le T, \tag{2.1}$$

where $S_t(\theta, X) \in \mathcal{P}$ and $\varepsilon \in (0,1]$. Let us denote by $\mathbf{P}_\theta^{(\varepsilon)}$ as the measure induced by the process (2.1) in the measurable space $(\mathcal{C}_T, \mathcal{B}_T)$ and by $x_t = x_t(\theta)$ the solution of the differential equation

$$\frac{dx_t}{dt} = S_t(\theta, x),\ \ x_0,\ \ 0 \le t \le T.$$

Our problem is how to estimate the unknown parameter θ by the observations (2.1) and to describe the asymptotic behavior of the estimator as $\varepsilon \longrightarrow 0$.

In this section we introduce the low boundary on the risk on any estimator of parameter θ and then we will use this bound to define asymptotically efficient estimators.

Theorem 2.1. *Suppose that the following holds:*

I. The functionals $S_t(\theta, X)$ satisfy the conditions $\mathcal{L}$.

II. There exists a d-dimensional vector–function

$$q(\theta) = \{q_t(\theta),\ 0 \le t \le T\} \in L_2[0,T]$$

such, that for any $v \in R^d$

$$\mathbf{P}_\theta - \lim_{\varepsilon \to 0} \left\| \left(S(\theta + \varepsilon v, X) - S(\theta, X)\right)\varepsilon^{-1} - (v, q(\theta)) \right\| = 0$$

and the matrix

$$I(\theta) = \int_0^T q_t(\theta) q_t(\theta)^T dt$$

is positively defined.

Then the family of measures $\left\{\mathbf{P}_\theta^{(\varepsilon)}, \theta \in \Theta\right\}$ is locally asymptotically normal in Θ with the normalizing matrix

$$\varphi_\varepsilon(\theta) = \varepsilon I(\theta)^{-\frac{1}{2}},$$

the vector

$$\Delta_\varepsilon(\theta, X) = I(\theta)^{-\frac{1}{2}} \varepsilon^{-1} \int_0^T q_t(\theta)[dX_t - S_t(\theta, X)dt]$$

in the representation (1.67) with $\mathbf{P}_\theta^{(\varepsilon)}$ probability 1, is equal to the vector

$$\Delta = I(\theta)^{-\frac{1}{2}} \int_0^T q_t(\theta) dW_t$$

which is Gaussian with the parameters $(0, I)$, I is a unit $d \times d$ matrix and for any estimator θ_ε^* of parameter θ and any $\theta_0 \in \Theta$, the following inequality holds:

$$\lim_{\delta \to 0} \liminf_{\varepsilon \to 0} \sup_{|\theta - \theta_0| \le \delta} \mathbf{E}_\theta l\left(\varepsilon^{-1} I(\theta_0)^{\frac{1}{2}} (\theta_\varepsilon^* - \theta)\right) \ge \mathbf{E} l(\Delta),$$

where $l(\cdot) \in \mathbf{W}_{e,2}$.

Proof. First we establish the LAN of this family and then cite Theorem 1.3.

By condition I, all the measures $\mathbf{P}_\theta^{(\varepsilon)}$, $\theta \in \Theta$ are equivalent and a normed likelihood ratio

$$\tilde{Z}_\varepsilon(v) \equiv L(\theta + \varepsilon v, \theta; X) = \frac{d\mathbf{P}_{\theta+\varepsilon v}^{(\varepsilon)}}{\mathbf{P}_\theta^{(\varepsilon)}}(X), \quad \theta + \varepsilon v \in \Theta,$$

on the observations (2.1) is (see Theorem 1.2, (1.56))

$$\begin{aligned} \ln \tilde{Z}_\varepsilon(v) &= \int_0^T \Delta f_t(v) dW_t - \frac{1}{2}\|\Delta f(v)\|^2 = \\ &= \left(v, \int_0^T q_t(\theta) dW_t\right) - \frac{1}{2}(I(\theta)v, v) + r_1(v) + r_2(v), \end{aligned}$$

where

$$\begin{aligned}\Delta f_t(v) &= [S_t(\theta+\varepsilon v,X)-S_t(\theta,X)]\,\varepsilon^{-1},\\ r_1(v) &= \int_0^T [\Delta f_t(v)-(v,q_t(\theta))]\,dW_t,\\ r_2(v) &= \frac{1}{2}\int_0^T \left[(\Delta f_t(v))^2-(v,q_t(\theta))^2\right]dt.\end{aligned}$$

For $r_1(v)$ using (1.7), we immediately obtain

$$\begin{aligned}&\mathbf{P}_\theta^{(\varepsilon)}\left\{\left|\int_0^T [\Delta f_t(v)-(v,q_t(\theta))]\,dW_t\right|>\delta_1\right\}\leq\\ &\leq \frac{\gamma}{\delta_1^2}+\mathbf{P}_\theta^{(\varepsilon)}\left\{\|\Delta f(v)-(v,q(\theta))\,\|^2>\gamma\right\}.\end{aligned}$$

Hence, choosing γ and δ_1 small but $\gamma\delta_1^{-2}$ also small we prove

$$\mathbf{P}_\theta-\lim_{\varepsilon\to 0} r_1(v)=0$$

for any fixed $v \in R^d$.

For the second term $r_2(v)$ we have

$$\begin{aligned}|r_2(v)|^2 &= \left(\int_0^T [\Delta f_t(v)-(v,q_t(\theta))]\,[\Delta f_t(v)+(v,q_t(\theta))]\,dt\right)^2\leq\\ &\leq 2\|\Delta f(v)-(v,q(\theta))\,\|^2\left(\|\Delta f(v)\|^2+\|\,(v,q(\theta))\,\|^2\right)\end{aligned}$$

and

$$\mathbf{P}_\theta-\lim_{\varepsilon\to 0} r_2(v)=0$$

because

$$\|\Delta f(v)\|^2\leq \|\Delta f(v)-(v,q(\theta))\,\|^2+(I(\theta)v,v)\,.$$

If we now put $v=I(\theta)^{-1/2}u$, then for the process $Z_\varepsilon(u)=\tilde{Z}_\varepsilon(I(\theta)^{-1/2}u)$ we obtain the representation (1.67) with the Gaussian vector $\Delta_\varepsilon(\theta,X)=\Delta$ and $\mathcal{L}_\theta\{\Delta\}=\mathcal{N}(0,I)$ for all $\varepsilon\in(0,1]$. So the family $\{\mathbf{P}_\theta^{(\varepsilon)},\theta\in\Theta\}$ is LAN in Θ and by Hajek-Le Cam's Theorem 1.3, the mentioned minimax bound on the risks of estimators is valid.

The matrix $I(\theta)$ will play the role of *Fisher's amount of information* and we shall call it also *Fisher's information matrix.*

REMARK 2.1 The proof of Hajek-Le Cam's theorem given in [53] could be slightly simplified here because the random vector Δ_ε in our problem is not only asymptotically normal, but also normal for all $\varepsilon\in(0,1]$. We give below this version of the proof from [53] also supposing for simplicity, that the loss function $l(u)=l_0(|u|), u\in\mathbf{R}^d$, where the function $l_0(x), x\in\mathbf{R}_+$ is monotone increasing, $l_0(x)<\exp(\gamma x^2)$ as $x\to\infty$ for any $\gamma>0$, $l(0)=0$ and continuous in point $x=0$.

Fix some $\theta_0\in\Theta$. By Theorem 2.1, the likelihood ratio

$$Z_\varepsilon(u)=L(\theta_0+\varphi_\varepsilon(\theta_0)u,\theta_0;X)$$

has representation

$$Z_\varepsilon(u) = \exp\{(u,\Delta) - \frac{1}{2}|u|^2 + \psi_\varepsilon(u,\theta_0)\}$$

for $u \in U_{\theta_0,\varepsilon} = \{u : \theta_0 + \varphi_\varepsilon(\theta_0)u \in \Theta\}$ with Gaussian Δ and (1.69). Introduce also the random field

$$Z(u) = \exp\{(u,\Delta) - \frac{1}{2}|u|^2\}, \quad u \in \mathbf{R}^d$$

and denote $V(M) = \pi^{d/2}\Gamma^{-1}(d/2+1)M^d$ the volume of a d-dimensional sphere of radius M. Below $l(\cdot)^a = l(\cdot) \wedge a$.

For any $\delta > 0$ and $M > 0$ from the nondegeneracy of the matrix $I(\theta_0, x)$ it follows that there exists $\varepsilon_1 = \varepsilon_1(\delta, M, I(\theta_0, x))$, such that for all $\varepsilon < \varepsilon_1$

$$S_M = \{u : |u| \le M\} \subset \{u : |I(\theta_0,x)^{-1/2}u| \le \delta\varepsilon^{-1}\}.$$

So we can write

$$\sup_{|\theta-\theta_0|<\delta} \mathbf{E}_\theta\, l(\varphi_\varepsilon(\theta_0)^{-1}(\theta_\varepsilon - \theta)) =$$

$$= \sup_{|I(\theta_0,x)^{-1/2}u|<\delta\varepsilon^{-1}} \mathbf{E}_{\theta_0} Z_\varepsilon(u)\, l(u_\varepsilon^* - u) \ge \sup_{|u|<M} \mathbf{E}_{\theta_0} Z_\varepsilon(u) l(u_\varepsilon^* - u)^a \ge$$

$$\ge V(M)^{-1} \int_{S_M} \mathbf{E}_{\theta_0} Z_\varepsilon(u)\, l(u_\varepsilon^* - u)^a\, du \equiv R_\varepsilon(u_\varepsilon^*, M),$$

where $u_\varepsilon^* = \varphi_\varepsilon(\theta_0)^{-1}(\theta_\varepsilon - \theta_0)$.

We now need the following known lemma (see [10], Theorem 1.5.4)

Lemma 2.1. *If the function $\xi(\varepsilon) \ge 0, \varepsilon \in [0,1]$ is continuous in probability at the point $\varepsilon = 0$ and $\mathbf{E}\xi(\varepsilon) = C$, then it is also continuous in the mean, i.e.*

$$\lim_{\varepsilon\to 0} \mathbf{E}|\xi(\varepsilon) - \xi(0)| = 0.$$

Proof. We have to prove the uniform integrability of $|\xi_n - \xi|$, where $\xi_n = \xi(\varepsilon_n), \xi = \xi(0)$ for any $\varepsilon_n \to 0$. Let $N > 0$, then

$$\mathbf{E}|\xi_n - \xi|\, \chi_{\{|\xi_n - \xi|>N\}} \le \mathbf{E}\xi_n\, \chi_{\{\xi_n>N\}} + \mathbf{E}\xi\, \chi_{\{\xi>N\}} =$$

$$= 2C - \mathbf{E}\xi_n\, \chi_{\{\xi_n\le N\}} - \mathbf{E}\xi\, \chi_{\{\xi\le N\}}.$$

The random variables $\xi_n\, \chi_{\{\xi_n\le N\}}$ are bounded by N , so

$$\lim_{n\to\infty} \mathbf{E}\xi_n\, \chi_{\{\xi_n\le N\}} = \mathbf{E}\xi\, \chi_{\{\xi\le N\}}.$$

Hence

$$\lim_{n\to\infty} \mathbf{E}|\xi_n - \xi|\, \chi_{\{|\xi_n-\xi|>N\}} \le 2\mathbf{E}\,\xi\, \chi_{\{\xi>\mathrm{N}\}}$$

and the last mathematical expectation tends to zero as $N \to \infty$.

Remember, that $\mathbf{E}_{\theta_0} Z_\varepsilon(u) = \mathbf{E}_{\theta_0} Z(u) = 1,\ Z_\varepsilon(u) > 0,\ Z(u) > 0$ and by (1.69) $\lim_{\varepsilon\to 0} Z_\varepsilon(u) = Z(u)$. This convergence, together with the boundedness of $l(\cdot)^a$ and inequality $\mathbf{E}|Z_\varepsilon(u) - Z(u)| \le 2$, allows us to write

$$R_\varepsilon(u_\varepsilon^*, M) = V(M)^{-1} \int_{S_M} \mathbf{E}_{\theta_0} Z(u)\, l(u_\varepsilon^* - u)^a\, du + o(1).$$

Denote the first term by $R(u_\varepsilon, M)$. We have

$$\begin{aligned} R(u_\varepsilon, M) &= V(M)^{-1} \int_{S_M} \mathbf{E}_{\theta_0} e^{(u,\Delta) - \frac{1}{2}|\Delta|^2} l(u_\varepsilon^* - u)^a\, du = \\ &= V(M)^{-1}\, \mathbf{E}_{\theta_0}\, e^{\frac{1}{2}|\Delta|^2} \int_{S_M} e^{-\frac{1}{2}|u-\Delta|^2} l(u_\varepsilon^* - u)^a\, du \end{aligned}$$

After changing the variables, $v = u - \Delta$, $v_\varepsilon^* = u_\varepsilon^* - \Delta$, we can write

$$\begin{aligned} V(M)\, R(v_\varepsilon^*, M) &= \mathbf{E}_{\theta_0}\, e^{\frac{1}{2}|\Delta|^2} \int_{|v+\Delta|<M} e^{-\frac{1}{2}|v|^2}\, l(v_\varepsilon^* - v)^a\, dv \ge \\ &\ge \mathbf{E}_{\theta_0}\, e^{\frac{1}{2}|\Delta|^2}\, \chi_{\{|\Delta|<M-\sqrt{M}\}} \int_{|v+\Delta|<M} e^{-\frac{1}{2}|v|^2} l(v_\varepsilon^* - v)^a\, dv \ge \\ &\ge \mathbf{E}_{\theta_0}\, e^{\frac{1}{2}|\Delta|^2}\, \chi_{\{|\Delta|<M-\sqrt{M}\}} \int_{|v|<\sqrt{M}} e^{-\frac{1}{2}|v|^2} l(v_\varepsilon^* - v)^a\, dv. \end{aligned}$$

Consider now minimizing the integral

$$J_M(\mathbf{b}) = \int_{S_{\sqrt{M}}} e^{-\frac{1}{2}|v|^2}\, l(\mathbf{b} - v)^a\, dv$$

with respect to $\mathbf{b} \in \mathbf{R}^d$. First we write the estimate

$$\begin{aligned} J_\infty(\mathbf{b}) - J_M(\mathbf{b}) &= \int_{|v|>\sqrt{M}} e^{-\frac{1}{2}|v|^2}\, l(\mathbf{b} - v)^a\, dv \le \\ &\le a(2\pi)^{d/2}\, \mathbf{P}_{\theta_0}^{(\varepsilon)}\{|\Delta| > \sqrt{M}\} < a\, d(2\pi)^{d/2}\, e^{-M/2}. \end{aligned}$$

So

$$J_M(\mathbf{b}) \ge J_\infty(\mathbf{b}) - a\, d\, (2\pi)^{d/2}\, e^{-M/2}.$$

Then to minimize

$$\begin{aligned} J_\infty(\mathbf{b}) &= \int e^{-\frac{1}{2}|v|^2}\, l(\mathbf{b} - v)^a\, dv = \\ &= \int e^{-\frac{1}{2}|y-\mathbf{b}|^2}\, l(y)^a\, dy \end{aligned}$$

we note that the problem has a spherical symmetry and $J_\infty(\mathbf{b}) = J_\infty(b)$, where $b = |\mathbf{b}|$. Hence, choosing the Cartesian coordinates, we can write

$$J_\infty(b) = \int_{\mathbf{R}^d} \exp\left\{-\frac{1}{2}(y_1 - b)^2 - \frac{1}{2}\sum_{j=2}^{d} y_j^2\right\}\, l(y)^a\, dy.$$

Direct calculation

$$\frac{\partial J_\infty(b)}{\partial b}\Big|_{b=0} = 0$$

and

$$\frac{\partial^2 J_\infty(b)}{\partial b^2} = \int [(y_1 - b)^2 + 1]\, e^{-\frac{1}{2}|y-b|^2}\, l(y)^a\, dy > 0$$

shows that the point of minimum is $b = 0$.

Therefore,

$$J_M(\mathbf{b}) \geq (2\pi)^{d/2}\mathbf{E}_{\theta_0}\, l(\Delta)^a \;-\; (2\pi)^{d/2}\, a\, d\, e^{-M/2}.$$

Finally, we have

$$R(v_\varepsilon^*, M) \geq$$

$$\geq (2\pi)^{d/2}[\mathbf{E}_{\theta_0} l(\Delta)^a - a d\, e^{-M/2}]\, V(M)^{-1} \int_{|x|<M-\sqrt{M}} e^{\frac{1}{2}|x|^2}\, e^{-\frac{1}{2}|x|^2}\, (2\pi)^{-d/2}\, dx =$$

$$= [\mathbf{E}_{\theta_0}\, l(\Delta)^a - a\, d\, e^{-M/2}] \frac{V(M - \sqrt{M})}{V(M)} \longrightarrow \mathbf{E}_{\theta_0}\, l(\Delta)^a$$

as $M \to \infty$. A second transition to the limit $a \to \infty$ proves the assertion.

REMARK 2.2. If the observed process is diffusion

$$dX_t = S_t(\theta, X_t) \;+\; \varepsilon\, dW_t, \quad X_0 = x_0,\; 0 \leq t \leq T,$$

then the following conditions are sufficient for the LAN of the family $\{\mathbf{P}_\theta^{(\varepsilon)},\ \theta \in \Theta\}$.

(R) *Function $S_t(\theta, x)$ has continuous bounded derivatives $\dot{S}_t(\theta, x)$, $S_t'(\theta, x)$, $\ddot{S}_t(\theta, x)$, $\dot{S}_t'(\theta, x)$ and the matrix*

$$I(\theta) = \int_0^T \dot{S}_t(\theta, x_t)\, \dot{S}_t(\theta, x_t)^T\, dt$$

is positive definite.

Here $x_t = x_t(\theta)$ is solution of the equation

$$\frac{dx_t}{dt} = S_t(\theta, x_t), \qquad x_0, \quad 0 \leq t \leq T,$$

point over the function corresponds to the differentiating with respect to θ, prim corresponds to the differentiating with respect to x and the index T - transposing. In this case we can put $q_t(\theta) = \dot{S}_t(\theta, x_t)$.

REMARK 2.3. Suppose that the observed process X_t, $0 \leq t \leq T$ belongs to the family of processes

$$dX_t = [S_t(X) \;+\; \theta\, \varepsilon\, g_t(X)]\, dt \;+\; \varepsilon\, dW_t, \quad X_0 = x_0,\; 0 \leq t \leq T,$$

where $\theta \in \Theta = \{\theta \;:\; |\theta - \theta_0| < m\} \subset R^1$, the functions $S_t(\cdot)$, $g_t(\cdot)$ satisfy the condition $\mathcal{L}$ and

$$I_0 = \int_0^T g_t(x)^2\, dt > 0.$$

Then the likelihood ratio admits the representation (θ_0 is a true value)

$$\begin{aligned} Z_\varepsilon(u) \;&\equiv\; \frac{\mathbf{P}^{(\varepsilon)}_{\theta_0+u/\sqrt{I_0}}}{\mathbf{P}^{(\varepsilon)}_{\theta_0}} = \\ &=\; \exp\left\{u\,\Delta \;-\; \frac{u^2}{2} + \;\psi_\varepsilon(u, \theta_0)\right\} = \\ &=\; Z(u)\, \exp\{\psi_\varepsilon(u, \theta_0)\}, \end{aligned}$$

where

$$\Delta = I_0^{-1/2} \int_0^T g_t(x)\, dW_t \sim \mathcal{N}(0,1)$$

and $\mathbf{P}_{\theta_0} - \lim_{\varepsilon\to 0} \psi_\varepsilon(u,\theta_0) = 0$.

Therefore, the above presented proof allows us to write the inequality

$$\sup_{|\theta-\theta_0|<m} \mathbf{E}_\theta l\left(I_0^{1/2}(\theta_\varepsilon - \theta)\right) \geq$$
$$\geq [\mathbf{E}_{\theta_0} l(\Delta)^a - a\, d\, e^{-M/2}]\left(1 - 1/\sqrt{M}\right) + o(1) \to \mathbf{E}_{\theta_0} l(\Delta)^a$$

where $M = m\, I_0^{1/2}$. So, if m is a free parameter (we can choose it), then for some m_0, ε_0 and a, we shall have the estimate

$$\sup_{|\theta-\theta_0|<m_0} \mathbf{E}_\theta l\left(I_0^{1/2}(\theta_\varepsilon - \theta)\right) \geq \frac{1}{2}\, \mathbf{E} l(\Delta)$$

for all $\varepsilon < \varepsilon_0$. Remember also that on the bounded interval, the function $l(\cdot)$ is also bounded.

2.2 Asymptotic behavior of estimators in the regular case

Below, we prove a "traditional" for the mathematical statistics theorem: if the problem is regular, then the maximum likelihood and a wide class of Bayes estimators are consistent, asymptotically normal, and asymptotically efficient.

We are given the same model of observations

$$dX_t = S_t(\theta, X)dt + \varepsilon dW_t, \quad X_0 = x_0, \quad 0 \leq t \leq T,$$

where $\theta \in \Theta \subset \mathbf{R}^d$, Θ is an open, bounded, convex set. Parameter ε is an "index of series", $\varepsilon \in (0,1]$ and the limit corresponds $\varepsilon \to 0$. The trend coefficient $S_t(\cdot, X)$ is a known function from the class $\mathcal{P}$. We have to estimate θ by the observations $X = \{X_t, 0 \leq t \leq T\}$. Introduce the notation : $x = \{x_t, 0 \leq t \leq T\}$, which is a solution of the equations

$$\frac{dx_t}{dt} = S_t(\theta, x), \quad x_0, \quad 0 \leq t \leq T,$$

the matrix

$$I(\theta, X) = \int_0^T \dot{S}_t(\theta, X)\dot{S}_t(\theta, X)^T dt,$$

where the vector $\dot{S}_t(\theta, X) = \frac{\partial}{\partial\theta} S_t(\theta, X)$. So the point over the function corresponds to the differentiability with respect to θ and the index T – transposing.

We study the properties of estimators under the following conditions.

I. *The coefficients* $\{S_t(\theta, X), 0 \leq t \leq T, \theta \in \Theta\}$ *satisfy the conditions* $\mathcal{L}$ *and the* *constants of these conditions do not depend on θ.*

II. *The random function $\{S_t(\theta, X), 0 \le t \le T\}$, $\theta \in \Theta$ is derivable on θ with $\mathbf{P}_\theta^{(\varepsilon)}$ probability 1 in $L_2[0,T]$ and the Fisher's information matrix $I(\theta) = I(\theta, x)$ is positive definite uniformly in $\theta \in \Theta$, i.e.*

$$0 < \inf_{\theta\in\Theta} \inf_{|l|=1} (I(\theta)l, l), \quad \sup_{\theta\in\Theta} \sup_{|l|=1} (I(\theta)l, l) < \infty.$$

III. *The derivative vector-process $\{\dot S_t(\theta, X), 0 \le t \le T\}$ is continuous in probability uniformly on $\theta \in \Theta$ in $L_2[0,T]$ on θ and on X in the point $\theta = \theta_0$ and* $X = \{x_t(\theta_0), 0 \le t \le T\}$, i.e.

(a) for any $\delta > 0$

$$\lim_{\varepsilon\to 0} \sup_{\theta_0\in\Theta} \mathbf{P}_{\theta_0}^{(\varepsilon)} \left\{ \sup_{\theta\in\Theta} \|\dot S(\theta, X) - \dot S(\theta, x)\| > \delta \right\} = 0;$$

(b)

$$\lim_{h\to 0} \sup_{\theta_0\in\Theta} \sup_{|\theta-\theta_0|<h} \|\dot S(\theta, x) - \dot S(\theta_0, x)\| = 0.$$

IV. *For some $m > \frac{d}{2}$,*

$$\sup_{\theta,\theta_1\in\Theta} \mathbf{E}_{\theta_1} \| |\dot S(\theta, X)|^m \|^2 < C. \tag{2.2}$$

V. *For any $\nu > 0$ and a compact $\mathbf{K} \subset \Theta$,*

$$\inf_{\theta\in\mathbf{K}} \inf_{|u|>\nu, \theta+u\in\Theta} \|S(\theta + u, x) - S(\theta, x)\| > 0.$$

The maximum likelihood estimator $\hat\theta_\varepsilon$ is defined by the equation

$$\frac{d\mathbf{P}_{\hat\theta_\varepsilon}^{(\varepsilon)}}{d\mathbf{P}_{\theta_1}^{(\varepsilon)}}(X) = \sup_{\theta\in\bar\Theta} \frac{d\mathbf{P}_\theta^{(\varepsilon)}}{d\mathbf{P}_{\theta_1}^{(\varepsilon)}}(X),$$

where θ_1 is some fixed value of $\theta, \theta_1 \in \Theta$. The likelihood ratio random field in our considerations has always a continuous modification.

Theorem 2.2. *Let conditions I–V be fulfilled, then uniformly on compacts $\mathbf{K} \subset \Theta$, the MLE $\hat\theta_\varepsilon$ as $\varepsilon \to 0$*

(a) is consistent

$$\mathbf{P}_\theta - \lim_{\varepsilon\to 0} \hat\theta_\varepsilon = \theta;$$

(b) is asymptotically normal

$$\mathcal{L}_\theta\{\varepsilon^{-1}I(\theta)^{\frac{1}{2}}(\hat{\theta}_\varepsilon-\theta)\}\longrightarrow\mathcal{N}(0,I);$$

(c) has converging moments for any $p>0$

$$\lim_{\varepsilon\to 0}\sup_{\theta\in\mathbf{K}}\left|\mathbf{E}_\theta|I(\theta)^{\frac{1}{2}}(\hat{\theta}_\varepsilon-\theta)|^p\varepsilon^{-p}-\mathbf{E}|\Delta|^p\right|=0,$$

where $\mathcal{L}\{\Delta\}\simeq\mathcal{N}(0,I)$, I – a unit $d\times d$ matrix.

(d) is LAM for loss functions $l(\cdot)\in\mathbf{W}_p$.

Proof. Below, we check the conditions of Theorem 1.6 using the following three lemmas.

Lemma 2.2. *Let conditions I–III be fulfilled, then the family $\{\mathbf{P}_\theta^{(\varepsilon)},\theta\in\Theta\}$ is uniformly asymptotically normal with the normalizing matrix $\varphi_\varepsilon(\theta)=\varepsilon I(\theta)^{-\frac{1}{2}}$ and the vector*

$$\Delta(\theta,X)=\varepsilon^{-1}\int_0^T\dot{S}_t(\theta,x)[dX_t-S_t(\theta,X)dt]$$

is under $\mathbf{P}_\theta^{(\varepsilon)}$ Gaussian, $\mathcal{L}_\theta\{\Delta(\theta,X)\}=\mathcal{N}(0,I(\theta))$.

Proof. Denote $\theta_{\varepsilon,u}=\theta_\varepsilon+\varphi_\varepsilon(\theta_\varepsilon)u_\varepsilon$, and suppose that $\theta_\varepsilon\in\mathbf{K}$ and $u_\varepsilon\to u$ as $\varepsilon\to 0$. All the measures $\mathbf{P}_\theta^{(\varepsilon)},\theta\in\Theta$ are equivalent (by I) and the normed likelihood ratio

$$Z_\varepsilon(u_\varepsilon)=L(\theta_{\varepsilon,u},\theta_\varepsilon;X)$$

could be written in the form

$$\begin{aligned}\ln Z_\varepsilon(u_\varepsilon) &= \varepsilon^{-1}\int_0^T[S_t(\theta_{\varepsilon,u},X)-S_t(\theta_\varepsilon,X)-(\varphi_\varepsilon(\theta_\varepsilon)u_\varepsilon,\dot{S}_t(\theta_\varepsilon,X))]dW_t+\\ &\quad+\left(I(\theta_\varepsilon)^{-\frac{1}{2}}u_\varepsilon,\int_0^T\dot{S}_t(\theta_\varepsilon,X)dW_t\right)-\frac{1}{2\varepsilon^2}\|S(\theta_{\varepsilon,u},X)-S(\theta_\varepsilon,X)\|^2.\end{aligned}$$

We now prove the uniform analog of Theorem 2.1. Put $q_t(\theta)=\dot{S}_t(\theta,x)$ and use the formula

$$S_t(\theta+h,X)-S_t(\theta,X)=\int_0^1(h,\dot{S}_t(\theta+hs,X))ds$$

in the following verifying

$$\begin{aligned}&\varepsilon^{-2}\|S(\theta_{\varepsilon,u},X)-S(\theta_\varepsilon,X)-(\varphi_\varepsilon(\theta_\varepsilon)u_\varepsilon,\dot{S}_t(\theta_\varepsilon,X))\|^2=\\ &=\varepsilon^{-2}\left\|\int_0^1(\varphi_\varepsilon(\theta_\varepsilon)u_\varepsilon,\dot{S}_t(\theta_\varepsilon+s\varphi_\varepsilon(\theta_\varepsilon)u_\varepsilon,X)-\dot{S}(\theta_\varepsilon,X))ds\right\|^2\le\\ &\le|I(\theta_\varepsilon)^{-\frac{1}{2}}u_\varepsilon|^2\sup_{\theta\in\mathbf{K}}\sup_{|v|<C\varepsilon}\|\dot{S}(\theta+v,X)-\dot{S}(\theta,X)\|^2\le\\ &\le C_1\sup_{\theta\in\mathbf{K}}\|\dot{S}(\theta,X)-\dot{S}(\theta,x)\|^2+\\ &\quad+C_1\sup_{\theta_0\in\mathbf{K}}\sup_{|\theta-\theta_0|<C\varepsilon}\|\dot{S}(\theta,x)-\dot{S}(\theta_0,x)\|^2,\end{aligned}$$

where for any $C > 0$ the last two norms as $\varepsilon \to 0$ tend to zero in probability by condition III.

The continuity of the matrix $I(\theta, X)$ on θ and on X at point $X = x$ also follows from III and the representation

$$\begin{aligned}
&I(\theta_\varepsilon, X) - I(\theta, x) = \\
&\quad = \int_0^T [\dot{S}_t(\theta_\varepsilon, X)\dot{S}_t(\theta_\varepsilon, X)^T - \dot{S}_t(\theta, x)\dot{S}_t(\theta, x)^T]dt = \\
&\quad = \int_0^T [(\dot{S}_t(\theta_\varepsilon, X) - \dot{S}_t(\theta, X))\dot{S}_t(\theta_\varepsilon, X)^T + (\dot{S}_t(\theta, X) - \dot{S}_t(\theta, x))\dot{S}_t(\theta_\varepsilon, X)^T + \\
&\quad + \dot{S}_t(\theta, x)(\dot{S}_t(\theta_\varepsilon, X) - \dot{S}_t(\theta_\varepsilon, x))^T + \dot{S}_t(\theta, x)(\dot{S}_t(\theta_\varepsilon, x) - \dot{S}_t(\theta, x)^T)]dt.
\end{aligned}$$

So, uniformly on $\theta \in \mathbf{K}$, we have the convergence

$$\mathbf{P}_\theta - \lim_{\varepsilon \to 0} I(\theta_\varepsilon, x)^{-\frac{1}{2}} I(\theta_\varepsilon, X) I(\theta_\varepsilon, x)^{-\frac{1}{2}} = I.$$

Hence, by Lemma 1.8, the stochastic integral

$$I(\theta)^{-\frac{1}{2}} \int_0^T \dot{S}_t(\theta, X) dW_t$$

is, uniformly on $\theta \in \mathbf{K}$, asymptotically normal with parameters $(0, I)$. This asymptotic normality also follows from the equality

$$\mathcal{L}\left\{ I(\theta)^{-\frac{1}{2}} \int_0^T \dot{S}_t(\theta, x) dW_t \right\} = \mathcal{N}(0, I)$$

and the uniform on $\theta \in \mathbf{K}$ convergence

$$\mathbf{P}_\theta - \lim_{\varepsilon \to 0} \|\dot{S}_t(\theta, X) - \dot{S}_t(\theta, x)\| = 0,$$

because for any $\delta > 0$ and $\gamma > 0$

$$\begin{aligned}
&\mathbf{P}_\theta^{(\varepsilon)} \left\{ \left| \int_0^T \dot{S}_t(\theta, X) dW_t - \int_0^T \dot{S}_t(\theta, x) dW_t \right| > \delta \right\} \leq \\
&\quad \leq \frac{\gamma}{\delta^2} + \mathbf{P}_\theta^{(\varepsilon)} \left\{ \|\dot{S}(\theta, X) - \dot{S}(\theta, x)\|^2 > \gamma \right\}.
\end{aligned}$$

Lemma 2.3. *Let conditions I-IV be fulfilled, then*

$$\sup_{\theta \in \mathbf{K}} \sup_{|u_i| < R, i=1,2} |u_2 - u_1|^{-2m} \mathbf{E}_\theta |Z_\varepsilon^{\frac{1}{2m}}(u_2) - Z_\varepsilon^{\frac{1}{2m}}(u_1)|^{2m} \leq C.$$

Proof. Put $\theta_i = \theta + \varphi_\varepsilon(\theta) u_i, i = 1, 2$;

$$\theta(l) = \theta_1 + (\theta_2 - \theta_1) l, \; l \in [0, 1], \quad \Delta S = \varepsilon^{-1}[S_t(\theta_2, X) - S_t(\theta_1, X)], \quad h = \theta_2 - \theta_1$$

and note that by changing the measure, we can write

$$\mathbf{E}_\theta|Z_\varepsilon^{\frac{1}{2m}}(u_2) - Z_\varepsilon^{\frac{1}{2m}}(u_1)|^{2m} = \mathbf{E}_{\theta_1}|L^{\frac{1}{2m}}(\theta_2,\theta_1;X) - 1|^{2m}.$$

The exponent index in $L(\cdot)$ by the Fubini theorem ([97], theorem 5.15) could be rewritten as

$$\begin{aligned}
&\int_0^T \Delta S dW_t - \frac{1}{2}\|\Delta S\|^2 = \\
&\quad = \varepsilon^{-1}\int_0^1\int_0^T (\dot{S}_t(\theta(l),X),h)dW_t dl - \varepsilon^{-1}\int_0^1\int_0^T \Delta S(\dot{S}_t(\theta(l),X),h)dtdl = \\
&\quad = 2m\int_0^1 (\psi(\theta(l)),h)dl,
\end{aligned}$$

where the vector-function $\psi(\cdot)$ is defined by the last equality.

Hence

$$\begin{aligned}
&\mathbf{E}_{\theta_1}|L^{\frac{1}{2m}}(\theta_2,\theta_1;X) - 1|^{2m} = \\
&\quad = \mathbf{E}_{\theta_1}\left|\exp\left\{\int_0^1 (\psi(\theta(l)),h)dl\right\} - 1\right|^2 \leq \\
&\quad \leq \mathbf{E}_{\theta_1}\left|\int_0^1 (\psi(\theta(l)),h)\exp\left\{\int_0^l (\psi(\theta(v)),h)dv\right\}dl\right|^{2m} \leq \\
&\quad \leq \int_0^1 \mathbf{E}_{\theta_1} L(\theta_2,\theta_1;X)(\psi(\theta(l)),h)^{2m} dl = \\
&\quad = (2m\varepsilon)^{-2m}\int_0^1 \mathbf{E}_{\theta_2}\left|\int_0^T \dot{S}_t(\theta(l),X),h)dW_t\right|^{2m} dl \leq \\
&\quad \leq \varepsilon^{-2m}\left(\frac{2m-1}{4m}\right)^m T^{m-1}\int_0^1 \mathbf{E}_{\theta_2}\int_0^T (\dot{S}_t(\theta(l),X),h)^{2m} dtdl \leq \\
&\quad \leq C(m,T)|u_2 - u_1|^{2m} \sup_{\theta,\theta_1\in\Theta} \mathbf{E}_{\theta_1}\||\dot{S}(\theta,x)|^m\|^2.
\end{aligned}$$

We use the condition IV to finish the proof.

Lemma 2.4. *Let conditions I–IV be fulfilled, then for any $p \in (0,1)$ and $\mathbf{K} \subset \Theta$, there exists a function $g(\mathbf{K},p,u) = g(|u|) \in \mathbf{G}$ such that*

$$\sup_{\theta\in\mathbf{K}} \mathbf{E}_\theta Z_{\varepsilon,\theta}^p(u) \leq \exp\{-g(|u|)\}.$$

Proof. An elementary inequality

$$a^2 \geq b^2 - 2|b(a-b)|$$

allows us to write

$$\begin{aligned}
&[S_t(\theta+u,X) - S_t(\theta,X)]^2 \geq \\
&\quad \geq [S_t(\theta+u,x) - S_t(\theta,x)]^2 - \\
&\quad -2|(S_t(\theta+u,x) - S_t(\theta,x))(|S_t(\theta+u,X) - S_t(\theta+u,x)| + \\
&\quad +|S_t(\theta,X) - S_t(\theta,x)|).
\end{aligned}$$

By condition II, as $|u| \to 0$, we have

$$\begin{aligned}
&\int_0^T [S_t(\theta+u,x)-S_t(\theta,x)]^2 dt = \\
&= \int_0^T \left| \int_0^1 (\dot{S}_t(\theta+lu,x),u)dl \right|^2 dt = \\
&= \int_0^T (\dot{S}_t(\theta,x);u)^2 dt + \int_0^T \left| \int_0^1 (\dot{S}_t(\theta+lu,x)-\dot{S}_t(\theta,x),u)dl \right|^2 + o(|u|^2) = \\
&= (I(\theta)u,u) + o(|u|^2) \geq |u|^2 (\inf_{|\lambda|=1} (I(\theta)\lambda,\lambda) + o(1)).
\end{aligned}$$

For all $\theta \in \mathbf{K}$, condition II and this inequality provide the two-sided inequality

$$\kappa|u|^2 \leq \int_0^T [S_t(\theta+u,x)-S_t(\theta,x)]^2 dt \leq A^2|u|^2 \tag{2.3}$$

with some constants $A > 0$ and $\kappa = \kappa(\mathbf{K}) > 0$.

Below, we use the Schwarz inequality

$$\begin{aligned}
&\left(\int_0^T [S_t(\theta+u,x)-S_t(\theta,x)][S_t(\theta,X)-S_t(\theta,x)]dt \right)^2 \leq \\
&\leq \int_0^T [S_t(\theta+u,x)-S_t(\theta,x)]^2 dt \int_0^T [S_t(\theta,X)-S_t(\theta,x)]^2 dt \leq \\
&\leq A^2|u|^2 \int_0^T [S_t(\theta,X)-S_t(\theta,x)]^2 dt.
\end{aligned}$$

By conditions I and V, we have

$$\int_0^T [S_t(\theta,X)-S_t(\theta,x)]^2 dt \leq C_1 \sup_{0\leq t\leq T} |X_t - x_t|^2 \leq C^2\varepsilon^2 \sup_{0\leq t\leq T} |W_t|^2.$$

Hence,

$$\int_0^T |S_t(\theta+u,x)-S_t(\theta,x)||S_t(\theta,X)-S_t(\theta,x)|dt \leq AC|u|\varepsilon \sup_{0\leq t\leq T} |W_t|. \tag{2.4}$$

Below, we denote $\Delta S_t = [S_t(\theta+\varphi_\varepsilon(\theta)u,X)-S_t(\theta,X)]\varepsilon^{-1}$ and use Hölder's inequality

$$\mathbf{E}(\xi\eta) \leq (\mathbf{E}|\xi|^{p_1})^{\frac{1}{p_1}} (\mathbf{E}|\eta|^{p_2})^{\frac{1}{p_2}}, \quad \frac{1}{p_1}+\frac{1}{p_2}=1, \quad p_1>1.$$

So, for any $q \in (0,p)$, we have

$$\begin{aligned}
\mathbf{E}Z^p_{\varepsilon,\theta}(u) &= \mathbf{E}\exp\left\{ -\frac{(p-q)}{2}\|\Delta S\|^2 \right\} \times \\
&\qquad \times \exp\left\{ p\int_0^T \Delta S_t dW_t - \frac{q}{2}\|\Delta S\|^2 \right\} \leq \\
&\leq \left(\mathbf{E}\exp\left\{ -\frac{(p-q)}{2}p_1\|\Delta S\|^2 \right\} \right)^{\frac{1}{p_1}} \times \\
&\qquad \times \left(\mathbf{E}\exp\left\{ p\,p_2 \int_0^T \Delta S_t dW_t - \frac{qp_2}{2}\|\Delta S\|^2 \right\} \right)^{\frac{1}{p_2}}.
\end{aligned}$$

Choose q and p_2 such that

$$p_2 = \frac{q}{p^2} > 1,$$

then by (1.9)

$$\mathbf{E}\exp\left\{\frac{q}{p}\int_0^T \Delta S_t dW_t - \frac{1}{2}\left\|\frac{q}{p}\Delta S\right\|^2\right\} \leq 1$$

and

$$\mathbf{E}_\theta Z_{\varepsilon,\theta}^p \leq \left(\mathbf{E}\exp\left\{-\frac{q(p-q)}{2(q-p^2)}\|\Delta S\|^2\right\}\right)^{\frac{q-p^2}{q}}.$$

Denote by

$$\gamma = \frac{q(p-q)}{2(q-p^2)} > 0.$$

Then, according to (2.3) and (2.4),

$$\begin{aligned}
&\mathbf{E}_\theta \exp\left\{-\gamma\|\Delta S\|^2\right\} \leq \\
&\leq \exp\left\{-\gamma\varepsilon^{-2}\int_0^T [S_t(\theta+\varphi_\varepsilon(\theta)u,x) - S_t(\theta,x)]^2 dt\right\} \times \\
&\times \mathbf{E}_\theta \exp\left\{2\gamma\varepsilon^{-2}\int_0^T (|S_t(\theta+\varphi_\varepsilon(\theta)u,X) - S_t(\theta+\varphi_\varepsilon(\theta)u,x)| + \right. \\
&\left. + |S_t(\theta,X) - S_t(\theta,x)|)|S_t(\theta+\varphi_\varepsilon(\theta)u,x) - S_t(\theta,x)|dt\right\} \leq \\
&\leq \exp\left\{-\gamma\kappa|I(\theta)^{-\frac{1}{2}}u|^2\right\}\mathbf{E}\exp\left\{4\gamma CA|u|\sup_{0\leq t\leq T}|W_t|\right\} \leq \\
&\leq \exp\left\{-\gamma\kappa_1|u|^2\right\}\left\{1 + 4\gamma CA|u|\sqrt{8\pi T}\exp\left(8\gamma^2C^2A^2T^2|u|^2\right)\right\}.
\end{aligned}$$

Choose now q to be sufficiently close to p in such a way that

$$\gamma = \frac{\kappa_1}{16C^2A^2T^2}.$$

Then

$$\begin{aligned}
\mathbf{E}_\theta Z_{\varepsilon,\theta}^p(u) &\leq \exp\left(-\frac{\kappa_1}{2}|u|^2\right)\left(1 + 4\gamma CA|u|\sqrt{8\pi T}\right)^{\frac{q-p^2}{q}} \leq \\
&\leq \exp\left(-\frac{\kappa_1}{2}|u|^2 + \left(\frac{q-p^2}{q}\right)4\gamma CA|u|\sqrt{8\pi T}|u|\right) = \exp(-g(|u|))
\end{aligned}$$

with $g(\cdot) \in \mathbf{G}$. Here we used the inequality $1 + x \leq e^x$.

The matrix B in (1.81) is

$$B(\theta_1,\theta_2) = I(\theta_2)^{\frac{1}{2}}I(\theta_1)^{-\frac{1}{2}}$$

and, by conditions II and III, this matrix is continuous in θ_1. So the normalizing matrix $\varphi_\varepsilon(\theta) \in \Phi(\mathbf{K})$ and by Remark 1.2, the MLE is LAN for any $l(\cdot) \in \mathbf{W}_p$.

Bayesian estimators. In the Bayesian approach, it is assumed that θ is a random vector. Denote by $\pi(y), y \in \Theta$ its prior density with respect to Lebesgue measure and suppose that $\pi(\cdot)$ is a positive, continuous, bounded function and the set Θ is an open bounded subset of $\mathbf{R}^d$. The loss function we choose as $l(\varphi_\varepsilon(\theta)^{-1}(v-y)), l(\cdot) \in \tilde{\mathbf{W}}$, so Equation (1.61) defining the BE $\tilde{\theta}_\varepsilon$ could be rewritten as

$$\int l(\varphi_\varepsilon(\theta)^{-1}(\tilde{\theta}_\varepsilon - y))p(y|X_t, 0 \le t \le T)dy =$$
$$= \inf_{v\in\Theta} \int l(\varphi_\varepsilon(\theta)^{-1}(v - y))p(y|X_t, 0 \le t \le T)dy.$$

So we have a class of BE $\tilde{\theta}_\varepsilon$ defined by a class of functions $l(\cdot) \in \tilde{\mathbf{W}}$.

Theorem 2.3. *Let conditions I–V be fulfilled but in condition IV $m \ge 1$ only, then, uniformly on $\theta \in \mathbf{K}$, the BE $\tilde{\theta}_\varepsilon$*

(a) *is consistent*

$$\mathbf{P}_\theta - \lim_{\varepsilon\to 0} \tilde{\theta}_\varepsilon = \theta,$$

(b) *is asymptotically normal*

$$\mathcal{L}_\theta\{\varphi_\varepsilon(\theta)^{-1}(\tilde{\theta}_\varepsilon - \theta)\} \Longrightarrow \mathcal{N}(0, I),$$

(c) *has converging moments: for any $p > 0$*

$$\lim_{\varepsilon\to 0} \sup_{\theta\in\mathbf{K}} \left|\mathbf{E}_\theta \left|I(\theta)^{\frac{1}{2}}(\tilde{\theta}_\varepsilon - \theta)\right|^p \varepsilon^{-p} - \mathbf{E}|\Delta|^p\right| = 0,$$

(d) *is LAM for loss functions $l(\cdot) \in \mathbf{W}_p$.*

Proof consists of verifying the conditions of Theorem 1.7. These conditions are fulfilled according to Lemmas 2.1 – 2.3.

Conditions of regularity. The meaning of "conditions of regularity" used in direct or implicit form in the above mentioned theorems could be clarified for the diffusion observations

$$dX_t = S(\theta, X_t)dt + \varepsilon dW_t, \quad X_0 = x_0, \quad 0 \le t \le T$$

with $\theta \in \Theta \subset \mathbf{R}^1$, as the following. Roughly speaking,

(a) The function $S(\theta, x)$ has two continuous bounded derivatives w.r.t. θ and x.

(b) The Fisher information $I(\theta)$ is positive.

(c) The initial value x_0 is nonrandom.

(d) The trend coefficient of observed process is really a member of the family $\{S(\theta, X_t), 0 \le t \le T\}, \theta \in \Theta$.

(e) The function

$$G(y,\theta)=\int_0^T[S(y,x_t)-S(\theta,x_t)]^2dt$$

has a unique minimum on $y\in\Theta$ at point $y=\theta$.

(f) The true value θ is an interior point of Θ.

The above-mentioned conditions (a) – (f) do not coincide with I – V and are only sufficient for the validity of the properties of the MLE and BE described in Theorems 2.2 and 2.3. For a better understanding of their role in the behavior of estimators below, we shall study the properties of the MLE and BE in the situations when they are violated (§2.4 – §2.8), but first we describe their properties for linear systems (§2.3).

2.3 Parameter estimation for linear systems

Let us consider several problems of parameter estimation which are included and excluded in the previous scheme.

Example 2.1. *(Ornstein-Uhlenbeck process)* Let $\theta\in(\alpha,\beta),|\alpha|+|\beta|<\infty$ and the observed process be

$$dX_t=\theta X_tdt+\varepsilon dW_t,\quad X_0=x_0,\quad 0\le t\le T,$$

then, for $x_0\neq 0$, all the conditions of Theorems 2.2 and 2.3 are fulfilled. Hence, the MLE and BE have all the properties mentioned in those theorems.
If we denote

$$\eta_\varepsilon=\left(\int_0^T X_t^2dt\right)^{-1}\int_0^T X_tdX_t$$

and introduce the sets

$$H_1=\{\omega:\ \eta_\varepsilon\in(\alpha,\beta)\},\ \ H_2=\{\omega:\ \eta_\varepsilon\le\alpha\},\ \ H_3=\{\omega:\ \eta_\varepsilon\ge\beta\},$$

then the MLE with $\mathbf{P}_\theta^{(\varepsilon)}$ probability 1 is

$$\hat\theta_\varepsilon=\eta_\varepsilon\chi_{\{H_1\}}+\alpha\chi_{\{H_2\}}+\beta\chi_{\{H_3\}}.$$

In the case $\alpha=-\infty,\beta=+\infty$ we have, of course, $\hat\theta_\varepsilon=\eta_\varepsilon$, but the constant $L=\theta$ in condition $\mathcal{L}$.

The stochastic process X_t converges to $x_t=x_0\exp\{\theta t\}$ uniformly on t and on $\theta\in\Theta$ The Fisher's information

$$I(\theta)=\frac{x^2}{2\theta}(e^{2\theta T}-1)$$

and so is uniformly on $\theta\in\mathbf{K}$, where $\mathbf{K}$ is arbitrarily compact in Θ

$$\mathcal{L}_\theta\{I(\theta)^{\frac12}\}\varepsilon^{-1}(\hat\theta_\varepsilon-\theta)\Longrightarrow\mathcal{N}(0,1).$$

Note that if $\theta \to 0$, then $I(\theta) \to x_0^2 T$ and for $x_0 \neq 0, \theta = 0$ the MLE is still asymptotically normal

$$\mathcal{L}_0\{\varepsilon^{-1}\hat{\theta}_\varepsilon\} \Longrightarrow \mathcal{N}\left(0, \frac{1}{x_0^2 T}\right).$$

The case $x_0 = 0$ and $\Theta = (-\infty, +\infty)$ corresponds to the representation

$$\hat{\theta}_\varepsilon = \theta + \frac{\int_0^T e^{\theta t} Y_t dW_t}{\int_0^T e^{2\theta t} Y_t^2 dt},$$

where the process $dY_t = e^{-\theta t} dW_t$, $Y_0 = 0$ does not depend on ε and the MLE is not even consistent (violation of conditions II and V).

Example 2.2. For a *linear diffusion-type process*

$$dX_t = \theta\left(\int_0^t X_s ds\right) dt + \varepsilon dW_t, \quad X_0 = x_0 \neq 0, 0 \leq t \leq T$$

with $\theta = (\alpha, \beta)$, we have the limit equation as

$$\frac{d^2 x_t}{dt^2} - \theta x_t = 0, \quad x_0, \quad \left.\frac{dx_t}{dt}\right|_{t=0} = 0.$$

Suppose, that $\theta < 0$, α and β are finite and denote by $\omega^2 = -\theta$. Then the solution of the limit equation is $x_t = x_0 \cos \omega t$ and

$$I(\theta) = \frac{x_0^2 T}{2\omega^2}\left(1 - \frac{\sin(2\omega T)}{2\omega T}\right).$$

All the conditions of Theorems 2.2 and 2.3 are fulfilled and the MLE $\hat{\theta}_\varepsilon$ and BE $\tilde{\theta}_\varepsilon$ are consistent and asymptotically normal:

$$\mathcal{L}\left\{\varepsilon^{-1}(\hat{\theta}_\varepsilon - \varepsilon)\right\} \Longrightarrow \mathcal{N}(0, I(\theta)^{-1}).$$

We also have the convergence of the moments and LAM of these estimators .

Example 2.3. Let the process $X_t, 0 \leq t \leq T$ be

$$dX_t = \frac{\theta}{\varepsilon} X_t dt + dW_t, \quad X_0 = x_0, \quad 0 \leq t \leq T, \tag{2.5}$$

where $\theta \in \Theta = (-\infty, +\infty)$. This model is not included in the scheme (2.1) but is interesting because it provides consistent estimation and corresponds to the large values of the "signal-to-noise ratio" with a large "signal".

The properties of MLE $\hat{\theta}_\varepsilon$ strongly depend on the values of θ and x_0.

The solution of Equation (2.5) is

$$X_t = x_0 e^{\frac{\theta}{\varepsilon} t} + \int_0^t \exp\left\{\frac{\theta}{\varepsilon}(t - s)\right\} dW_s. \tag{2.6}$$

First, let $\theta < 0$. We divide Equation (2.5) by $\sqrt{\varepsilon}$ and change the variable $s = t/\varepsilon$. So $s \in [0, T_\varepsilon]$ with $T_\varepsilon = T\varepsilon^{-1} \to \infty$. Denote also $Y_s = X_{s\varepsilon}\varepsilon^{-\frac{1}{2}}$. Then, for this process, we have

$$dY_s = \theta Y_s ds + dw_s, \quad Y_0 = \frac{x_0}{\sqrt{\varepsilon}}, \quad 0 \le s \le T_\varepsilon,$$

where the realizations of $w_s = W_{s\varepsilon}\varepsilon^{-\frac{1}{2}}$ depend on ε.

We also introduce the equation

$$d\tilde{Y}_s = \theta\tilde{Y}_s ds + dw_s, \quad \tilde{Y}_0 = 0, \quad 0 \le s \le T_\varepsilon. \tag{2.7}$$

For this linear system, we have an ergodic theorem (see [46] or [49]) : with $\mathbf{P}_\theta^{(\varepsilon)}$ probability 1

$$\frac{1}{T_\varepsilon}\int_0^{T_\varepsilon} \tilde{Y}_s^2 ds \longrightarrow -\frac{1}{2\theta}$$

and for the difference $Y_s - \tilde{Y}_s$ using the representation (2.6) we obtain

$$\int_0^{T_\varepsilon} (Y_s - \tilde{Y}_s)^2 ds = \frac{x_0^2}{\varepsilon}\int_0^{T_\varepsilon} e^{2\theta t} dt = \frac{x_0^2}{2\theta\varepsilon}(e^{2\theta T_\varepsilon} - 1).$$

Hence

$$\begin{aligned}
\frac{1}{T_\varepsilon}\int_0^{T_\varepsilon} Y_s^2 ds &= \frac{1}{T_\varepsilon}\int_0^{T_\varepsilon} (\tilde{Y}_s + Y_s - \tilde{Y}_s)^2 dt = \\
&= \frac{1}{T_\varepsilon}\int_0^{T_\varepsilon} \tilde{Y}_s^2 ds + \frac{2}{T_\varepsilon}\int_0^{T_\varepsilon} \tilde{Y}_s(Y_s - \tilde{Y}_s) ds + \frac{1}{T_\varepsilon}\int_0^{T_\varepsilon} (Y_s - \tilde{Y}_s)^2 ds = \\
&= \frac{1}{T_\varepsilon}\int_0^{T_\varepsilon} \tilde{Y}_s^2 ds + \frac{2x_0}{T_\varepsilon\sqrt{\varepsilon}}\int_0^{T_\varepsilon} \tilde{Y}_s e^{\theta s} ds + \frac{x_0^2}{2\theta T}(e^{2\theta T_\varepsilon} - 1) \longrightarrow \\
&\longrightarrow -\frac{1}{2\theta} - \frac{x_0^2}{2\theta T} = -\frac{1}{2\theta T}(x_0^2 + T).
\end{aligned} \tag{2.8}$$

By the central limit theorem for the stochastic integrals (Lemma 1.8), the convergence (2.8) provides

$$\mathcal{L}\left\{\frac{1}{\sqrt{T_\varepsilon}}\int_0^{T_\varepsilon} Y_s dw_s\right\} \Longrightarrow \mathcal{N}\left(0, -\frac{x_0^2 + T}{2\theta T}\right).$$

So we have established the asymptotic normality of the MLE $\hat{\theta}_\varepsilon$ in the case $\theta < 0$ because

$$\begin{aligned}
\varepsilon^{-1}(\hat{\theta}_\varepsilon - \theta) &= \frac{\int_0^T X_t dW_t}{\int_0^T X_t^2 dt} = \frac{1}{\varepsilon}\frac{\int_0^{T_\varepsilon} Y_s dw_s}{\int_0^{T_\varepsilon} Y_s^2 ds} = \\
&= \frac{1}{\varepsilon\sqrt{T_\varepsilon}}\frac{\frac{1}{\sqrt{T_\varepsilon}}\int_0^{T_\varepsilon} Y_s dw_s}{\frac{1}{T_\varepsilon}\int_0^{T_\varepsilon} Y_s^2 ds}
\end{aligned}$$

and

$$\mathcal{L}_\theta\left\{\varepsilon^{-\frac{1}{2}}(\hat{\theta}_\varepsilon - \theta)\right\} \Longrightarrow \mathcal{N}\left(0, -\frac{2\theta}{x_0^2 + T}\right).$$

Consider now the case $\theta > 0$ and at the beginning we suppose that $x_0 \neq 0$. Note that

$$\int_0^t e^{\frac{2\theta(t-s)}{\varepsilon}} ds = \frac{\varepsilon}{2\theta}(e^{\frac{2\theta t}{\varepsilon}} - 1).$$

Therefore

$$\mathbf{E}\left(\int_0^t e^{\frac{\theta(t-s)}{\varepsilon}} dw_s\right)^2 = \frac{\varepsilon}{2\theta}(e^{\frac{2\theta t}{\varepsilon}} - 1)$$

and from the representation (2.6), we obtain

$$\varepsilon^{-1} e^{-\frac{2\theta T}{\varepsilon}} \mathbf{E}_\theta \int_0^T (X_t - x_0 e^{\frac{\theta t}{\varepsilon}})^2 dt \longrightarrow 0.$$

Hence

$$\mathbf{P}_\theta - \lim_{\varepsilon \to 0} \varepsilon^{-1} e^{-\frac{2\theta T}{\varepsilon}} \int_0^T X_t^2 dt = \frac{x_0^2}{2\theta}$$

and by Lemma 1.8, we have

$$\mathcal{L}_\theta\{\varepsilon^{-\frac{1}{2}} e^{-\frac{\theta T}{\varepsilon}}(\hat{\theta}_\varepsilon - \theta)\} \Longrightarrow \mathcal{N}\left(0, \frac{2\theta}{x_0^2}\right).$$

It now remains to consider the case $\theta > 0$ and $x_0 = 0$. Once more, we change the variables $s = \varepsilon^{-1} t$, $w_s = \varepsilon^{-1/2} W_{s\varepsilon}$, $Y_s = \varepsilon^{-1/2} X_{s\varepsilon}$. Then for the process Y_s, we have the same equation

$$dY_s = \theta Y_s ds + dw_s, \quad Y_0 = 0, \quad 0 \le s \le T_\varepsilon = T\varepsilon^{-1} \longrightarrow \infty,$$

but its solution with probability 1 goes to infinity.

Introduce a sequence ($\varepsilon \to 0$) of processes

$$\xi_t = \int_0^t e^{-\theta s} dw_s, \quad \eta_t = \int_0^t e^{\theta s} dw_s.$$

We have

$$\mathcal{L}\{\xi_t\} \Longrightarrow \mathcal{L}\{\xi\} = \mathcal{N}\left(0, \frac{1}{2\theta}\right),$$

where

$$\xi = \int_0^\infty e^{-\theta s} d\tilde{w}_s$$

with some Wiener process $\tilde{w}_s, s \ge 0$.

Integrating by parts, we have

$$\begin{aligned} e^{-2\theta T} \int_0^T Y_s^2 ds &= e^{-2\theta T} \int_0^T e^{2\theta s} \xi_s^2 ds = \\ &= \frac{e^{-2\theta T} e^{2\theta T}}{2\theta} \xi_T^2 - \frac{e^{-2\theta T}}{\theta} \int_0^T e^{2\theta s} \xi_s d\xi_s - \frac{e^{-2\theta T}}{\theta} \int_0^T e^{2\theta t} e^{-2\theta t} dt = \\ &= \frac{\xi_T^2}{2\theta} - \frac{1}{\theta} e^{-2\theta T} \int_0^T e^{\theta s} \xi_s dw_s - \frac{T e^{-2\theta T}}{\theta}. \end{aligned}$$

This equality together with

$$\mathbf{E}\left(\int_0^T e^{\theta s}\xi_s dw_s\right)^2 = \int_0^T e^{2\theta s}\mathbf{E}\xi_s^2 ds = \frac{1}{2\theta}\int_0^T e^{2\theta s}(1-e^{-2\theta s})ds$$

by Lemma 1.8 provides

$$\mathcal{L}_\theta\left\{e^{-2\theta T}\int_0^T Y_s^2 ds\right\} \Longrightarrow \mathcal{L}\left\{\frac{\xi^2}{2\theta}\right\}.$$

For the stochastic integral in the assumptions made by Itô formula, we have

$$\begin{aligned}\int_0^T Y_s dw_s &= \int_0^T e^{\theta s}\xi_s dw_s = \int_0^T \xi_s d\eta_s = \\ &= \xi_T\eta_T - T - \int_0^T \eta_s d\xi_s = \xi_T\eta_T - T - \int_0^T \eta_s e^{-\theta s} dw_s.\end{aligned}$$

Direct calculation gives

$$\mathcal{L}\{e^{-\theta T}\eta_T\} \longrightarrow \mathcal{L}\{\eta\} - \mathcal{N}\left(0, \frac{1}{2\theta}\right)$$

and at the same time,

$$\mathbf{E}(\xi\eta) = \lim_{T\to\infty}\mathbf{E}(\xi_T\eta_T e^{-\theta T}) = \lim_{T\to\infty} Te^{-\theta T} = 0.$$

Hence, the limit distribution of the pair $(\xi_T, e^{-\theta T}\eta_T)$ is a Gaussian distribution of two independent variables. So

$$\mathcal{L}\left\{\frac{\xi_T}{\eta_T}e^{\theta T}\right\} \Longrightarrow \mathcal{L}\{\zeta\},$$

where ζ has a Cauchy distribution.

Finally

$$\mathcal{L}_\theta\{2\theta e^{\frac{\theta T}{\varepsilon}}(\hat\theta_\varepsilon - \theta)\} \Longrightarrow \mathcal{L}\{\zeta\}.$$

Collecting all these properties of the MLE, we formulate the following theorem.

Theorem 2.4. *The MLE $\hat\theta_\varepsilon$ as $\varepsilon \to 0$ has the properties:*

1. *For $\theta < 0$*

$$\mathcal{L}_\theta\{\varepsilon^{-\frac{1}{2}}(\hat\theta_\varepsilon - \theta)\} \Longrightarrow \mathcal{N}\left(0, -\frac{2\theta}{x_0^2 + T}\right).$$

2. *For $\theta > 0$ and $x_0 \neq 0$*

$$\mathcal{L}_\theta\{\varepsilon^{-\frac{1}{2}}e^{\frac{\theta T}{\varepsilon}}(\hat\theta_\varepsilon - \theta)\} \Longrightarrow \mathcal{N}\left(0, \frac{2\theta}{x_0^2}\right).$$

3. *For $\theta > 0$ and $x_0 = 0$*

$$\mathcal{L}_\theta\{2\theta e^{\frac{\theta T}{\varepsilon}}(\hat\theta_\varepsilon - \theta)\} \Longrightarrow \mathit{Cauchy}.$$

Example 2.4. We can also have any polynomial rate of convergence of the MLE to a true value, as in the case $T \to \infty$ (see, for example, [65]).

Let us consider the process

$$dX_t = \frac{\theta}{\varepsilon}\,|X_t|^\beta\,dt + dW_t,\ \ X_0 = x_0,\ 0 \le t \le T,$$

where $\beta \in (0,1)$ and θ is an unknown parameter, $\theta \in (0,\infty)$. The MLE is

$$\hat{\theta}_\varepsilon = \max\left(0,\ \frac{\int_0^T |X_t|^\beta\,dX_t}{\int_0^T |X_t|^{2\beta}\,dt}\right).$$

Denote

$$\sigma^2(\theta) = \theta^{\frac{2}{1-\beta}} \left(\frac{1-\beta}{3-\beta}\,T\right)^{\frac{3-\beta}{1-\beta}}.$$

Then we have the convergence

$$\mathcal{L}\{\varepsilon^{-\frac{\beta}{1-\beta}}\,(\hat{\theta}_\varepsilon - \theta)\ \} \Longrightarrow \mathcal{N}(0,\sigma^2(\theta)).$$

To prove this, we introduce the stochastic process $Y_t = X_t\,\varepsilon^{\frac{1}{1-\beta}}$ which admits the stochastic differential

$$dY_t = \theta|Y_t|^\beta dt + \varepsilon^{\frac{1}{1-\beta}}\,dW_t, Y_0 = 0,\ 0 \le t \le T.$$

It can be shown that this process converges to the function

$$y_t = (\theta(1-\beta)t)^{\frac{1}{1-\beta}}\,,\ 0 \le t \le T,$$

which is a solution of the limit equation ($\varepsilon = 0$). Hence,

$$\mathbf{P}_\theta - \lim_{\varepsilon\to 0} \varepsilon^{-\frac{2\beta}{1-\beta}} \int_0^T |X_t|^{2\beta} dt = \sigma^2(\theta)$$

and this convergence provides the above-mentioned asymptotic normality of the MLE.

Now, for any $N > 0$, we can put $\beta = N/(1+N)$ and have the convergence

$$\mathcal{L}\{\varepsilon^N(\hat{\theta}_\varepsilon - \theta)\} \Longrightarrow \mathcal{N}(0,\sigma^2(\theta)).$$

2.4 Nondifferentiable and "too differentiable" trends

Conditions III–V of Theorem 2.1 are formulated in terms of the first derivative of the function $S_t(\theta, X)$ with respect to θ and this derivative is used in the proof of Lemmas 2.1 and 2.2. The derivative $\dot{S}_t(\cdot)$ is also used in

$$\mathcal{L}_\theta\{(\hat{\theta}_\varepsilon - \theta)\varepsilon^{-1}\} \Longrightarrow \mathcal{N}\left(0, (\int_0^T \dot{S}_t(\theta, x)^2 dt)^{-1}\right).$$

So the following questions arise: is it necessary to have a first derivative of the trend for the asymptotic normality of the MLE and BE and what changes in the properties of estimators if Fisher information $I(\theta) = 0$?

Below we propose an example of parameter estimation by observations of the diffusion-type process with nondifferentiable drift, and study the properties of estimators when $I(\theta) = 0$.

Nondifferentiable trend. Let the observed process be

$$dX_t = X_{\theta t}\, dt + \varepsilon\, dW_t, \quad X_0 = x_0 \neq 0, \quad 0 \leq t \leq T, \tag{2.9}$$

where $\theta \in (\alpha, \beta) = \Theta$, $0 < \alpha < \beta < 1$ and we have to estimate θ by the observations $\{X_t, 0 \leq t \leq T\}$.

From the representation

$$X_t = x_0 + \int_0^t X_{\theta s} ds + \varepsilon W_t \tag{2.10}$$

we find that the process X_t has the same smoothness as the Wiener process, so the derivative of $X_{\theta t}$ by θ does not exist in any sense for $t \in (0, T]$.

Denote by x_t, the solution of

$$\frac{dx_t}{dt} = x_{\theta t}, \quad x_0, \quad 0 \leq t \leq T$$

and put

$$\sigma(\theta)^2 = \int_0^T t^2 x_{\theta^2 t}^2 dt.$$

Below $l(\cdot)$ is a loss function from the class $\mathbf{W}_{e,2}$ and $\xi \sim \mathcal{N}(0,1)$. As before, $\mathbf{P}_\theta^{(\varepsilon)}$ denotes the probability measure induced in $(\mathcal{C}_T, \mathcal{B}_T)$ by the observed process.

The Hajek-Le Cam's inequality is given by the following theorem.

Theorem 2.5. *For any estimator $\bar{\theta}_\varepsilon$ of parameter θ, we have*

$$\lim_{\delta \to 0} \liminf_{\varepsilon \to 0} \sup_{|\theta - \theta_0| < \delta} \mathbf{E}_\theta l(\sigma(\theta_0)\varepsilon^{-1}(\bar{\theta}_\varepsilon - \theta)) \geq \mathbf{E} l(\xi). \tag{2.11}$$

Proof of this theorem consist of the verification of LAN of the family $\{\mathbf{P}_\theta^{(\varepsilon)}, \theta \in \Theta\}$ and then applying it to Theorem 2.1. Below, we establish, the uniform on $\theta \in \mathbf{K}$ LAN for this model, where $\mathbf{K}$ is an arbitrary compact in Θ.

Theorem 2.6. *The MLE $\hat{\theta}_\varepsilon$ as $\varepsilon \to 0$ is uniformly on $\theta \in \mathbf{K}$*

(a) consistent:

$$\mathbf{P}_\theta - \lim_{\varepsilon \to 0} \hat{\theta}_\varepsilon = \theta;$$

(b) asymptotically normal:

$$\mathcal{L}_\theta\{\varepsilon^{-1}\sigma(\theta)(\hat{\theta}_\varepsilon - \theta)\} \Longrightarrow \mathcal{N}(0,1);$$

(c) LAM for $l(\cdot) \in \mathbf{W}_p$.

Proof is similar to the proof of Theorem 2.2 and is based on the corresponding three lemmas and Theorem 1.6.

Lemma 2.5. *The family of measures* $\{\mathbf{P}_\theta^{(\varepsilon)}, \theta \in \Theta\}$ *is uniformly on* $\theta \in \mathbf{K}$ *LAN with the normalizing function* $\varphi_\varepsilon(\theta) = \sigma(\theta)^{-1}\varepsilon$ *and*

$$\Delta(\theta, X) = \sigma(\theta)^{-1}\varepsilon^{-1}\int_0^T t x_{\theta^2 t}[dX_t - X_{\theta t}dt].$$

Proof. Denote $\theta_u = \theta + \varphi_\varepsilon(\theta)u$ and represent the normed likelihood ratio $Z_\varepsilon(u_\varepsilon) = L(\theta + \varphi_\varepsilon(\theta)u_\varepsilon, \theta, X^\cdot)$ as

$$Z_\varepsilon(u_\varepsilon) = \exp\left\{u\Delta(\theta, X) - \frac{1}{2}u^2 + r_1 + r_2\right\},$$

where $r_1 = r_1(\theta, u, u_\varepsilon, X)$ and $r_2 = r_2(\theta, u, u_\varepsilon, X)$ are defined by

$$\begin{aligned} r_1(\theta, u_\varepsilon, X) &= (u_\varepsilon - u)\sigma(\theta)^{-1}\int_0^T t x_{\theta^2 t} dW_t + \\ &\quad + \varepsilon^{-1}\int_0^T (X_{\theta_{u_\varepsilon} t} - X_{\theta t} - t u_\varepsilon \varphi_\varepsilon(\theta) x_{\theta^2 t})dW_t, \\ r_2(\theta, u_\varepsilon, X) &= \varepsilon^{-2}\int_0^T [X_{\theta_{u_\varepsilon} t} - X_{\theta t}]^2 dt - u^2. \end{aligned}$$

Note also that

$$\mathcal{L}_\theta\{\Delta(\theta, X)\} = \mathcal{N}(0, 1).$$

Rewrite r_2 in the form

$$\begin{aligned} r_2 &= \varepsilon^{-2}\int_0^T (X_{\theta_{u_\varepsilon} t} - X_{\theta t})^2 - \\ &\quad - \varepsilon^{-2}\varphi_\varepsilon(\theta)^2 u_\varepsilon^2 \int_0^T x_{\theta^2 t}^2 t^2 dt + u_\varepsilon^2 - u^2 \end{aligned}$$

and consider separately the first two terms. We have

$$\begin{aligned} &\varepsilon^{-2}\mathbf{E}_\theta \int_0^T [X_{\theta_{u_\varepsilon} t} - X_{\theta t} - u_\varepsilon \varepsilon t \sigma(\theta)^{-1} x_{\theta^2 t}]^2 dt = \\ &= \varepsilon^{-2}\mathbf{E}_\theta \int_0^T \left\{\int_{\theta t}^{\theta_{u_\varepsilon} t} [X_{\theta s} - x_{\theta^2 t}]ds + \varepsilon[W_{\theta_{u_\varepsilon} t} - W_{\theta t}]\right\}^2 dt \le \\ &\le \frac{2\sigma(\theta)^{-1}|u_\varepsilon|T}{\varepsilon}\int_0^T\int_{\theta t}^{\theta_{u_\varepsilon} t} \mathbf{E}_\theta[X_{\theta s} - x_{\theta^2 t}]^2 ds dt + 2T|u_\varepsilon|\sigma(\theta)^{-1}\varepsilon. \qquad (2.12) \end{aligned}$$

For the last mathematical expectation, we obtain the estimate

$$\begin{aligned} \sup_{\theta t \le s \le \theta_{u_\varepsilon} t} \mathbf{E}_\theta |X_{\theta s} - x_{\theta^2 t}|^2 &\le 2 \sup_{\theta t \le s \le \theta_{u_\varepsilon} t} \mathbf{E}_\theta |X_{\theta s} - x_{\theta s}|^2 + \\ &\quad + 2 \sup_{\theta t \le s \le \theta_{u_\varepsilon} t} |x_{\theta s} - x_{\theta^2 t}|. \end{aligned}$$

Convergence to zero of the last term follows from the uniform continuity of the function x_t and for the first one we have

$$\begin{aligned}\mathbf{E}_\theta|X_t - x_t|^2 &\leq 2\mathbf{E}_\theta\left(\int_0^t (X_{\theta s} - x_{\theta s})ds\right)^2 + 2\varepsilon^2\mathbf{E}W_t^2 \leq \\ &\leq 2T\int_0^t \mathbf{E}_\theta(X_{\theta s} - x_{\theta s})^2 ds + 2\varepsilon^2 t \leq \\ &\leq \frac{2T}{\theta}\int_0^{\theta t} \mathbf{E}_\theta(X_v - x_v)^2 dv + 2\varepsilon^2 T < \\ &< \frac{2T}{\theta}\int_0^t \mathbf{E}_\theta(X_v - x_v)^2 dv + 2\varepsilon^2 T,\end{aligned}$$

so we can apply Lemma 1.11 to prove convergence to zero.

Collecting all these estimates, we obtain

$$\lim_{\varepsilon\to 0} \varepsilon^{-2} \sup_{\theta\in\mathbf{K}} \mathbf{E}_\theta\|X_{\theta_{u_\varepsilon} t} - X_{\theta t} - u_\varepsilon\varphi_\varepsilon(\theta)tx_{\theta^2 t}\|^2 = 0$$

for any sequence $u_\varepsilon \to u$. So we have established the convergence

$$\lim_{\varepsilon\to 0}\sup_{\theta\in\mathbf{K}} \mathbf{E}_\theta|r_2| = 0.$$

Tchebyshev inequality provides

$$\begin{aligned}\mathbf{P}_\theta^{(\varepsilon)}\{|r_1| > \delta\} &\leq \frac{\mathbf{E}r_1^2}{\delta^2} \leq 2\frac{(u_\varepsilon - u)^2}{\delta^2} + \\ &+ \frac{2}{\delta^2}\mathbf{E}_\theta\|X_{\theta_{u_\varepsilon} t} - X_{\theta t} - u_\varepsilon\varphi_\varepsilon(\theta)tx_{\theta^2 t}\|^2 \longrightarrow 0\end{aligned}$$

also uniformly on $\theta \in \mathbf{K}$.

Lemma 2.6. *For any $R > 0$ and $\mathbf{K} \subset \Theta$ we have*

$$\sup_{\theta\in\mathbf{K}}\sup_{|u_i|<R} |u_2 - u_1|^{-2}\mathbf{E}_\theta|Z_\varepsilon^{\frac{1}{4}}(u_2) - Z_\varepsilon^{\frac{1}{4}}(u_1)|^4 \leq B(1 + R^6)$$

with some constant $B > 0$.

Proof. Denote

$$\theta_i = \theta + \varphi_\varepsilon(\theta)u_i,\ i = 1,2; \quad \Delta X_t = X_{\theta_1 t} - X_{\theta_2 t};$$

$$Y_t = \exp\left\{\frac{1}{4\varepsilon}\int_0^t \Delta X_s dW_s - \frac{1}{8\varepsilon^2}\int_0^t (\Delta X_s)^2 dt\right\}.$$

The process Y_t satisfies with $\mathbf{P}_{\theta_2}^{(\varepsilon)}$ probability 1 to equality

$$Y_T = \left(\frac{d\mathbf{P}_{\theta_1}^{(\varepsilon)}}{d\mathbf{P}_{\theta_2}^{(\varepsilon)}}(X)\right)^{\frac{1}{4}}$$

and, by the Itô formula (1.30), admits the representation

$$dY_t = -\frac{3}{32\varepsilon^2}(\Delta X_t)^2 Y_t dt + \frac{1}{4\varepsilon}\Delta X_t Y_t dW_t, \quad Y_0 = 1.$$

Hence,

$$Y_T = 1 - \frac{3}{32\varepsilon^2}\int_0^T (\Delta X_t)^2 Y_t dt + \frac{1}{4\varepsilon}\int_0^T \Delta X_t Y_t \, dW_t.$$

Below, we use the inequality $|a+b|^4 \le 8a^4 + 8b^4$ and the Schwarz inequality.

$$\begin{aligned}
&\mathbf{E}_\theta |Z_\varepsilon^{\frac14}(u_2) - Z_\varepsilon^{\frac14}(u_1)|^4 = \\
&= \mathbf{E}_\theta \frac{d\mathbf{P}_{\theta_2}^{(\varepsilon)}}{d\mathbf{P}_\theta^{(\varepsilon)}} |1 - Y_T|^4 = \mathbf{E}_{\theta_2}|1 - Y_T|^4 \le \\
&\le 8\mathbf{E}_{\theta_2}\left|\frac{3}{32\varepsilon^2}\int_0^T (\Delta X_t)^2 Y_t dt\right|^4 + 8\mathbf{E}_{\theta_2}\left|\frac{1}{4\varepsilon}\int_0^T \Delta X_t Y_t dW_t\right|^4 \le \\
&\le \frac{8\cdot 3^4 \cdot T^3}{32^4 \varepsilon^8}\int_0^T \mathbf{E}_{\theta_2} Y_t^4 (\Delta X_t)^8 dt + \frac{9T}{8\varepsilon^4}\int_0^T \mathbf{E}_{\theta_2} Y_t^4 (\Delta X_t)^4 dt = \\
&= \frac{C_1}{\varepsilon^8}\int_0^T \mathbf{E}_{\theta_1}(\Delta X_t)^8 dt + \frac{C_2}{\varepsilon^4}\int_0^T \mathbf{E}_{\theta_1}(\Delta X_t)^4 dt.
\end{aligned} \tag{2.13}$$

Further,

$$\begin{aligned}
\mathbf{E}_{\theta_1}(\Delta X_t)^8 &= \left\{\int_{\theta_1 t}^{\theta_2 t} X_{\theta_1 s} ds + \varepsilon[W_{\theta_2 t} - W_{\theta_1 t}]\right\}^8 \le \\
&\le C_3[(\theta_2 - \theta_1)t]^7 \int_{\theta_1 t}^{\theta_2 t} \mathbf{E}_{\theta_1}(X_{\theta_1 s})^8 ds + C_4 \varepsilon^8 t^4 (\theta_2 - \theta_1)^4.
\end{aligned}$$

Using the representation (2.10) we derive the inequality

$$\mathbf{E}_{\theta_1}(X_t)^8 \le a + b\int_0^t \mathbf{E}_{\theta_1}(X_{\theta_1 s})^8 ds \le a + \frac{b}{\theta_1}\int_0^t \mathbf{E}_{\theta_1}(X_s)^8 ds$$

with some constants $a = a(x_0, T) > 0$ and $b = b(T) > 0$. So the mathematical expectation $\sup_{t,\theta} \mathbf{E}_\theta (X_t)^8$ is bounded by Lemma 1.11.

Hence,

$$\mathbf{E}_{\theta_1}(\Delta X_t)^8 \le C_1|u_2 - u_1|^8 \varepsilon^8 + C_2|u_2 - u_1|^4 \varepsilon^{12}$$

and analogously

$$\mathbf{E}_{\theta_1}(\Delta X_t)^4 \le C_3|u_2 - u_1|^4 \varepsilon^4 + C_4|u_2 - u_1|^2 \varepsilon^6.$$

with some constants $C_i > 0, i = 1, \dots, 4$.

By substituting these inequalities in (2.13), we obtain

$$\begin{aligned}
\mathbf{E}_\theta |Z_\varepsilon^{\frac14}(u_2) - Z_\varepsilon^{\frac14}(u_1)|^4 &\le C_1|u_2 - u_1|^8 + \\
&+ C_2|u_2 - u_1|^4 + C_3|u_2 - u_1|^2,
\end{aligned}$$

where the constants C_i do not depend on θ and ε. Finally, we deduce

$$\sup_{|u_i|<R} |u_2-u_1|^{-2}\mathbf{E}_\theta|Z_\varepsilon^{\frac{1}{4}}(u_2)-Z_\varepsilon^{\frac{1}{4}}(u_1)|^4 \le B(1+R^6)$$

with some constant $B>0$ which does not depend on θ.

Lemma 2.7. *For any $\mathbf{K}\subset\Theta$ and $p\in(0,1)$, there exists $g(\cdot)\in\mathbf{G}$ such that*

$$\sup_{\theta\in\mathbf{K}} \mathbf{E}_\theta Z_\varepsilon^p(u)\le e^{-g(u)}.$$

Proof. The equality

$$x_t = x_0+\int_0^t x_{\theta s}ds,\quad x_0>0$$

provides the following two inequalities

$$\begin{aligned} x_t &\ge x_0,\\ x_t &= x_0+\theta^{-1}\int_0^{\theta t} x_s ds\le x_0+\theta^{-1}\int_0^t x_s ds,\end{aligned}$$

so, by Lemma 1.11, we have

$$x_t\le x_0\exp\left(\frac{t}{\theta}\right).$$

Hence, denoting $\theta_u=\theta+\varepsilon\sigma(\theta)^{-1}u$, we can write

$$\begin{aligned} |x_{\theta_u t}-x_{\theta t}| &= |\int_{\theta t}^{\theta_u t} x_{\theta s}ds|\ge x_0|\theta_u-\theta|t = x_0|u|\varepsilon\sigma^{-1}(\theta)t,\\ |x_{\theta_u t}-x_{\theta t}| &\le \int_{\theta t}^{\theta_u t} x_0 e^{\frac{T}{\theta}}ds = x_0 e^{\frac{T}{\theta}}\,|u|\varepsilon\sigma(\theta)^{-1}t.\end{aligned}$$

Further, with $\mathbf{P}_\theta^{(\varepsilon)}$ probability 1, we have

$$\begin{aligned} |X_t-x_t| &\le \int_0^t |X_{\theta s}-x_{\theta s}|ds+\varepsilon|W_t|\le\\ &\le \frac{1}{\theta}\int_0^{\theta t}|X_s-x_s|ds+\varepsilon\sup_{0\le t\le T}|W_t|\le\\ &\le \frac{1}{\theta}\int_0^t |X_s-x_s|ds+\varepsilon\eta\end{aligned}$$

and by Lemma 1.11

$$\sup_{0\le t\le T}|X_t-x_t|\le\varepsilon\exp\left(\frac{T}{\theta}\right)\eta.$$

Note also that

$$\int_0^T (x_{(\theta+u)t}-x_{\theta t})^2dt\ge x_0^2|u|^2\frac{T^3}{3}$$

and if $\mathbf{K}=[\alpha_1,\beta_1]\subset(\alpha,\beta)$, then

$$\inf_{\theta\in\mathbf{K}}\int_0^T t^2x_{\theta^2 t}^2dt=\int_0^T t^2x_{\alpha_1^2 t}dt>0.$$

Now the required inequality follows from the proof of Lemma 2.3.

Lemmas 2.1–2.3 allow us to use Theorem 1.6 and this completes the proof of Theorem 2.6. We have similar result for a wide class of Bayesian estimators of the parameter θ.

REMARK 2.4. This result admits several generalizations.

First, we can consider the nonlinear process

$$dX_t = S(X_{\theta t}, t)dt + \varepsilon\sigma(X_t, t)dW_t, \quad X_0 = x_0, \quad 0 \le t \le T$$

with some smooth on x function $S(x,t)$ and using a similar argument to prove the asymptotic normality

$$\mathcal{L}_\theta\{\varepsilon^{-1}(\hat{\theta}_\varepsilon - \theta)\} \Longrightarrow \mathcal{N}(0, \sigma(\theta)^{-2}),$$

where

$$\sigma(\theta)^2 = \int_0^T t^2 S_x'(x_{\theta t})^2 \frac{x_{\theta^2 t}^2}{\sigma(t, x_t)^2} dt.$$

(see [1]).

REMARK 2.5. The scheme (2.9) can be considered as an information transmission by a channel with a feedback (see [97], §16.3). The message is a duration of delay defined by θ. So it is interesting also to consider the model

$$dX_t = X_{t-\theta}dt + \varepsilon dW_t, \quad X_s = x_0, \quad s \le 0, \quad 0 \le t \le T,$$

where $\theta \in (\alpha, \beta), \alpha > 0$ and to estimate the *delay* θ.

Theorem 2.7. *Let $x_0 > 0$, then the MLE $\hat{\theta}_\varepsilon$ uniformly on $\theta \in \mathbf{K}$ is*
(a) consistent,
(b) asymptotically normal:

$$\mathcal{L}_\theta\{(\hat{\theta}_\varepsilon - \theta)\varepsilon^{-1}\} \Longrightarrow \mathcal{N}(0, \sigma(\theta)^{-2}),$$

where

$$\sigma(\theta)^2 = \int_0^T x_{t-2\theta}^2 dt,$$

(c) LAM.

Proof of this theorem is similar to the proof of Theorem 2.4 and even simpler.

REMARK 2.6. We present one generalization with the observed process having a different smoothness with respect to different components of the estimated parameter.

We observe the diffusion-type process

$$dX_t = \lambda X_{\theta t}dt + \varepsilon dW_t, \quad X_0 = x_0 > 0, \quad 0 \le t \le T,$$

where $\theta \in (0,1), \lambda \in (\alpha, \beta), \alpha > 0$ and we have to estimate the vector (λ, θ). Denote $(\hat{\lambda}_\varepsilon, \hat{\theta}_\varepsilon)$ as the MLE of this vector and x_t as the solution of the equation

$$\frac{dx_t}{dt} = \lambda x_{\theta t}, \quad x_0, \quad 0 \le t \le T.$$

We introduce the matrix

$$I(\lambda,\theta)=\begin{pmatrix} \int_0^T x_{\theta t}^2 dt & \lambda\int_0^T t x_{\theta^2 t} x_{\theta t} dt \\ \lambda\int_0^T t x_{\theta t} x_{\theta^2 t} dt & \lambda^2\int_0^T t^2 x_{\theta^2 t}^2 dt \end{pmatrix}$$

and put $\varphi_\varepsilon(\theta)=\varepsilon I(\lambda,\theta)^{-\frac{1}{2}}$.

Theorem 2.8. *The MLE* $(\hat\lambda_\varepsilon,\hat\theta_\varepsilon)$ *as* $\varepsilon\to 0$ *is uniformly consistent and asymptotically normal:*

$$\mathcal{L}_{\lambda,\theta}\left\{\varepsilon^{-1}I(\lambda,\theta)^{\frac{1}{2}}\begin{pmatrix}\hat\lambda_\varepsilon-\lambda\\ \hat\theta_\varepsilon-\theta\end{pmatrix}\right\}\Longrightarrow\mathcal{N}(0,I),$$

where I *is a unit* 2×2 *matrix.*

Proof. The uniform LAN is based on the uniform convergence

$$\mathbf{P}_\theta-\lim_{\varepsilon\to 0}\varepsilon^{-2}\|(\lambda+\varepsilon u_1)X_{(\theta+\varepsilon u_2)t}-\lambda X_{\theta t}-\varepsilon u_1 x_{\theta t}-u_2 t\varepsilon x_{\theta^2 t}\|=0$$

and the verification of this convergence has no additional difficulties compared with Lemma 2.1.

For the quantity

$$r=\mathbf{E}_\theta\left[Z_\varepsilon^{\frac{1}{6}}(\mathbf{u})-Z_\varepsilon^{\frac{1}{6}}(\mathbf{v})\right]^6,$$

where $Z_\varepsilon(\mathbf{u})=L(\lambda+\varepsilon u_1,\theta+\varepsilon u_2;\ \lambda,\theta;\ X)$, $\mathbf{u}=(u_1,u_2)$ and repeating the derivation in the Lemma 2.5, we obtain the estimate

$$r\le C_p\{|\mathbf{u}-\mathbf{v}|^{12}+|\mathbf{u}-\mathbf{v}|^6+\varepsilon^3|\mathbf{u}-\mathbf{v}|^3\}.$$

Hence

$$\sup_{\theta\in\mathbf{K}}\ \sup_{|\mathbf{u}|<R,|\mathbf{v}|<R}|\mathbf{u}-\mathbf{v}|^{-3}\mathbf{E}_\theta\left[Z_\varepsilon^{\frac{1}{6}}(\mathbf{u})-Z_\varepsilon^{\frac{1}{6}}(\mathbf{v})\right]^6\le B_1(1+R^9).$$

Further, let the condition

$$\inf_{\theta\in\mathbf{K}}\inf_{|\mathbf{u}|>h}\int_0^T[\lambda+u_1)x_{(\theta+u_2)t}-\lambda x_{\theta t}]^2dt>0 \tag{2.14}$$

be violated, then there exist such $\theta\in\mathbf{K}$ and $u=(u_1,u_2)$, so that

$$\int_0^T[(\lambda+u_1)x_{(\theta+u_2)t}-\lambda x_{\theta t}]^2dt=0,$$

i.e. for all $t\in[0,T]$,

$$(\lambda+u_2)x_{(\theta+u_2)t}-\lambda x_{\theta t}=0.$$

Hence, we have for all $t\in[0,T]$

$$\frac{x_{(\theta+u_2)t}}{x_{\theta t}}=\frac{\lambda}{\lambda+u_1},$$

but for $t = 0$, we obtain $\lambda = \lambda + u_1$ and so $u_1 = 0$. Further

$$\frac{x_{(\theta+u_2)t}}{x_{\theta t}} - 1 = \int_{\theta t}^{(\theta+u_2)t} \frac{x_{\theta s}}{x_{\theta t}} ds \geq u_2 t$$

for $u_2 > 0$. This contradiction proves the inequality (2.14) and we can apply the proof of Lemma 2.6 to derive an analogous inequality.

REMARK 2.7. The last result could be generalized to a multi-dimensional case of the following type:

$$dX_t = \sum_{j=1}^{k} \lambda_j X_{t-b_j} dt + \varepsilon dw_t, \quad X_0 = x_0, \quad 0 \leq t \leq T,$$

where the vector parameter $(\lambda_1, \ldots, \lambda_k, b_1, \ldots, b_k)$ is unknown. The properties of the MLE of this parameter are described in [78], where its consistence and asymptotic normality are proved.

"Too smooth" trend ($I(\theta) = 0$). Let the observed process be

$$dX_t = S(\theta, X_t)dt + \varepsilon dW_t, \quad X_0 = x_0, \quad 0 \leq t \leq T,$$

where $\theta \in (\alpha, \beta)$ is one-dimensional unknown parameter. In the regular case, the MLE $\hat{\theta}_\varepsilon$ and BE $\tilde{\theta}_\varepsilon$ are consistent and asymptotically normal (Theorems 2.2 and 2.3):

$$\mathcal{L}_\theta\{\varepsilon^{-1}(\hat{\theta}_\varepsilon - \theta)\} \Longrightarrow \mathcal{N}(0, I(\theta)^{-1}).$$

So the condition $I(\theta) > 0$ is necessary one for this asymptotic normality.

Below, we study the properties of the MLE and BE when

$$I(\theta) = \int_0^T \dot{S}(\theta, x_t)^2 dt = 0.$$

Of course, we can not admit that $I(\theta) = 0$ on some interval because, in this case, the function $S(\theta, x)$ does not depend on θ and even a consistent estimation is impossible. So we suppose that there exists one point (true value) θ_0 such that $I(\theta_0) = 0$ and we shall study the asymptotic of $\hat{\theta}_\varepsilon - \theta_0$ and $\tilde{\theta}_\varepsilon - \theta_0$ as $\varepsilon \to 0$. This condition is equivalent to the condition

$$\dot{S}(\theta_0, x) = 0, \quad a \leq x \leq b,$$

where

$$a = \min_{0 \leq t \leq T} x_t(\theta_0), \quad b = \max_{0 \leq t \leq T} x_t(\theta_0).$$

Moreover, to immediately consider a more general case, we suppose that

I. *The first $k - 1$ derivatives of $S(\theta, x)$ w.r.t. θ at point θ_0 are equal to zero:*

$$S^{(j)}(\theta_0, x) = 0, \quad j = 1, \ldots, k-1, \quad x \in [a^*, b^*],$$

where

$$a^* = \inf_{\theta\in\Theta} \min_{0\leq t\leq T} x_t(\theta), \quad b^* = \sup_{\theta\in\Theta} \max_{0\leq t\leq T} x_t(\theta).$$

II. *The* k-th *derivative* $S^{(k)}(\theta_0, x)$ *is such that*

$$I_k(\theta_0) = \frac{1}{(k!)^2}\int_0^T S^{(k)}(\theta_0, x_t)^2 dt > 0.$$

Here x_t is a solution of the equation

$$\frac{x_t}{dt} = S(\theta_0, x_t), \quad x_0, \quad 0 \leq t \leq T.$$

We also assume that the other conditions of regularity are satisfied.

III. *For any* $\nu > 0$, *the function*

$$g(\nu) = \inf_{|\theta-\theta_0|>\nu} \int_0^T [S(\theta, x_t) - S(\theta_0, x_t)]^2 dt > 0 \tag{2.15}$$

and the functions $S(\theta_0, x)$ *and* $\dot{S}(\theta, x)$ *satisfy the Lipschitz condition*

$$|S(\theta_0, x) - S(\theta_0, y)| + |\dot{S}(\theta, x) - \dot{S}(\theta, y)| \leq L|x - y|.$$

The properties of estimators essentially depend on the number k. First , we consider the more simple case of k odd.

k **odd.** Denote by

$$\xi_k(\theta_0) = \frac{1}{k!}\int_0^T S^{(k)}(\theta_0, x_t) dW_t \sim \mathcal{N}(0, I_k(\theta_0)),$$

then we have

Theorem 2.9. *Let the function* $S(\theta, x)$ *have* $k+1$ *continuous bounded derivatives on* θ *and conditions I–III be fulfilled, then the MLE* $\hat{\theta}_\varepsilon$ *is consistent:*

$$\mathbf{P}_{\theta_0} - \lim_{\varepsilon\to 0} \hat{\theta}_\varepsilon = \theta_0$$

and the normalized difference

$$\mathcal{L}_{\theta_0}\left\{\varepsilon^{-\frac{1}{k}}(\hat{\theta}_\varepsilon - \theta_0)\right\} \Longrightarrow \mathcal{L}\left\{\left(\frac{\xi_k(\theta_0)}{I_k(\theta_0)}\right)^{\frac{1}{k}}\right\}.$$

The moments also converge.

Proof. We shall check the conditions of Theorem 1.4. First recall that by the Taylor formula

$$\begin{aligned} S(\theta_0+h,x_t) &= S(\theta_0,x_t)+\sum_{j=1}^{k}\frac{S^{(j)}(\theta_0,x_t)}{j!}h^j+\frac{S^{(k+1)}(\tilde\theta,x_t)}{(k+1)!}h^{k+1}= \\ &= S(\theta_0,x_t)+\frac{h^k}{k!}S^{(k)}(\theta_0,x_t)+\frac{S^{(k+1)}(\tilde\theta,x_t)}{(k+1)!}h^{k+1},\quad \tilde\theta\in[\theta_0,\theta_0+h). \end{aligned}$$

We put $\varphi_\varepsilon=\varepsilon^{\frac{1}{k}}$, then we can write

$$\begin{aligned} &\int_0^T[S(\theta_0+\varphi_\varepsilon u,X_t)-S(\theta_0,X_t)-\frac{u^k\varepsilon}{k!}S^{(k)}(\theta_0,x_t)]^2dt\le \\ &\le 2\int_0^T[S(\theta_0+\varphi_\varepsilon u,x_t)-S(\theta_0,x_t)-\frac{u^k\varepsilon}{k!}S^{(k)}(\theta_0,x_t)]^2dt+ \\ &\quad +2\int_0^T[S(\theta_0+\varphi_\varepsilon u,X_t)-S(\theta_0+\varphi_\varepsilon,x_t)+S(\theta_0,x_t)-S(\theta_0,X_t)]^2dt\le \\ &\le \frac{2}{((k+1)!)^2}\varepsilon^{2+\frac{2}{k}}u^{2k+2}\int_0^T S^{(k+1)}(\tilde\theta,x_t)^2dt+ \\ &\quad +2\int_0^T[S'(\theta_0+\varphi_\varepsilon u,\tilde X_t)-S'(\theta_0,\tilde{\tilde X}_t)]^2(X_t-x_t)^2dt\le \\ &\le Cu^{2k+2}\varepsilon^{2+\frac{2}{k}}+6\left(\int_0^T[S'(\theta_0+\varphi_\varepsilon u,\tilde X_t)-S'(\theta_0+\varphi_\varepsilon u,x_t)]^2dt+\right. \\ &\quad \left.+\int_0^T[S'(\theta_0+\varphi_\varepsilon u,x_t)-S'(\theta_0,x_t)]^2dt+\int_0^T[S'(\theta_0,\tilde{\tilde X}_t)-S'(\theta_0,x_t)]^2dt\right)\times \\ &\quad \times\sup_{0\le t\le T}|X_t-x_t|^2\le \\ &\le Cu^{2k+2}\varepsilon^{2+\frac{2}{k}}+C_1\sup_{0\le t\le T}|X_t-x_t|^4+C_2u^2\varepsilon^{\frac{2}{k}}\sup_{0\le t\le T}|X_t-x_t|^2. \end{aligned}$$

Here $\tilde X_t,\tilde{\tilde X}_t\in(x_t,X_t)$. Thus, for any bounded sequence $u_\varepsilon, u_\varepsilon\to u$,

$$\mathbf{P}_{\theta_0}-\lim_{\varepsilon\to0}\varepsilon^{-2}\int_0^T\left[S(\theta_0+\varphi_\varepsilon u_\varepsilon,X_t)-S(\theta_0,X_t)-\frac{u_\varepsilon^k}{k!}\varepsilon S^{(k)}(\theta_0,x_t)\right]^2dt=0.$$

This convergence provides the limits

$$\mathbf{P}_{\theta_0}-\lim_{\varepsilon\to0}\varepsilon^{-2}\int_0^T[S(\theta_0+\varphi_\varepsilon u,X_t)-S(\theta_0,X_t)]^2dt=u^{2k}I_k(\theta_0)$$

and

$$\mathbf{P}_{\theta_0}-\lim_{\varepsilon\to0}\varepsilon^{-1}\int_0^T[S(\theta_0+\varphi_\varepsilon u,X_t)-S(\theta_0,X_t)]dW_t=u^k\xi_k.$$

So the normalized likelihood ratio

$$Z_\varepsilon(u_\varepsilon)\equiv\frac{d\mathbf{P}^{(\varepsilon)}_{\theta_0+\varphi_\varepsilon u_\varepsilon}}{d\mathbf{P}^{(\varepsilon)}_{\theta_0}},\quad u_\varepsilon\in((\alpha-\theta_0)\varepsilon^{-1},(\beta-\theta_0)\varepsilon^{-1})$$

for any $u \in \mathbf{R}^1$ and $u_\varepsilon \to u$ converges to

$$Z(u) = \exp\left\{u^k \xi_k(\theta_0) - \frac{u^{2k}}{2} I_k(\theta_0)\right\}$$

and we have also convergence of all finite-dimensional distributions of $Z_\varepsilon(u)$.

Consider now the increments of $Z_\varepsilon(u)$, as was done in Lemma 2.2

$$\begin{aligned}
&\mathbf{E}_{\theta_0}\left[Z_\varepsilon^{\frac{1}{2}}(u_2) - Z_\varepsilon^{\frac{1}{2}}(u_1)\right]^2 = \\
&= \mathbf{E}_{\theta_1}\left[\left(\frac{Z_\varepsilon(u_2)}{Z_\varepsilon(u_1)}\right)^{\frac{1}{2}} - 1\right]^2 \leq \\
&\leq \mathbf{E}_{\theta_1}\int_0^T [S(\theta_0 + \varphi_\varepsilon u_2, X_t) - S(\theta_0 + \varphi_\varepsilon u_1, X_t)]^2 dt + \\
&+\mathbf{E}_{\theta_2}\int_0^T [S(\theta_0 + \varphi_\varepsilon u_2, X_t) - S(\theta_0 + \varphi_\varepsilon u_1, X_t)]^2 dt. \qquad (2.16)
\end{aligned}$$

We have

$$\begin{aligned}
&\mathbf{E}_{\theta_1}\int_0^T [S(\theta_0 + \varphi_\varepsilon u_2, X_t) - S(\theta_0 + \varphi_\varepsilon u_1, X_t)]^2 dt - \\
&= (u_2 - u_1)^2 \varepsilon^{\frac{2}{k}} \int_0^T \mathbf{E}_{\theta_1}[\dot S(\tilde\theta, x_t(\theta_1)) + (\dot S(\tilde\theta, X_t) - \dot S(\tilde\theta, x_t(\theta_1)))]^2 dt \leq \\
&\leq 2(u_2 - u_1)^2 \varepsilon^{\frac{2}{k}} \int_0^T [\dot S(\theta_0, x_t(\theta_1)) + \sum_{j=1}^{k-2} \frac{S^{(j+1)}(\theta_0, x_t(\theta_1))}{j!}(\gamma\varphi_\varepsilon(u_2 - u_1))^j + \\
&+\frac{S^{(k)}(\tilde{\tilde\theta}, x_t(\theta_1))}{(k-1)!}(\gamma\varphi_\varepsilon(u_2 - u_1))^{k-1}]^2 dt + \\
&+(u_2 - u_1)^2 \varepsilon^{\frac{2}{k}} C \sup_{0\leq t\leq T} \mathbf{E}_{\theta_1}|X_t - x_t(\theta_1)|^2 \leq \\
&\leq C_1(u_2 - u_1)^{2k}\varepsilon^2 + C_2(u_2 - u_1)^2 \varepsilon^{2+\frac{2}{k}}.
\end{aligned}$$

We have the same estimates for the second integral in (2.16), so we can write

$$\sup_{|u_i|<R} \mathbf{E}_{\theta_0}\left[Z_\varepsilon^{\frac{1}{2}}(u_2) - Z_\varepsilon^{\frac{1}{2}}(u_1)\right]^2 \leq C(\varepsilon^{\frac{2}{k}} + R^{2k-2})(u_2 - u_1)^2.$$

To derive the last inequality, we first note that

$$\lim_{u\to 0} \frac{1}{u^{2k}} \int_0^T [S(\theta_0 + u, x_t(\theta_0)) - S(\theta_0, x_t(\theta_0))]^2 dt = I_k(\theta_0)$$

and this limit, together with (2.16) and condition III, provides the inequality

$$\int_0^T [S(\theta_0 + u, x_t(\theta_0)) - S(\theta_0, x_t(\theta_0))]^2 dt \geq \kappa u^{2k},$$

where $\kappa > 0$ and for all $u \in (\alpha - \theta_0, \beta - \theta_0)$. So by repeating the calculations of Lemma 2.3, we establish

$$\mathbf{E}_{\theta_0} Z_\varepsilon^{\frac{1}{2}}(u) \leq e^{-g(u)}$$

with $g(\cdot) \in \mathbf{G}$.

The conditions of (nonuniform on θ version) Theorem 1.4 are satisfied, hence

$$\mathcal{L}_{\theta_0}\left\{\varepsilon^{-\frac{1}{k}}(\hat{\theta}_\varepsilon - \theta_0)\right\} \longrightarrow \mathcal{L}\left\{\arg\sup_u Z(u)\right\} = \mathcal{L}\left\{\left(\frac{\xi_k(\theta_0)}{I_k(\theta_0)}\right)^{\frac{1}{k}}\right\}.$$

Suppose now that θ is a random variable with density $\pi(y)$ positive at point θ_0 and we have a quadratic loss function, then under the assumption made according to Theorem 1.5, the Bayesian estimator has the same asymptotic properties as the MLE because

$$\frac{\int_{-\infty}^{\infty} uZ(u)du}{\int_{-\infty}^{\infty} Z(u)du} = \left(\frac{\xi_k(\theta_0)}{I_k(\theta_0)}\right)^{\frac{1}{k}}.$$

Consider now the case of

k even. In this case, under the same conditions as in Theorem 2.9, the limit likelihood ratio process $Z(u)$ is a symmetric function of u so for $\xi_k(\theta_0) > 0$ it has two equal maximums at the points

$$u_+ = (\xi_k(\theta_0)I_k^{-1}(\theta_0))^{\frac{1}{k}}, \quad u_- = -u_+$$

and the "limit MLE" is not defined. Of course, if $\xi_k(\theta_0) < 0$, then $u_+ = u_- = 0$ and with probability $\frac{1}{2}$ the limit of $\varepsilon^{\frac{1}{k}}(\hat{\theta}_\varepsilon - \theta_0)$ is zero. To simplify the problem, we suppose that the first $k-1$ derivatives are as in condition I and that the k-th derivatives from the left and from the right at point θ_0 are different. So

$$\rho = \frac{|\int_0^T S^{(k)}(\theta_0+, x_t)S^{(k)}(\theta_0-, x_t)dt|}{(k!)^2(I_k(\theta_0+)I_k(\theta_0-))^{\frac{1}{2}}} < 1. \tag{2.17}$$

Denote this condition by II'. In this case, we have two limit processes

$$\begin{aligned} Z_+(u) &= \exp\left\{u^k\xi_k(\theta_0+) - \frac{u^{2k}}{2}I_k(\theta_0+)\right\}, \quad u \geq 0, \\ Z_-(u) &= \exp\left\{u^k\xi_k(\theta_0-) - \frac{u^{2k}}{2}I_k(\theta_0-)\right\}, \quad u < 0 \end{aligned}$$

and the random variable

$$\hat{u} = \arg\max\left(\sup_{u\geq 0} Z_+(u), \sup_{u\leq 0} Z_-(u)\right) = \left(\frac{\xi_k(\theta_0+)}{I_k(\theta_0+)} \bigvee \frac{\xi_k(\theta_0-)}{I_k(\theta_0-)} \bigvee 0\right)^{\frac{1}{k}}.$$

So if $\xi_k(\theta_0+) < 0$ and $\xi_k(\theta_0-) < 0$, then $\hat{u} = 0$.

Theorem 2.10. *Let k be even, the function $S(\theta, x)$ has $k-1$ continuous bounded derivatives w.r.t. θ, $S^{(k)}(\theta, x)$ is absolutely continuous on θ, and conditions I, II', III be satisfied, then the MLE $\hat{\theta}_\varepsilon$ is consistent:*

$$\mathbf{P}_{\theta_0} - \lim_{\varepsilon\to 0} \hat{\theta}_\varepsilon = \theta_0$$

and

$$\mathcal{L}_{\theta_0}\left\{\varepsilon^{-\frac{1}{k}}(\hat{\theta}_\varepsilon - \theta_0)\right\} \to \mathcal{L}\{\hat{u}\}.$$

The moments of this difference also converge.

Proof is similar to that given above.

We consider the weak convergence of the couple $\{(Z_\varepsilon(u), u \geq 0), (Z_\varepsilon(u), u \leq 0)\}$ to the couple $\{(Z_+(u), u \geq 0), (Z_-(u), u \leq 0)\}$, checking the conditions of Theorem 1.4.

For the Bayes estimator with a quadratic loss function, the limit random variable $\tilde{u}$, i.e.

$$\mathcal{L}_{\theta_0}\{\varepsilon^{-\frac{1}{k}}(\tilde{\theta}_\varepsilon - \theta_0)\} \longrightarrow \mathcal{L}\{\tilde{u}\},$$

is defined by the equality

$$\tilde{u} = \left(\int_{-\infty}^{0} uZ_-(u)du + \int_0^\infty uZ_+(u)du\right)\left(\int_{-\infty}^{0} Z_-(u)du + \int_0^\infty Z_+(u)du\right)^{-1}.$$

Note, that if $\rho = 1$, then $\tilde{u} = 0$.

The behavior of the MLE $\tilde{\theta}_\varepsilon$ in the case where $\rho = 1$ requires us to consider the next term in the expansion of $\ln Z_\varepsilon(u)$ by the powers of ε and to compare the left and right sides, as will be done in §2.8, in similar situation of coinciding limit maximums. Here it could be done using the same technique.

2.5 Random initial value

Let the *initial value* X_0 of the observed diffusion-type process

$$dX_t = S_t(\theta, X)\, dt + \varepsilon\, dW_t, \quad 0 \leq t \leq T \tag{2.18}$$

be a $\mathcal{F}_0$–measurable *random* variable and suppose that its distribution does not depend on θ. For simplicity, we suppose that the unknown parameter θ is one-dimensional, $\theta \in \Theta = (\alpha, \beta)$. Under conditions $\mathcal{L}$, the process $X_t, 0 \leq t \leq T$ converges to the solution of the equation

$$\frac{dx_t}{dt} = S_t(\theta, x), \quad x_0 = X_0, \quad 0 \leq t \leq T \tag{2.19}$$

uniformly on $t \in [0, T]$, but this solution $\{x_t, 0 \leq t \leq T\}$ is now an $\mathcal{F}_0$–measurable random function. So the Fisher's information

$$I(\theta) = \int_0^T \dot{S}_t(\theta, x)^2\, dt$$

is also an $\mathcal{F}_0$–measurable random variable .

Below, we describe the properties of the MLE of parameter θ constructed by the observation (2.18).

First, we introduce a corresponding minimax bound for the risks of estimators in this situation and then we prove the local asymptotic minimaxity of the MLE.

Theorem 2.11. *Let the following conditions be satisfied:*

(a) The function $S_t(\theta, x)$ satisfies the conditions $\mathcal{L}$;

(b) has two continuous bounded derivatives w.r.t. θ and the first derivative $\dot{S}_t(\theta, X)$ is continuous on X at the point $x = \{x_s, 0 \le s \le t\}$ in $L_2[0,T]$;

(c) $I(\theta) \ge \kappa$ with probability 1, where κ is a positive number.

Then the family of measures $\{\mathbf{P}_\theta^{(\varepsilon)}, \theta \in \Theta\}$ as $\varepsilon \to 0$ is conditionally LAN in Θ, i.e., the likelihood ratio

$$\frac{d\mathbf{P}_{\theta+\varepsilon u}^{(\varepsilon)}}{d\mathbf{P}_\theta^{(\varepsilon)}}(X) = \exp\left\{u\Delta(\theta, X) - \frac{u^2}{2}I(\theta) + \psi_\varepsilon(u,\theta)\right\}, \quad u \in U_{\theta,\varepsilon} = \{u : \theta + \varepsilon u \in \Theta\}, \tag{2.20}$$

conditional distribution

$$\mathcal{L}_\theta\{\Delta(\theta, X)|\mathcal{F}_0\} = \mathcal{N}(0, I(\theta)) \tag{2.21}$$

with probability 1 and for any $u \in \mathbf{R}$

$$\mathbf{P}_\theta - \lim_{\varepsilon\to 0} \psi_\varepsilon(u,\theta) = 0. \tag{2.22}$$

Moreover, for $\theta_0 \in \Theta$, any estimator θ_ε and a loss function $l(\cdot) \in \mathbf{W}_{e,2}$,

$$\lim_{\delta\to 0}\liminf_{\varepsilon\to 0}\sup_{|\theta-\theta_0|<\delta} \mathbf{E}_\theta l(\varepsilon^{-1}(\theta_\varepsilon - \theta)) \ge \mathbf{E}_{\theta_0} l(\xi), \tag{2.23}$$

where $\mathcal{L}_{\theta_0}\{\xi|\mathcal{F}_0\} = \mathcal{N}(0, I(\theta_0)^{-1})$. So

$$\mathbf{E}_{\theta_0} l(\xi) = \frac{1}{\sqrt{2\pi}}\int_{-\infty}^{\infty} e^{-\frac{x^2}{2}}\mathbf{E}\, l(x\, I(\theta_0)^{-1/2})\, dx$$

Proof. This theorem consist of two parts. First, we derive the representation (2.20) with (2.21) and (2.22) and then, using this representation, we establish the analog of Hajek-Le Cam's inequality for the minimax risks of estimators. Representation (2.20) - (2.22) is proved in Theorem 2.12 below.

Here we present the necessary modification of the proof of Theorem 2.1. As before, we denote by $Z_\varepsilon(u)$ the likelihood ratio (2.20) and

$$Z(u) = \exp\left\{u\Delta - \frac{u^2}{2}I(\theta)\right\}.$$

We put also $I(\theta_0) = I_0$. For any estimator θ_ε and a loss function $l(\cdot) \in \mathbf{W}_{e,2}$ such that $\mathbf{E}l(\xi) < \infty$ we have

$$\sup_{|\theta-\theta_0|<\delta} \mathbf{E}_\theta\, l(\varepsilon^{-1}(\theta_\varepsilon - \theta)) =$$
$$= \sup_{|u|<\delta\varepsilon^{-1}} \mathbf{E}_{\theta_0+\varepsilon u}\, l(\varepsilon^{-1}(\theta_\varepsilon - \theta) - u) \ge \sup_{|u|<M} \mathbf{E}_{\theta_0}\, Z_\varepsilon(u)\, l(u_\varepsilon - u) \ge$$

$$\geq \frac{1}{2M}\int_{-M}^{M}\mathbf{E}_{\theta_0}\, e^{u\Delta-\frac{u^2}{2}I_0}\, l(u_\varepsilon-u)^a\,du+o(1)=$$

$$=\frac{1}{2M}\mathbf{E}_{\theta_0}e^{\frac{\Delta^2}{2I_0}}\int_{-M}^{M} e^{-\frac{I_0}{2}(u-\frac{\Delta}{I_0})^2}\, l(u_\varepsilon-u)^a\,du+o(1)=$$

$$=\frac{1}{2M}\mathbf{E}_{\theta_0}\, e^{\frac{\Delta^2}{2I_0}}\int_{-M-\frac{\Delta}{I_0}}^{M-\frac{\Delta}{I_0}} e^{-\frac{I_0}{2}v^2}\, l(v_\varepsilon-v)^a\,dv+o(1)\geq$$

$$\geq\frac{1}{2M}\mathbf{E}_{\theta_0}\chi_{\{|\Delta I_0^{-1}|<M-\sqrt{M}\}}e^{\frac{\Delta^2}{2I_0}}\int_{-\sqrt{M}}^{\sqrt{M}} e^{-\frac{I_0}{2}v^2}\, l(v_\varepsilon-v)^a\,dv+o(1)$$

$$\geq\frac{1}{2M}\mathbf{E}_{\theta_0}\chi_{\{|\Delta I_0^{-1}|<M-\sqrt{M}\}}e^{\frac{\Delta^2}{2I_0}}\left\{\mathbf{E}[l(\xi)^a|\mathcal{F}_0]\sqrt{\frac{2\pi}{I_0}}-ae^{-\frac{I_0M}{2}}\right\}+o(1),$$

where ξ is conditionally normal : $\mathcal{L}(\xi|\mathcal{F}_0)=\mathcal{N}(0,I_0^{-1})$, so we can write

$$\int_{-\sqrt{M}}^{\sqrt{M}} e^{-\frac{I_0v^2}{2}}l(v_\varepsilon-v)^a dv\geq\sqrt{\frac{2\pi}{I_0}}\mathbf{E}[l(\xi)^a|\mathcal{F}_0]-ae^{-\frac{\kappa M}{2}}.$$

For the first integral, we have the following estimates

$$\frac{1}{2M}\mathbf{E}_{\theta_0}\chi_{\{|\frac{\Delta}{I_0}|<M-\sqrt{M}\}}\, e^{\frac{\Delta^2}{2I_0}}\sqrt{\frac{2\pi}{I_0}}\;\mathbf{E}[l(\xi)^a|\mathcal{F}_0]=$$

$$=\frac{1}{2M}\mathbf{E}_{\theta_0}\sqrt{\frac{2\pi}{I_0}}\;\mathbf{E}[l(\xi)^a|\mathcal{F}_0]\;\mathbf{E}\left(\chi_{\{|\frac{\Delta}{I_0}|<M-\sqrt{M}\}}e^{\frac{\Delta^2}{I_0}}|\mathcal{F}_0\right)=$$

$$=\frac{1}{2M}\mathbf{E}_{\theta_0}\sqrt{\frac{2\pi}{I_0}}\;\mathbf{E}[l(\xi)^a|\mathcal{F}_0]\;\frac{1}{\sqrt{2\pi I_0}}\int_{-(M-\sqrt{M})I_0}^{(M-\sqrt{M})I_0} e^{-\frac{x^2}{2I_0}}\,e^{\frac{x^2}{2I_0}}dx=$$

$$=\mathbf{E}_{\theta_0}l(\xi)^a(1-\frac{1}{\sqrt{M}})\longrightarrow\mathbf{E}_{\theta_0}l(\xi)^a.$$

Hence, the desired inequality is proved.

Let the family of measures $\{\mathbf{P}_\theta^{(\varepsilon)},\theta\in\Theta\}$ be conditionally LAN (i.e. satisfies (2.20) – (2.22)), the random variable $I(\theta)>\kappa>0$, and $l(\cdot)\in\mathbf{W}_{e,2}$. We call the estimator $\bar\theta_\varepsilon$ *local asymptotic minimax* if for any $\theta_0\in\Theta$

$$\lim_{\delta\to0}\lim_{\varepsilon\to0}\sup_{|\theta-\theta_0|<\delta}\mathbf{E}_\theta l(\varepsilon^{-1}(\bar\theta_\varepsilon-\theta))=\mathbf{E}l(\xi).$$

Theorem 2.12. *Let the function $S_t(\theta,X)$ have two bounded continuous derivatives w.r.t. θ, the first derivative $\dot S_t(\theta,X)$ is continuous in $L_2[0,T]$ at point $x=\{x_t,0\leq t\leq T\}$, and with probability 1*

$$\int_0^T[S_t(\theta+u,x)-S_t(\theta,x)]^2dt\geq\kappa|u|^2,$$

where $\kappa>0$ is nonrandom.

Then the MLE is

(a) consistent

$$\mathbf{P}_\theta-\lim_{\varepsilon\to0}\hat\theta_\varepsilon=\theta;$$

(b) asymptotically conditionally normal

$$\mathcal{L}_\theta\{\varepsilon^{-1}(\hat\theta_\varepsilon - \theta)\} \Longrightarrow \mathcal{L}_\theta\{\xi\};$$

(c) LAM for $l(\cdot) \in \mathbf{W}_p$.

Proof. By condition $\mathcal{L}$, we have

$$\int_0^T S_t(\theta, X)^2 dt \le 4L_1^2 \int_0^T \int_0^t (1 + X_v^2) dK(v) dt + 4L_2^2 \int_0^T (1 + X_t^2) dt < \infty$$

with probability 1 because the process $X_t, 0 \le t \le T$ is continuous and its trajectories are bounded functions. So, by Theorem 1.2, all the measures $\mathbf{P}_\theta^{(\varepsilon)}, \theta \in \Theta$ are equivalent and the likelihood ratio

$$Z_\varepsilon(u) = \frac{d\mathbf{P}_{\theta+\varepsilon u}^{(\varepsilon)}}{d\mathbf{P}_\theta^{(\varepsilon)}}(X)$$

with $\mathbf{P}_\theta^{(\varepsilon)}$ probability 1 admits the representation (recall , that X_0 does not depend on θ)

$$\begin{aligned} \ln Z_\varepsilon(u_\varepsilon) &= \varepsilon^{-1} \int_0^T S_t(\theta + \varepsilon u_\varepsilon, X) - S_t(\theta, X)] dW_t - \\ &- \frac{1}{2\varepsilon^2} \int_0^T [S_t(\theta + \varepsilon u_\varepsilon, X) - S_t(\theta, X)]^2 dt. \end{aligned}$$

By Lemma 1.3, we have the same inequality

$$\sup_{0 \le t \le T} |X_t - x_t| \le C\varepsilon \sup_{0 \le t \le T} |W_t|$$

and, hence, the same convergence of X_t to x_t.

So repeating the corresponding inference of Lemma 2.1, we establish a convergence which is uniform on $\theta \in \mathbf{K}$. So,

$$\mathbf{P}_\theta - \lim_{\varepsilon \to 0} \varepsilon^{-1} \| S(\theta + \varepsilon u_\varepsilon, X) - S(\theta, X) - u_\varepsilon \dot S(\theta, x) \| = 0,$$

where $u_\varepsilon \to u \in \mathbf{R}$. This convergence allows us to write the representation (2.20) for $Z_\varepsilon(u_\varepsilon)$ with (2.22) and

$$\Delta(\theta, X) = \int_0^T \dot S_t(\theta, x)\, dW_t.$$

The conditional normality (2.21) of $\Delta(\theta, X)$ follows from the following lemma.

Lemma 2.8. *If the random function* $f \in \mathcal{P}$ *and*

$$\mathbf{E}\left[\exp\left\{\frac{1}{2}\int_0^T f_t^2 dt\right\} | \mathcal{F}_0\right] < \infty. \tag{2.24}$$

Then with probability 1

$$\mathbf{E}\left[\exp\left\{i\int_0^T f_t dW_t + \frac{1}{2}\int_0^T f_t^2 dt\right\}|\mathcal{F}_0\right] = 1. \tag{2.25}$$

This lemma with nonconditional mathematical expectations was presented in [100]. This conditional version follows from a slight modification of the proof of lemma 3.3.1 in [69] as follows. For the stochastic process

$$V(T) = \exp\left\{i\int_0^T f_t dW_t + \frac{1}{2}\int_0^T f_t^2 dt\right\}$$

by Itô formula we can write

$$V(T) = 1 + i\int_0^T f_t\, V(t)\, dW_t$$

and

$$\mathbf{E}\left[\left(\int_0^T f_t^2\, |V(t)|^2\, dt\right)^{1/2} |\mathcal{F}_0\right] =$$

$$= \mathbf{E}\left[\left(\int_0^T f_t^2\, \exp\{\int_0^t f_s^2 ds\}\, dt\right)^{1/2} |\mathcal{F}_0\right] =$$

$$= \mathbf{E}\left[\left(\exp\{\int_0^T f_t^2 dt\} - 1\right)^{1/2} |\mathcal{F}_0\right] < \mathbf{E}\left[\exp\{\frac{1}{2}\int_0^T f_t^2 dt\}|\mathcal{F}_0\right] < \infty$$

So the mathematical expectation of the stochastic integral is zero (see [104]) and $\mathbf{E}V(T)|\mathcal{F}_0 = 1$.

Let $\lambda \in \mathbf{R}$ be an arbitrary number and $f_t \in \mathcal{P}$ be $\mathcal{F}_0$ measurable for all $t \in [0,T]$. By (2.23) we have

$$\mathbf{E}\left[\exp\left\{i\lambda\int_0^T f_t dW_t\right\}|\mathcal{F}_0\right] = \exp\left\{-\frac{\lambda^2}{2}\int_0^T f_t^2 dt\right\}.$$

So the conditional characteristic function of the integral $\int_0^T f_t dW_t$ corresponds to the characteristic function of the Gaussian law with parameters 0 and $\int_0^T f_t^2 dt$.

Hence

$$\mathbf{E}\left[\exp\{i\lambda\Delta(\theta, X))\}|\mathcal{F}_0\right] = \exp\left\{-\frac{1}{2}I(\theta)\lambda^2\right\}.$$

The boundedness of the first derivatives allows us (using the proof of Lemma 2.2) to derive the inequality

$$\sup_{\theta\in\mathbf{K}} \mathbf{E}|Z_\varepsilon^{\frac{1}{2}}(u_1) - Z_\varepsilon^{\frac{1}{2}}(u_2)|^2 \le C|u_2 - u_1|^2$$

and the derivation of the inequality

$$\sup_{\theta\in\mathbf{K}} \mathbf{E}_\theta Z_\varepsilon^{\frac{1}{2}}(u) \leq C\exp\{-\kappa_1|u|^2\}$$

is also a repetition of Lemma 2.3.

The limiting process $Z(u)$ attains its maximum at the point

$$\hat{u} = I(\theta)^{-1}\Delta(\theta, x).$$

The properties of the likelihood ratio established above allow us to use Theorem 1.4 to finish the proof of Theorem 2.5.

The wide class of Bayes estimators has, of course, the same properties as the MLE.

2.6 Misspecified models

We now consider the problem of parameter estimation in the situation when the observer does not know the true parametric family (*misspecified model*).

Let the observed process ("nature") be

$$dX_t = b_t(\mu, X)dt + \varepsilon dW_t, \quad X_0 = x_0, \quad o \leq t \leq T, \tag{2.26}$$

where $\mu \in \mathcal{M} \subset \mathbf{R}^{d_1}$ and we have to identify this process. We do not know the true drift $b_t(\mu, X)$ and we suppose that the observed process belongs to a parametric family

$$dX_t = S_t(\theta, X)dt + \varepsilon dW_t, \quad X_0 = x_0, \quad 0 \leq t \leq T \tag{2.27}$$

with $\theta \in \Theta \subset \mathbf{R}^{d_2}$, i.e. we suppose that the true model is (2.27) and we try to estimate the "unknown" parameter θ by the observations (2.26), which have nothing in common with the process (2.27). We would like to describe the asymptotic properties of the MLE $\hat{\theta}_\varepsilon$ and BE $\tilde{\theta}_\varepsilon$ in such a "stupid" problem. There are several reasons to consider such a type of problem. First of all, a gap always exists between the mathematical models ("ideal") and the real data. So, the behavior of estimators is also interesting to study in situations when the observations correspond to "close" parametric or nonparametrical vicinities of the prescribed model. Sometimes, we can know the true drift, but the exact expression for $b_t(\cdot)$ is too complicated and depends on too many parameters or has a form which couldn't be used to provide a special solution of some problems. So we would like to find a correspondence between $b(\mu, X), \mu \in \mathcal{M}$ and a more simple and tractable family $S(\theta, X), \theta \in \Theta$, approximating $b(\mu, X), \mu \in \mathcal{M}$ in some sense. It will be shown below that the above-mentioned approach gives a solution for this problem.

First, we introduce a likelihood ratio for the model (2.27)

$$\begin{aligned}\frac{d\mathbf{P}_\theta^{(\varepsilon)}}{d\mathbf{P}_{\theta_0}^{(\varepsilon)}}(X) &= \exp\left\{\frac{1}{\varepsilon^2}\int_0^T [S_t(\theta, X) - S_t(\theta_0, x)]dX_t\right.\\ &\left. -\frac{1}{2\varepsilon^2}\int_0^T \left[S_t(\theta, X)^2 - S_t(\theta_0, X)^2\right]dt\right\}\end{aligned} \tag{2.28}$$

and then we substitute the observations (2.26) in (2.28). So the "MLE" is defined by

$$\hat{\theta}_\varepsilon = \arg\sup_{\theta\in\Theta} \frac{d\mathbf{P}_\theta^{(\varepsilon)}}{d\mathbf{P}_{\theta_0}^{(\varepsilon)}}(X)$$

with some fixed $\theta_0 \in \Theta$. We are interested in the asymptotic properties of $\hat{\theta}_\varepsilon$ as $\varepsilon \to 0$.

Note that the method of investigation applied in §2.2 could not be directly used here , because

$$\mathbf{E}_\mu \frac{d\mathbf{P}_\theta^{(\varepsilon)}}{d\mathbf{P}_{\theta_0}^{(\varepsilon)}}(X) \neq 1$$

for all $\theta_0 \in \Theta, \theta \neq \theta_0$. We propose another approach concerned with the "nonasymptotic" expansions and "good" sets.

Denote $x = \{x_t, 0 \leq t \leq T\}$ as the solution of the equation

$$\frac{dx_t}{dt} = b_t(\mu, x),\ x_0,\ 0 \leq t \leq T \tag{2.29}$$

(the limit function for the process (2.26)), introduce the function

$$G(\theta, \mu, x) - \int_0^T [S_t(\theta, x) - b_t(\mu, x)]^2 dt,$$

and define a "true" value

$$\theta^* = \arg\inf_{\theta\in\Theta} G(\theta, \mu, x). \tag{2.30}$$

Of course, $\theta^* = \theta^*(\mu)$ and $x_t = x_t(\mu)$.

We consider the one-dimensional bounded set $\Theta = (\alpha, \beta)$ and the integral-type trend coefficients

$$\begin{aligned} S_t(\theta, X) &= \int_0^t K(\theta, t, s, X_s)ds + V(\theta, t, X_t), \\ b_t(\mu, X) &= \int_0^t K_b(\mu, t, s, X_s)ds + V_b(\mu, t, X_t). \end{aligned}$$

The nature of the set $\mathcal{M}$ does not interested us at first. As a particular case of our problem, we have an ordinary estimation problem, if $\mathcal{M} = \Theta$ and $S_t(\theta, X) = b_t(\theta, X)$ for all $t \in [0, T]$ and $X \in \mathcal{C}_T$.

Maximum likelihood estimator. The main result of this section is the "consistence" of MLE:

$$\mathbf{P}_\mu - \lim_{\varepsilon\to 0} \hat{\theta}_\varepsilon = \theta^* \tag{2.31}$$

and its asymptotic normality

$$\mathcal{L}_\mu \left\{\varepsilon^{-1}(\hat{\theta}_\varepsilon - \theta^*)\right\} \Longrightarrow \mathcal{N}(0, \sigma^2)$$

under some regularity conditions. The limit variance σ^2 should be described below.

As follows from (2.31) and (2.30), the MLE $\hat{\theta}_\varepsilon$ chooses in the family $\{S(\theta,\cdot),\theta \in \Theta\}$ an element $S(\theta^*,\cdot)$ closest in $L_2[0,T]$ to the unknown function $b(\mu,\cdot)$ along the trajectory of (2.29). So this maximum likelihood approach solves the above-mentioned problem of fitting the model. The same limits also have the Bayes estimators.

This example shows that the MLE and BE are "robust" in a certain sense.

The limit (2.31) also allows us to consider the following problem of consistent estimation of μ for the model (2.26) with a "complicated" $b(\cdot)$.

Suppose, that $\theta^* \in \Theta$ and the equation

$$\int_0^T [S_t(\theta^*,x) - b_t(\mu,x)]\dot{S}_t(\theta^*,x)dt = 0, \quad \mu \in \mathcal{M} \tag{2.32}$$

has a unique solution $\mu = \mu(\theta^*)$ ($\mathcal{M}$–one-dimensional). Then we can construct an estimator $\mu^*_\varepsilon = \mu(\theta^*_\varepsilon)$ and, supposing the smoothness of $S(\cdot)$ and $b(\cdot)$ on θ, x and μ, we can prove the real consistency

$$\mathbf{P}_\mu - \lim_{\varepsilon\to 0} \mu^*_\varepsilon = \mu.$$

The advantage of this approach is that we solve once the deterministic equation (2.32) with a "complicate" $b(\cdot,\cdot)$ to find the function $\mu^*(\theta^*)$ and then maximize a "simple" function (2.28).

Note also another advantage of this approach. If $\Theta = \mathcal{M}$ and the functions $S(\cdot)$ and $b_t(\cdot)$ coincide along the trajectory of (2.29), then the MLE $\hat{\theta}_\varepsilon$ is consistent in the usual sense even if these two functions are different outside the set $C(\mathcal{M}) = \{(x_t(\mu), 0 \le t \le T), \mu \in \mathcal{M}\}$. Below, we consider this situation separately.

We need the following conditions.

I. The measurable functions $V(\theta,t,x), V_b(\mu,t,x), K(\theta,t,s,x), K_b(\mu,t,s,x)$ *are bounded and have four bounded continuous derivatives w.r.t.* θ *and* x*;*

II. The function $G(\theta,\mu,x)$ *for every* $\mu \in \mathcal{M}$ *has a unique minimum at point* $\theta^* \in \Theta$ *and for any compact* $\mathbf{K}_\mu \subset \mathcal{M}$ *and any* $\nu > 0$

$$\inf_{\mu\in\mathbf{K}_\mu} \ddot{G}(\theta^*,\mu,x) > 0,$$

$$\inf_{\mu\in\mathbf{K}_\mu} \inf_{|\theta-\theta^*|>\nu} [G(\theta,\mu,x) - G(\theta^*,\mu,x)] > 0.$$

Denote also $\Theta_\mu = \{\theta^*(\mu), \mu \in \mathcal{M}\}$ and suppose, that $\Theta_\mu \subset \Theta$. As before, the point $\dot{S}$ correspond to differentiating w.r.t. θ and prim S' w.r.t. x.

Theorem 2.13. *Let conditions I–II be fulfilled, then MLE* $\hat{\theta}_\varepsilon$ *is uniformly on* $\mu \in \mathbf{K}_\mu$

(a) "consistent"

$$\mathbf{P}_\mu - \lim_{\varepsilon\to 0} \hat{\theta}^*_\varepsilon = \theta^*;$$

(b) asymptotically normal

$$\mathcal{L}_\mu\left\{\varepsilon^{-1}(\hat{\theta}_\varepsilon - \theta^*)\right\} \Longrightarrow \mathcal{N}(0,\sigma^2)$$

with the variance $\sigma^2 = \sigma^2(\theta^*,\mu)$ *described below.*

Proof. By condition I, Equation (2.26) has a unique solution X and all the measures $\mathbf{P}_\theta^{(\varepsilon)}, \theta \in \Theta$ are equivalent. We take the likelihood ratio formula for the model (2.26) and substitute there the observations (2.28). Denote

$$\begin{aligned} L(\theta^* + u, \theta^*, X) &= \\ &= \ln \frac{d\mathbf{P}_{\theta^*+u}^{(\varepsilon)}}{d\mathbf{P}_{\theta^*}^{(\varepsilon)}}(X) = \frac{1}{\varepsilon}\int_0^T [S_t(\theta^* + u, X) - \\ &- S_t(\theta^*, X)]dW_t - \frac{1}{2\varepsilon^2}\int_0^T [S_t(\theta^* + u, X) - b_t(\mu, X)]^2 dt + \\ &+ \frac{1}{2\varepsilon^2}\int_0^T [S_t(\theta^*, X) - b_t(\mu, X)]^2 dt. \end{aligned}$$

Introduce a "good" set

$$\mathcal{A}_1 = \left\{\omega : \sup_{|u|>\nu_\varepsilon} L(\theta^* + u, \theta^*, X) < 0\right\}$$

which localizes the problem in the following sense: for all $\omega \in \mathcal{A}_1$ we have $\hat\theta_\varepsilon \in (\theta^* - \nu_\varepsilon, \theta^* + \nu_\varepsilon)$ because $L(\theta^*, \theta^*, X) = 0$. Here $\nu_\varepsilon = \varepsilon^\delta, \delta \in (0, \frac{1}{2})$.

By condition I and Lemma 1.1, the random process $L(\theta^* + u, \theta^*; X), u \in (\alpha - \theta^*, \beta - \theta^*)$ is differentiable on u and with $\mathbf{P}_\mu^{(\varepsilon)}$ probability 1, the random variable $\hat u_\varepsilon = \theta_\varepsilon^* - \theta^*$ is one of the solutions of the maximum likelihood equation

$$\varepsilon \int_0^T \dot S_t(\theta^* + u, X)dW_t - \int_0^T \dot S_t(\theta^* + u, X)[S_t(\theta^* + u, X) b_t(\mu, X)]dt = 0. \qquad (2.33)$$

Rewrite this equation in the form

$$\varepsilon\dot\eta(u) - \frac{1}{2}\dot G(\theta^* + u, \mu, X) = 0, \quad u \in (-\nu_\varepsilon, \nu_\varepsilon)$$

and denote it as

$$F(u, \varepsilon) = 0, \quad u \in (-\nu_\varepsilon, \nu_\varepsilon). \qquad (2.34)$$

Introduce also the second "good" set

$$A_2 = \left\{ \omega : \begin{array}{c} \sup_{0\le t\le T} |W_t| < C_2 C_1^{-1}\varepsilon^{-1+\delta} \\ \sup_{|u|<\nu_\varepsilon} \int_0^T \ddot S_t(\theta^* + u, X)dW_t < \frac{1}{4}\varepsilon^{-1}\ddot G(\theta^*, x) \\ \left|\int_0^T \dot S_t(\theta^*, X)dW_t\right| < \frac{1}{4}\ddot G(\theta^*, x)\nu_\varepsilon\varepsilon^{-1} \end{array} \right.$$

where $G(\theta, x) = G(\theta, \mu, x)$ and the constants C_1 and C_2 will be defined below.

Lemma 2.9. *Let conditions I–II be fulfilled and $\omega \in \mathcal{A}_1 \bigcap \mathcal{A}_2$, then there exists a number $\varepsilon_1 > 0$ such that for all $\varepsilon < \varepsilon_1$, Equation (2.33) has on interval $(-\nu_\varepsilon, \nu_\varepsilon)$ a unique solution $\hat u_\varepsilon$ and $\hat\theta_\varepsilon = \theta^* + \hat u_\varepsilon$.*

Proof. First, we differentiate Equation (2.33) w.r.t. u

$$\frac{\partial F(u,\varepsilon)}{\partial u} = \varepsilon\ddot{\eta}(u) - \ddot{G}(\theta^* + u, X)$$

and note that

$$\sup_{|u|\le\nu_\varepsilon} \frac{\partial F(u,\varepsilon)}{\partial u} \le \varepsilon \sup_{|u|\le\nu_\varepsilon} \ddot{\eta}(u) - \inf_{|u|\le\nu_\varepsilon} \ddot{G}(\theta^* + u, X). \tag{2.35}$$

By Lemma 1.13, we have

$$\sup_{0\le t\le T} |X_t - x_t| \le \varepsilon C_b T \sup_{0\le t\le T} |W_t|. \tag{2.36}$$

For the difference, we have

$$\sup_{|u|\le\nu_\varepsilon} |\ddot{G}(\theta^* + u, X) - \ddot{G}(\theta^*, x)| \le \sup_{|u|\le\nu_\varepsilon} |\ddot{G}(\theta^* + u, X) -$$
$$-\ddot{G}(\theta^* + u, x)| + \sup_{|u|\le\nu_\varepsilon} |\ddot{G}(\theta^* + u, x) - \ddot{G}(\theta^*, x)|.$$

For the first item of the last expression, we can write

$$\begin{aligned}
&|\ddot{G}(\theta, X) - \ddot{G}(\theta, x)| \le \\
&\le \int_0^T |\ddot{S}(\theta, X)[S_t(\theta, X) - b_t(\mu, X)]|dt + \\
&\quad + \int_0^T |\ddot{S}_t(\theta, x)[S_t(\theta, x) - b_t(\mu, x)]|dt + \int_0^T |\dot{S}_t(\theta, X)^2 - \dot{S}_t(\theta, x)^2|dt \le \\
&\le \int_0^T |\ddot{S}_t(\theta, X) - \ddot{S}_t(\theta, x)||S_t(\theta, X) - b_t(\mu, X)|dt + \\
&\quad + \int_0^T |\ddot{S}(\theta, x)||S_t(\theta, X) - S_t(\theta, x) + b_t(\mu, x) - b_t(\mu, X)|dt + \\
&\quad + \int_0^T |\dot{S}_t(\theta, X) - \dot{S}_t(\theta, x)||\dot{S}_t(\theta, X) + \dot{S}_t(\theta, x)|dt \le \\
&\le C_1\varepsilon \sup_{0\le t\le T} |W_t|.
\end{aligned}$$

For example,

$$\begin{aligned}
&|\dot{S}_t(\theta, X) - \dot{S}_t(\theta, x)| \le \\
&\le \int_0^t |\dot{K}(\theta, t, s, X_s) - \dot{K}(\theta, t, s, x_s)|ds + \\
&\quad + |\dot{V}(\theta, t, X_t) - \dot{V}(\theta, t, x_t)| \le C\varepsilon \sup_{0\le t\le T} |W_t|,
\end{aligned}$$

where we have used the boundedness of the derivatives $\dot{K}'(\cdot), \dot{V}'(\cdot)$.

Hence

$$|\ddot{G}(\theta, X) - \ddot{G}(\theta, x)| \le C\varepsilon \sup_{0\le t\le T} |W_t|. \tag{2.37}$$

For the second item, we have

$$
\begin{aligned}
&|\ddot{G}(\theta^* + u, x) - \ddot{G}(\theta^*, x)| \leq \\
&\leq \int_0^T |\ddot{S}_t(\theta^* + u, x)[S_t(\theta^* + u, x) - b_t(\mu, x)] - \\
&\quad -\ddot{S}_t(\theta^*, x)[S_t(\theta^*, x) - b_t(\mu, x)]|dt \int_0^T |\dot{S}_t(\theta, x)^2 - \dot{S}_t(\theta^* + u, x)^2|dt \leq \\
&\leq \int_0^T |\ddot{S}_t(\theta^* + u, x) - \ddot{S}_t(\theta^*, x)||\dot{S}_t(\theta^* + u, x) - b_t(\mu, x)|dt + \\
&\quad + \int_0^T |\ddot{S}(\theta^*, x)||S_t(\theta^* + u, x) - S_t(\theta^*, x)dt + \\
&\quad + \int_0^T |\dot{S}_t(\theta^* + u, x) - \dot{S}_t(\theta^*, x)||\dot{S}_t(\theta^*, x) + \dot{S}_t(\theta^*, x)|dt \leq C|u|,
\end{aligned}
$$

because we can always write

$$
\begin{aligned}
&|\ddot{S}_t(\theta^* + u, x) - \ddot{S}_t(\theta^*, x)| \leq \\
&\leq \int_0^t |\ddot{K}(\theta^* + u, t, s, x_s) - \ddot{K}(\theta^*, t, s, x_s)|ds + \\
&\quad + |\ddot{V}(\theta^* + u, t, x_t) - \ddot{V}(\theta^*, t, x_t)| \leq C|u|.
\end{aligned}
$$

Joining all the inequalities above, we obtain

$$
\sup_{|u| \leq \nu_\varepsilon} |\ddot{S}_t(\theta^* + u, X) - \ddot{S}_t(\theta^*, x)| \leq C_1 \varepsilon \sup_{0 \leq t \leq T} |W_t| + C_2 \nu_\varepsilon.
$$

Referring to (2.35), we write

$$
\begin{aligned}
&\sup_{|u| < \nu_\varepsilon} \frac{\partial F(u, \varepsilon)}{\partial u} \leq \\
&\leq \varepsilon \sup_{|u| < \nu_\varepsilon} \int_0^T \ddot{S}_t(\theta^* + u, X) dW_t - \ddot{G}(\theta^*, x) + \\
&\quad + C_1 \varepsilon \sup_{0 \leq t \leq T} |W_t| + C_2 \varepsilon^\delta < -\frac{1}{2} \ddot{G}(\theta^*, x)
\end{aligned}
\tag{2.38}
$$

for all $\varepsilon < \varepsilon_1$, where

$$
\varepsilon_1 = \left(\frac{\ddot{G}(\theta^*, x)}{8C_2} \right)^{\frac{1}{\delta}}.
$$

Note that

$$
\begin{aligned}
F(-\nu_\varepsilon, \varepsilon) &= F(0, \varepsilon) + \int_0^{-\nu_\varepsilon} \frac{\partial F}{\partial u} du \geq \varepsilon \int_0^T \dot{S}_t(\theta^*, X) dW_t + \\
&\quad + \frac{\nu_\varepsilon}{2} \ddot{G}(\theta^*, x) \geq \frac{1}{4} \nu_\varepsilon \ddot{G}(\theta^*, x) > 0
\end{aligned}
$$

and, analogously,

$$
F(\nu_\varepsilon, \varepsilon) \leq -\frac{1}{4} \nu_\varepsilon \ddot{G}(\theta^*, x) < 0.
$$

These two inequalities, together with (2.38), provide the existence and uniqueness of the solution of Equation (2.33).

We also need two technical lemmas.

Lemma 2.10. *Let the functions $K_b(\mu,t,s,x)$ and $V_b(\mu,t,x)$ have two continuous bounded derivatives w.r.t. x, then the random process (2.26) with $\mathbf{P}_\mu^{(\varepsilon)}$ probability 1 is continuously differentiable w.r.t. ε and the derivative $X_t^{(1)}$ satisfies the equation*

$$dX_t^{(1)} = \left[\int_0^t K_b'(\mu,t,s,X_s)X_s^{(1)}ds + V_b'(\mu,t,X_t)X_t^{(1)}\right]dt + dW_t,\ X_0^{(1)} = 0. \qquad (2.39)$$

Proof. First, we consider the difference $V_t(h) = |X_t(\varepsilon+h) - X_t(\varepsilon)|$. We can write

$$\begin{aligned} V_t(h) &\leq \int_0^t\int_0^s |K_b(\mu,s,v,X_v(\varepsilon+h)) - K_b(\mu,s,v,X(\varepsilon))|dvds + \\ &\quad + \int_0^t |V_b(\mu,s,X_s(\varepsilon+h) - V_b(\mu,s,X_s(\varepsilon))|ds + h|W_t| \leq \\ &\leq T\int_0^t |K_b'(\mu,s,v,\tilde{X}_s)|V_s(h)ds + \int_0^t |V_b'(\mu,s,\tilde{\tilde{X}}_s)|V_s(h)ds + |h||W_t| \leq \\ &\leq C_1T\int_0^t V_s(h)ds + |h|\sup_{0\leq t\leq T}|W_t|, \end{aligned}$$

so by Lemma 1.11 we obtain the uniform convergence

$$\lim_{h\to 0}\sup_{0\leq t\leq T}|X_t(\varepsilon+h) - X_t(\varepsilon)| \leq e^{C_1T^2}\sup_{0\leq t\leq T}|W_t|\lim_{h\to 0}|h| = 0. \qquad (2.40)$$

Denote now

$$U_t(h) = \left|[X_t(\varepsilon+h) - X_t(\varepsilon)]h^{-1} - X_t^{(1)}(\varepsilon)\right|.$$

The similar inequalities for this process are as follows

$$\begin{aligned} U_t(h) &\leq \int_0^t\int_0^s |[K_b(\mu,s,v,X_v(\varepsilon+h)) - K_b(\mu,s,v,X(\varepsilon))]h^{-1} - \\ &\quad -K_b'(\mu,s,v,X_v(\varepsilon))X_v^{(1)}|dvds \leq \\ &\leq \int_0^t \left|[V_b(\mu,s,X_s(\varepsilon+h)) - V_b(\mu,s,X_s(\varepsilon))]h^{-1} - V_b'(\mu,s,X_s)X_s^{(1)}\right|ds \leq \\ &\leq \int_0^t\int_0^s |K_b'(\mu,s,v,\tilde{X}_v)[X_v(\varepsilon+h) - X_v(\varepsilon)]h^{-1} - K_b'(\mu,s,v,X_v)X_v^{(1)}|dvds + \\ &\quad + \int_0^t \left|V_b'(\mu,s,\tilde{X}_s)[X_s(\varepsilon+h) - X_s(\varepsilon)]h^{-1} - V_b'(\mu,s,X_s)X_s^{(1)}\right|ds \leq \\ &\leq \int_0^t\int_0^s |K_b'(\mu,s,v,\tilde{X}_v)|U_v(h)dvds + \\ &\quad + \int_0^t\int_0^s |K_b'(\mu,s,v,\tilde{X}_v) - K_b'(\mu,s,v,X_v)||X_v^{(1)}dvds + \\ &\quad + \int_0^t |V_b'(\mu,s,\tilde{X}_s)U_s(h)ds + \int_0^t |V_b'(\mu,s,X_s) - V_b'(\mu,s,\tilde{X}_s)||X_s^{(1)}|ds \leq \\ &\leq C_1\int_0^t\int_0^s U_v(h)dvds + C_2\int_0^T\int_0^s |X_v^{(1)}|dvds\sup_{0\leq t\leq T}|X_t(\varepsilon+h) - X_t(\varepsilon)| + \end{aligned}$$

$$+C_3\int_0^t U_s(h)ds + C_4\int_0^T |X_s^{(1)}|ds \sup_{0\le t\le T}|X_t(\varepsilon+h)-X_t(\varepsilon)|.$$

Once more using Lemma 1.11, we obtain the necessary limit: uniformly on $\mu\in\mathbf{K}_\mu$

$$\mathbf{P}_\mu^{(\varepsilon)} - \lim_{h\to 0}\sup_{0\le t\le T} U_t(h) = 0.$$

Lemma 2.11. *Let the functions $K(\theta,t,s,x)$, $K_b(\mu,t,s,x)$, $V(\theta,t,x)$, $V_b(\mu,t,x)$ have two continuous bounded derivatives w.r.t. x, then the stochastic process*

$$\eta(\varepsilon) = \int_0^T S_t(\theta,X)dW_t, \quad \varepsilon\in(0,1]$$

has, with probability 1, a derivative w.r.t. ε and

$$\frac{\partial}{\partial\varepsilon}\int_0^T S_t(\theta,X)dW_t = \int_0^T\left(\int_0^t K'(\theta,t,s,X_s)X_s^{(1)}ds\right)dW_t + \int_0^T V'(\theta,t,X_s)X_t^{(1)}dW_t. \tag{2.41}$$

Proof is a direct consequence of Lemmas 1.1 and 2.10.

Note also that under the same assumptions, we have, with probability 1,

$$\frac{\partial}{\partial\varepsilon}\int_0^T S_t(\theta,X)dt =$$
$$= \int_0^T\left(\int_0^t K'(\theta,t,s,X_s)X_s^{(1)}ds\right)dt + \int_0^T V'(\theta,t,X_t)X_t^{(1)}dt. \tag{2.42}$$

Equation (2.34) defines an implicit function $u=u(\varepsilon)$ and the first two derivatives of this function are obtained by the formulas

$$u'_\varepsilon = -\frac{F'_\varepsilon}{\dot F_u}, \quad u''_{\varepsilon\varepsilon} = \frac{2\dot F'_{u\varepsilon}F'_\varepsilon\dot F_u - \ddot F_{uu}(F'_\varepsilon)^2 - F''_{\varepsilon\varepsilon}(\dot F_u)}{(\dot F_u)^3}. \tag{2.43}$$

Using Lemmas 2.10 and 2.11, we obtain from (2.33) the expressions for the first derivatives:

$$\begin{aligned}
F'_\varepsilon \;=\; & \int_0^T \dot S_t(\theta^*+u,X)dW_t + \\
& +\varepsilon\int_0^T\left[\int_0^t \dot K'(\theta^*+u,t,s,X_s)X_s^{(1)}ds + \dot V'(\theta^*+u,t,X_t)X_t^{(1)}\right]dW_t - \\
& -\int_0^T [S_t(\theta^*+u,X)-b_t(\mu,X)]\times \\
& \quad\times\left(\int_0^t \dot K'(\theta^*+u,t,s,X_s)X_s^{(1)}ds + \dot V'(\theta^*+u,t,X_t)X_t^{(1)}\right)dt - \\
& -\int_0^T \dot S_t(\theta^*+u,X)\times \\
& \quad\times\left(\int_0^t [K'(\theta^*+u,t,s,X_s)-K'_b(\mu,t,s,X_s)]X_s^{(1)}ds\right)dt + \\
& +[V'(\theta^*+u,t,X_t)-V'_b(\mu,t,X_t)]X_t^{(1)};
\end{aligned} \tag{2.44}$$

$$\dot{F}_u = \varepsilon\int_0^T \ddot{S}_t(\theta^*+u,X)dW_t - \int_0^T \dot{S}_t(\theta^*+u,X)^2dt - \\ - \int_0^T \ddot{S}_t(\theta^*+u,X)[S_t(\theta^*+u,X)-b_t(\mu,X)]dt. \tag{2.45}$$

The direct differentiating also gives us the exact expressions for the second derivatives $F''_{\varepsilon\varepsilon}, F''_{u\varepsilon}, F''_{uu}$, but they are cumbersome and, taking in account that we shall need only some upper estimates for them, we restrict ourselves to the calculation of these estimates only. So we can write

$$\sup_{|u|<\nu_\varepsilon} |\ddot{F}_{uu}| \le \varepsilon \sup_{|u|\le\nu_\varepsilon}\left|\int_0^T S_t^{(3)}(\theta^*+u,X)dW_t\right| + C_1, \tag{2.46}$$

$$\sup_{|u|<\nu_\varepsilon} |\dot{F}'_{u\varepsilon}| \le \sup_{|u|<\nu_\varepsilon}\left|\int_0^T \ddot{S}_t(\theta^*+u,X)dW_t\right| + \\ +\varepsilon \sup_{|u|\le\nu_\varepsilon}\left|\int_0^T\left(\int_0^t \ddot{K}'(\theta^*+u,t,s,X_s)X_s^{(1)}ds\right)dW_t\right| + \\ +\varepsilon \sup_{|u|<\nu_\varepsilon}\left|\int_0^T \ddot{V}'(\theta^*+u,t,X_t)X_t^{(1)}dW_t\right| + C_2\sup_{0\le t\le T}|X_t^{(1)}|, \tag{2.47}$$

$$\sup_{|u|<\nu_\varepsilon} |F''_{\varepsilon\varepsilon}| \le \sup_{|u|\le\nu_\varepsilon}\left|\int_0^T\left(\int_0^t \dot{K}'(\theta^*+u,t,s,X_s)X_s^{(1)}ds+\right.\right. \\ \left.\left.+\dot{V}'(\theta^*+u,t,X_t)X_t^{(1)}\right)dW_t\right| + \\ +\varepsilon \sup_{|u|\le\nu_\varepsilon}\left|\int_0^T\left(\int_0^t \dot{K}''(\theta^*+u,t,s,X_s)(X_s^{(1)})^2ds+\right.\right. \\ \left.\left.+\dot{V}''(\theta^*+u,t,X_t)(X_t^{(1)})^2\right)dW_t\right| + \\ +\varepsilon \sup_{|u|\le\nu_\varepsilon}\left|\int_0^T\left(\int_0^t \dot{K}'(\theta^*+u,t,s,X_s)X_s^{(2)}ds+\right.\right. \\ \left.\left.+\dot{V}'(\theta^*+u,t,X_t)X_t^{(2)}\right)dW_t\right| + \\ +C_3\sup_{0\le t\le T}|X_t^{(1)}|^2 + C_4\sup_{0\le t\le T}|X_t^{(2)}|.$$

Introduce also the third "good" set

$$\mathcal{A}_3 = \left\{\omega:\ \sup_{|u|<\nu_\varepsilon}|u''_{\varepsilon\varepsilon}(u)| \le 2\varepsilon^{-\frac{1}{2}}\right\}$$

and we can say that the following lemma is proved.

Lemma 2.12. *Let conditions I–II be fulfilled, then the MLE $\hat{\theta}_\varepsilon$ admits the representation*

$$\hat{\theta}_\varepsilon\chi_{\{\mathcal{A}\}} = (\theta^*+\varepsilon\xi+\varepsilon^{\frac{3}{2}}\eta)\chi_{\{\mathcal{A}\}},$$

where $\mathcal{A} = \cap_{j=1}^{3} \mathcal{A}_j$, *the random variable* $|\eta| < 1$ *and* ξ *is the following Gaussian random variable:*

$$\begin{aligned}
\xi &= \ddot{G}(\theta^*, x)^{-1} \Big\{ \int_0^T \dot{S}_t(\theta^*, x) dW_t - \\
&- \int_0^T [S_t(\theta^*, x) - b_t(\mu, x)] \left(\int_0^t \dot{K}'(\theta^*, t, s, x_s) x_s^{(1)} ds + \dot{V}'(\theta^*, t, x_t) x_t^{(-1)} \right) dt - \\
&- \int_0^T \dot{S}_t(\theta^*, x) \left(\int_0^t [K'(\theta^*, t, s, x_s) - K_b'(\mu, t, s, x_s)] x_s^{(1)} ds + \right. \\
&+ \left. [V''(\theta^*, x_t) - V_b''(\mu, t, x_t)] x_t^{(1)} \right) dt \Big\}
\end{aligned}$$

where $x_t^{(1)}$ *is a solution of the linear equation*

$$dx_t^{(1)} = \left(\int_0^t K_b'(\mu, t, s, x_s) x_s^{(1)} ds + V_b'(\mu, t, x_t) x_t^{(1)} \right) dt + dW_t, \quad x_0^{(1)} = 0$$

with x_t *from (2.29).*

We obtain the representation for ξ by a direct computation of (2.44) , (2.45) for $-F_\varepsilon'/F_u'$ at point $\varepsilon = 0$.

We also need the following lemma.

Lemma 2.13. *Let the process* $\{y_t, 0 \le t \le T\}$ *be a solution of the linear equation*

$$dy_t = [\int_0^t A(t, s) y_s ds + B(t) y_t] dt + dW_t, \quad y_0 = 0,$$

where $A(t,s)$ *and* $B(t)$ *are measurable bounded functions, then* $\{y_t, 0 \le t \le T\}$ *is a Gaussian process with zero mean.*

Proof. Note that if the process $\tilde{y}_t, 0 \le t \le T$ satisfies the system of equations

$$\begin{aligned}
d\tilde{y}_t &= [Z_t + B(t)\tilde{y}_t] dt + dW_t, \quad \tilde{y}_0, \\
dZ_t &= A(t)\tilde{y}_t dt, \qquad\qquad Z_0
\end{aligned} \tag{2.48}$$

with some bounded measurable $A(\cdot), B(\cdot)$ and Gaussian $(\tilde{y}, Z_0)$ independent on W_t, $0 \le t \le T$, then it is a Gaussian process (see, for example, [4],[97]).

Consider the difference

$$\begin{aligned}
\mathbf{E}(y_t - \tilde{y}_t)^2 &\le 2\mathbf{E} \left(\int_0^t \int_0^s (A(s, v) y_v - A(s)\tilde{y}_v) dv ds \right)^2 + \\
&\quad + 2\mathbf{E} \left(\int_0^t B(s)(y_s - \tilde{y}_s) ds \right)^2 \le \\
&\le 2T \int_0^t \mathbf{E} \left(\int_0^s A(s, v)(y_v - \tilde{y}_v) dv + \int_0^s (A(s, v) - A(v)) y_v dv \right)^2 + \\
&\quad + 2T \int_0^t B(s)^2 \mathbf{E}(y_s - \tilde{y}_s)^2 ds \le \\
&\le C_1 \int_0^t \mathbf{E}(y_s - \tilde{y}_s)^2 ds + C_2 \int_0^T \int_0^T (A(t, s) - A(s))^2 ds dt
\end{aligned}$$

with some constants C_1, C_2, where we have used the boundedness of $A(t,s), A(s)$ and $B(t)$. By Lemma 1.11, we can write

$$\sup_{0\le t\le T} \mathbf{E}(y_t - \tilde{y}_t)^2 \le C \int_0^T \int_0^T [A(t,s) - A(s)]^2 ds dt.$$

First assume that the function $A(t,s)$ is continuous on $t \in [0,T]$. Divide the interval $[0,T]$ in N parts by the points $t(j) = TN^{-1}j, j = 0,1,\dots,N$ and put

$$A(t) = A(0,t)$$

for $t \in [0,t(1)]$, then the process $\tilde{y}_t, t \in [0,t(1)]$ satisfies the system (2.48) and so is Gaussian. For $t \in [t(1),t(2)]$, we put $A(t) = A(t(1),t)$ and consider the solution of the system (2.48) on $[t(1),t(2)]$ with initial Gaussian $(\tilde{y}_{t(1)}, Z_{t(1)})$, then $\tilde{y}_t, t \in [t(1),t(2)]$ is also Gaussian. Repeating this procedure until the interval $[t(N-1),T]$, we establish that $\{\tilde{y}_t, 0 \le t \le T\}$ with step wise $A(\cdot)$ is a Gaussian process. For continuous on t functions $A(t,s)$, we have

$$\lim_{N\to\infty} \int_0^T \int_0^T [A(t,s) - A(s)]^2 ds dt = 0.$$

Moreover, for any vector $\lambda \in \mathbf{R}^m$, we can prove

$$\lim_{N\to\infty} \mathbf{E}\left(\sum_{i=1}^m \lambda_i (y_{t_i} - \tilde{y}_{t_i})\right)^2 = 0,$$

in a similar way, so $y_t, 0 \le t \le T$ is also a Gaussian process.

For measurable bounded $A(t,s)$, we can introduce the functions

$$A_M(t,s) = M \int_{(t-\frac{1}{M})\vee 0} A(v,s) dv,$$

which are continuous on $t \in [0,T]$ and such that

$$\lim_{M\to\infty} \int_0^T \int_0^T [A_M(t,s) - A(t,s)]^2 dt ds = 0.$$

Using these $A_M(t,s)$, we construct similar approximation of y_t as above.

Below, we formulate a result which is more general than that of Theorem 2.13.

Theorem 2.14. *Let conditions I–II be fulfilled, then there exist $\varepsilon_1 > 0$ such that for all $\varepsilon < \varepsilon_1$ the MLE $\hat{\theta}_\varepsilon$ admits the representation*

$$\hat{\theta}_\varepsilon = \theta^* + \left\{\varepsilon\xi + \eta\varepsilon^{\frac{3}{2}}\right\}\chi_{\{\mathcal{A}\}} + \zeta\chi_{\{\mathcal{A}^c\}}, \tag{2.49}$$

where $|\eta| < 1$ with probability 1 and we also have

$$\sup_{\mu\in K_\mu} \mathbf{P}_\mu^{(\varepsilon)}\{\mathcal{A}^c\} \le C_1 \exp\left\{-\kappa_1\varepsilon^{-\gamma_1}\right\}, \tag{2.50}$$

$$\sup_{\mu\in K_\mu} \mathbf{P}_\mu^{(\varepsilon)}\{|\zeta| > \varepsilon^\delta\} \le C_2 \exp\left\{-\kappa_2\varepsilon^{-\gamma_2}\right\} \tag{2.51}$$

with some positive constants $\kappa_i, c_i, \gamma_i, i = 1,2$.

Proof. This representation was established in Lemma 2.12 For $\omega \in \mathcal{A}$, so it remains only to prove the inequalities (2.50), (2.51).

Note that for the compliment of $\mathcal{A}$, we have an estimate

$$\mathbf{P}_\mu^{(\varepsilon)}\{\mathcal{A}^c\} \le \sum_{j=1}^{3} \mathbf{P}_\mu^{(\varepsilon)}\{\mathcal{A}_j^c\}.$$

We begin from $\mathcal{A}_1^c$. Under conditions I–II, the function $G(\theta,\mu,x)$ satisfies the equality

$$\lim_{u\to 0}[G(\theta^*+u,\mu,x) - G(\theta^*,\mu,x)]u^{-2} = \ddot{G}(\theta^*,\mu,x) > 0$$

because $\dot{G}(\theta^*,\mu,x) = 0$ and the second derivative is separated from zero, so by condition II, for any $\mathbf{K}_\mu \subset \mathcal{M}$, we can find a $\kappa > 0$, such that

$$\inf_{\mu\in\mathbf{K}_\mu} [G(\theta^*+u,\mu,x) - G(\theta^*,\mu,x)] \ge \kappa u^2. \tag{2.52}$$

By analogy with (2.37), we derive the estimate

$$|G(\theta^*+u,X) - G(\theta^*+u,x)| \le C\varepsilon \sup_{0\le t\le T} |W_t| \tag{2.53}$$

with the constant $C > 0$. Denote $\hat{G}(\theta^*+u,\mu,x) = G(\theta^*+u,\mu,x) - G(\theta^*,\mu,x)$. The inequalities (2.52) and (2.53) allow us to write

$$\begin{aligned}
\mathbf{P}_\mu^{(\varepsilon)}\{\mathcal{A}_1^c\} &= \mathbf{P}_\mu^{(\varepsilon)}\left\{\sup_{|u|>\nu_\varepsilon}\left[\int_0^T \Delta S_t(u)dW_t - \frac{1}{2\varepsilon}\hat{G}(\theta^*+u,\mu,X)\right] > 0\right\} \le \\
&\le \mathbf{P}_\mu^{(\varepsilon)}\left\{\sup_{|u|>\nu_\varepsilon}\left[\int_0^T \Delta S_t(u)dW_t - \frac{1}{4\varepsilon}\hat{G}(\theta^*+u,\mu,x)\right] > 0\right\} + \\
&\quad +\mathbf{P}_\mu^{(\varepsilon)}\left\{\sup_{|u|>\nu_\varepsilon} |\hat{G}(\theta^*+u,\mu,X) - \hat{G}(\theta^*+u,\mu,x)| > \right. \\
&\quad \left. > \inf_{|u|>\nu_\varepsilon} \hat{G}(\theta^*+u,\mu,x)\right\} \le \\
&\le \mathbf{P}_\mu^{(\varepsilon)}\left\{\sup_{|u|>\nu_\varepsilon} [\int_0^T \Delta S_t(u)dW_t \ge \frac{\kappa}{4}\nu_\varepsilon^2\varepsilon^{-1}\right\} + \\
&\quad +\mathbf{P}\left\{C\varepsilon \sup_{0\le t\le T} |W_t| > \kappa\nu_\varepsilon^2\right\},
\end{aligned} \tag{2.54}$$

where we have denoted $\Delta S_t(u) = S_t(\theta^*+u,X) - S_t(\theta^*,X)$.

For the stochastic integral, by Lemma 1.5 we can write

$$\mathbf{P}_\mu^{(\varepsilon)}\left\{\sup_{|u|>\nu_\varepsilon} [\int_0^T \Delta S_t(u)dW_t \ge \frac{\kappa}{4}\varepsilon^{-1+2\delta}\right\} \le$$

$$\leq \mathbf{P}_\mu^{(\varepsilon)}\left\{\sup_{|u|>\nu_\varepsilon}[\int_0^T[\Delta S_t(u)-\Delta S_t(-\nu_\varepsilon)]dW_t \geq \frac{\kappa}{8}\varepsilon^{-1+2\delta}\right\}+$$
$$+\mathbf{P}_\mu^{(\varepsilon)}\left\{[\int_0^T \Delta S_t(-\nu_\varepsilon)dW_t > \frac{\kappa}{8}\varepsilon^{-1+2\delta}\right\} \leq C_1\exp\left\{-\kappa_1\varepsilon^{-1+2\delta}\right\}.$$

Remember that by (1.49)

$$\mathbf{P}\left\{\sup_{0\leq t\leq T}|W_t| > \frac{\kappa}{C}\varepsilon^{-1+2\delta}\right\} \leq 2\exp\left\{-\frac{\kappa^2}{2C^2T}\varepsilon^{-2+4\delta}\right\}.$$

Hence

$$\sup_{\mu\in\mathbf{K}_\mu}\mathbf{P}_\mu^{(\varepsilon)}\{\mathcal{A}_1^c\} \leq C\exp\{-\kappa_0\varepsilon^{-1+2\delta}\}. \tag{2.55}$$

Consider now the second probability of complement.

$$\begin{aligned}
\mathbf{P}_\mu^{(\varepsilon)}\{\mathcal{A}_2^c\} \;\leq\; & \mathbf{P}\{\sup_{0\leq t\leq T}|W_t| > C_2C_1^{-1}\varepsilon^{-1+\delta}\}+\\
&+\mathbf{P}_\mu^{(\varepsilon)}\left\{\sup_{|u|\leq\nu_\varepsilon}\int_0^t \ddot{S}_t(\theta^*+u,X)dW_t > \frac{1}{4}\varepsilon^{-1}\ddot{G}(\theta^*,x)\right\}+\\
&+\mathbf{P}_\mu^{(\varepsilon)}\left\{\left|\int_0^t \dot{S}_t(\theta^*,X)dW_t\right| > \frac{1}{4}\varepsilon^{-1+\delta}\ddot{G}(\theta^*,x)\right\} \leq\\
\leq\; & 2\exp\left\{-\frac{C_2^2}{2C_1^2T}\varepsilon^{-2+2\delta}\right\}+C\exp\left\{-\kappa_2\varepsilon^{-1-\delta}\right\}+\\
&+C\exp\left\{-\kappa_3\varepsilon^{-1+\delta}\right\} \leq C\exp\left\{-\tilde{\kappa}_0\varepsilon^{-1+\delta}\right\},
\end{aligned} \tag{2.56}$$

where we have used a derivation similar to that discussed in Remark 1.1.

To estimate the last probability $\mathbf{P}_\mu^{(\varepsilon)}\{\mathcal{A}_3^c\}$, we note that for $\omega\in\mathcal{A}_2$

$$\inf_{|u|<\nu_\varepsilon}\left|\frac{\partial F}{\partial u}\right| > \frac{1}{2}\ddot{G}(\theta^*,x),$$

so we can write

$$\mathbf{P}_\mu^{(\varepsilon)}\{\mathcal{A}_3^c\} \leq \mathbf{P}_\mu^{(\varepsilon)}\{\mathcal{A}_3^c\bigcap\mathcal{A}_2\}+\mathbf{P}\{\mathcal{A}_2^c\}$$

and it is sufficient to impose the conditions

$$\sup_{|u|\leq\nu_\varepsilon}|\ddot{F}_{uu}|(F_\varepsilon')^2 < \frac{2}{3}\varepsilon^{-\frac{1}{2}}\ddot{G}(\theta^*,x)^3, \tag{2.57}$$

$$\sup_{|u|\leq\nu_\varepsilon}|\dot{F}_{u\varepsilon}'||F_\varepsilon'||\dot{F}_u| < \frac{2}{3}\varepsilon^{-\frac{1}{2}}\ddot{G}(\theta^*,x)^3,$$

$$\sup_{|u|\leq\nu_\varepsilon}|F_{\varepsilon\varepsilon}''||\dot{F}_u|^2 < \frac{2}{3}\varepsilon^{-\frac{1}{2}}\ddot{G}(\theta^*,x)^3, \tag{2.58}$$

which could be replaced in turn by the conditions

$$\sup_{|u|\le\nu_\varepsilon} |\dot{F}'_{u\varepsilon}| < \varepsilon^{-\frac{1}{6}},$$
$$\sup_{|u|\le\nu_\varepsilon} |F'_\varepsilon| < \frac{2}{3}\varepsilon^{-\frac{1}{6}}\ddot{G}(\theta^*,x),$$
$$\sup_{|u|\le\nu_\varepsilon} |\dot{F}_u| < \varepsilon^{-\frac{1}{6}}.$$

Now the estimate of $\mathbf{P}_\mu^{(\varepsilon)}\{\mathcal{A}_3^c\}$ is reduced to estimates of the probabilities of each of them. Say

$$\begin{aligned}
&\mathbf{P}_\mu^{(\varepsilon)}\left\{\sup_{|u|<\nu_\varepsilon} |\dot{F}'_{u\varepsilon}| > \varepsilon^{-\frac{1}{6}}\right\} \le \\
&\le \mathbf{P}_\mu^{(\varepsilon)}\left\{\sup_{|u|<\nu_\varepsilon} |\int_0^T \ddot{S}_t(\theta^*+u,X)dW_t| > \frac{1}{3}\varepsilon^{-\frac{1}{6}}\right\} + \\
&+\mathbf{P}_\mu^{(\varepsilon)}\left\{\sup_{|u|<\nu_\varepsilon} \left|\int_0^T (\int_0^t \ddot{K}'(\theta^*+u,t,s,X_s)X_s^{(1)}ds + \right.\right.\\
&\left.\left.+\ \ddot{V}'(\theta^*+u,t,X_t)X_t^{(1)})dW_t\right| > \frac{1}{3}\varepsilon^{-\frac{7}{6}}\right\} + \\
&+\mathbf{P}_\mu^{(\varepsilon)}\left\{\sup_{0\le t\le T} |X_t^{(1)}| > \frac{C_2^{-1}}{3}\varepsilon^{-\frac{1}{6}}\right\}.
\end{aligned} \tag{2.59}$$

For the first probability at the right-hand side of (2.59) as in (2.56), using Lemmas 1.5, we obtain

$$\mathbf{P}_\mu^{(\varepsilon)}\left\{\sup_{|u|<\nu_\varepsilon} \left|\int_0^T \ddot{S}_t(\theta^*+u_0,X)dW_t\right| > \frac{1}{3}\varepsilon^{-\frac{1}{6}}\right\} \le C\exp\{-\kappa_3\varepsilon^{-\frac{1}{6}}\}.$$

Further, the process $X_t^{(1)}$ satisfies the inequality

$$|X_t^{(1)}| \le \int_0^t \left(\int_0^s |K'_b(\mu,s,v,X_v)\,X_v^{(1)}|dv + |V'_b(\mu,s,X_s)\,X_s^{(1)}|\right) ds + \sup_{0\le t\le T} |W_t|,$$

and so by Lemma 1.11, we can write

$$\sup_{0\le t\le T} |X_t^{(1)}| \le M_1 \sup_{0\le t\le T} |W_t| \tag{2.60}$$

with $\mathbf{P}_\mu^{(\varepsilon)}$ probability 1. An analogous inequality for $X_t^{(2)}$ provides

$$\sup_{0\le t\le T} |X_t^{(2)}| \le M_2 \sup_{0\le t\le T} |W_t|^2$$

with some positive constants M_1, M_2.

So the condition

$$\mathbf{E}_\mu \exp\left\{\gamma \sup_\theta \left[\int_0^T \left(\int_0^t \ddot{K}'(\theta^* + u, t, s, X_s) X_s^{(1)} ds \; + \right.\right.\right.$$

$$\left.\left.\left. + \ddot{V}'(\theta^* + u, t, X_t) X_t^{(1)}\right)^2 dt\right]^{\frac{1}{2}}\right\} \le \mathbf{E}_\mu \exp\{C \sup_{0\le t\le T} |W_t|\} = M_3$$

of Lemma 1.5 is fulfilled with $p = 2$ and, using this lemma, we can write

$$\mathbf{P}_\mu^{(\varepsilon)}\left\{\sup_{|u|<\nu_\varepsilon} \left|\int_0^T \left(\int_0^t \ddot{K}'(\theta^* + u, t, s, X_s) X_s^{(1)} ds \; + \right.\right.\right.$$

$$\left.\left.\left. + \ddot{V}'(\theta^* + u, t, X_t) X_t^{(1)}\right) dW_t\right| > \frac{1}{3}\varepsilon^{-\frac{7}{6}}\right\} \le C \exp\{-\kappa\varepsilon^{-\frac{7}{12}}\}.$$

Inequality (2.60) allows us to have an exponential estimate for the last term on the right-hand side of (2.59).

In a similar way, estimates on the probabilities of compliments of the sets defined by (2.57) and (2.58) could be obtained.

Joining all the estimates, we arrive at

$$\sup_{\mu \in bfK_\mu} \mathbf{P}_\mu^{(\varepsilon)}\{\mathcal{A}^c\} \le C \exp\{-\kappa\varepsilon^{-\gamma}\},$$

with some constants $C > 0, \kappa > 0$ and $\gamma > 0$.

The inequality (2.51) has already been proved above because

$$\begin{aligned}
\mathbf{P}_\mu^{(\varepsilon)}\{|\zeta| > \varepsilon^\delta\} &= \mathbf{P}_\mu^{(\varepsilon)}\left\{\sup_{|u|>\nu_\varepsilon} L(\theta^* + u, \theta^*, X) > \sup_{|u|\le\nu_\varepsilon} L(\theta^* + u, \theta^*, X)\right\} \le \\
&\le \mathbf{P}_\mu^{(\varepsilon)}\left\{\sup_{|u|>\nu_\varepsilon} L(\theta^* + u, \theta^*, X) > 0\right\} = \\
&= \mathbf{P}_\mu^{(\varepsilon)}\{\mathcal{A}_1^c\} \le C \exp\{-\kappa_0\varepsilon^{-1+2\delta}\}.
\end{aligned}$$

So Theorem 2.14 is proved.

The representation (2.49) with (2.50) and (2.51) allows us to prove the "two-sided" asymptotic normality of MLE in the following way:

$$\begin{aligned}
&\mathbf{P}_\mu^{(\varepsilon)}\{(\hat{\theta}_\varepsilon - \theta^*)\varepsilon^{-1} < x\} = \\
&= \mathbf{P}_\mu^{(\varepsilon)}\{(\hat{\theta}_\varepsilon - \theta^*)\varepsilon^{-1} < x, \mathcal{A}\} + \mathbf{P}_\mu^{(\varepsilon)}\{(\hat{\theta}_\varepsilon - \theta^*)\varepsilon^{-1} < x, \mathcal{A}^c\} \le \\
&\le \mathbf{P}_\mu^{(\varepsilon)}\{\xi + \eta\varepsilon^{\frac{1}{2}} < x, \mathcal{A}\} + \mathbf{P}_\mu^{(\varepsilon)}\{\mathcal{A}^c\} \le \mathbf{P}_\mu^{(\varepsilon)}\{\xi < x + \sqrt{\varepsilon}\} + C \exp\{-\kappa_1\varepsilon^{-\gamma_1}\}.
\end{aligned}$$

On the other hand

$$\begin{aligned}
&\mathbf{P}_\mu^{(\varepsilon)}\{(\hat{\theta}_\varepsilon - \theta^*)\varepsilon^{-1} < x\} \ge \\
&\ge \mathbf{P}_\mu^{(\varepsilon)}\{(\hat{\theta}_\varepsilon - \theta^*)\varepsilon^{-1} < x, \mathcal{A}\} = \mathbf{P}_\mu^{(\varepsilon)}\{\xi + \eta\sqrt{\varepsilon} < x, \mathcal{A}\} \ge \\
&\ge \mathbf{P}_\mu^{(\varepsilon)}\{\xi < x - \sqrt{\varepsilon}, \mathcal{A}\} \ge \mathbf{P}_\mu^{(\varepsilon)}\{\xi < x - \sqrt{\varepsilon}\} - \mathbf{P}_\mu^{(\varepsilon)}\{\mathcal{A}^c\}.
\end{aligned}$$

The asymptotic normality now follows from the normality of ξ and the inequalities

$$\mathbf{P}_\mu^{(\varepsilon)}\{\xi < x - \sqrt{\varepsilon}\} - C\exp\{-\kappa_1\varepsilon^{-\gamma_1}\} \leq$$
$$\leq \mathbf{P}_\mu^{(\varepsilon)}\{(\hat\theta_\varepsilon - \theta^*)\varepsilon^{-1} < x\} \leq \mathbf{P}_\mu^{(\varepsilon)}\{\xi < x + \sqrt{\varepsilon}\} + C\exp\{-\kappa_1\varepsilon^{-\gamma_1}\}.$$

REMARK 2.8. The problem of estimation θ^* could be also considered as follows. Suppose that between two sets $\mathcal{M}$ and Θ_μ there is a one-to-one correspondence defined by the equality

$$\int_0^T [S_t(\theta^*,x) - b_t(\mu,x)]^2 dt = \inf_{\theta\in\Theta_\mu}\int_0^T [S_t(\theta,x) - b_t(\mu,x)]^2 dt.$$

Then we can introduce the reparametrization

$$\tilde b_t(\theta^*,x) = b_t(\mu(\theta^*),x),$$

where $x = \{x_t(\theta^*), 0 \leq t \leq T\}$ is a solution of the equation

$$\frac{dx_t}{dt} = \tilde b_t(\theta^*,x), \quad x_0, \quad 0 \leq t \leq T$$

and consider the problem of estimation θ^* by the observations

$$dX_t = \tilde b_t(\theta^*,X)dt + \varepsilon dW_t, \quad X_0 = x_0, \quad 0 \leq t \leq T$$

with a known function $\tilde b(\cdot)$. Then Theorem 2.13 allows us to describe the properties of a "real" MLE of θ^* by the observations (2.26).

Theorem 2.15. *Let the functions $K(\theta,t,s,x)$, $K_b(\mu,t,s,x)$, $V(\theta,t,x)$, $V_b(\mu,t,x)$ be continuous and bounded with their four first derivatives w.r.t. θ,μ and x and*

$$\inf_{|u|>h}\int_0^T [\tilde b_t(\theta^*+u,x) - \tilde b_t(\theta^*,x)]^2 dt > 0,$$

$$\inf_{\theta^*\in\Theta_\mu} I(\theta^*) > 0,$$

where

$$I(\theta^*) = \int_0^T \dot{\tilde b}_t(\theta^*,x)^2 dt = \int_0^T \dot b_t(\mu,x)^2 \dot\mu_{\theta^*}^2(\theta^*)dt.$$

Then the MLE $\hat\theta_\varepsilon^*$ is consistent uniformly on $\theta^* \in \mathbf{K}_\theta$, $\varepsilon^{-1}(\hat\theta_\varepsilon^* - \theta^*)$ is asymptotically normal with parameters $(0, I(\theta^*)^{-1})$, and $\hat\theta_\varepsilon^*$ is LAM.

REMARK 2.9. It is also interesting to consider the situation when the limit solutions of systems coincide. Let the observed process be

$$dX_t = b_t(\theta_0,X_t)dt + \varepsilon dW_t, \quad X_0 = x_0, \quad 0 \leq t \leq T \tag{2.61}$$

and the supposed parametric model

$$dX_t = S_t(\theta_0,X_t)dt + \varepsilon dW_t, \quad X_0 = x_0, \quad 0 \leq t \leq T \tag{2.62}$$

in the limit as $\varepsilon \to 0$ converge to the same function $x = (x_t, 0 \le t \le T)$ which is a solution of

$$\frac{dx_t}{dt} = S_t(\theta_0, x_t), \quad x_0, \quad 0 \le t \le T.$$

So, two functions $b_t(\theta_0, x_t)$ and $S_t(\theta_0, x_t), 0 \le t \le T$ coincide along the trajectory x, but outside x they can differ. Of course, the measures induced by the processes (2.61) and (2.62) need not coincide and we suppose that

$$\mathbf{P}^{(\varepsilon)}_{b(\theta_0)}\left\{\int_0^T [b_t(\theta_0, X_t) - S_t(\theta_0, X_t)]^2 dt = 0\right\} = 0$$

for $\varepsilon \neq 0$.

Denote $g_t(\theta, x) = S_t(\theta, x) - b_t(\theta, x)$ and $g_t'(\theta, x)$ is the derivative of $g_t(\theta, x)$ w.r.t. x. As before, $\dot{S}_t(\theta, x)$ is a derivative of $S_t(\theta, x)$ w.r.t. θ.

Also introduce a Gaussian process $x_t^{(1)}, 0 \le t \le T$ by the equation

$$dx_t^{(1)} = b_t'(\theta_0, x_t) x_t^{(1)} dt + dW_t, \quad x_0^{(1)} = 0, \quad 0 \le t \le T$$

and a Gaussian random variable

$$\xi = I(\theta_0)^{-1}\left\{\int_0^T \dot{S}_t(\theta_0, x_t) dW_t - \int_0^T \dot{S}_t(\theta_0, x_t) x_t^{(1)} g_t'(\theta_0, x_t) dt\right\}$$

with

$$I(\theta_0) = \int_0^T \dot{S}_t(\theta_0, x_t)^2 dt.$$

Theorem 2.16. *Let the function $S_t(\theta, x)$ have four bounded continuous derivatives on θ and x and $b_t(\theta_0, x)$ have three bounded continuous derivatives on x, the quantity $I(\theta_0) > 0$ and for any $\nu > 0$*

$$\inf_{|\theta - \theta_0| > \nu} \|S(\theta, x) - S(\theta_0, x)\| > 0.$$

Then the MLE $\hat{\theta}_\varepsilon$ is consistent and asymptotically normal:

$$\mathcal{L}_{b(\theta_0)}\{\varepsilon^{-1}(\hat{\theta}_\varepsilon - \theta_0)\} \Longrightarrow \mathcal{L}\{\xi\}.$$

Proof. The loglikelihood ratio for the model (2.62) on observations (2.61) has the representation

$$\begin{aligned} L(u, X) &= \frac{1}{\varepsilon}\int_0^T [S_t(\theta_0 + u, X_t) - S_t(\theta_0, X_t)] dW_t - \\ &- \frac{1}{2\varepsilon^2}\int_0^T [S_t(\theta_0 + u, X_t)^2 - S_t(\theta_0, X_t)^2 - \\ &- 2b_t(\theta_0, X_t)(S_t(\theta_0 + u, X_t) - S_t(\theta_0, X_t))] dt. \end{aligned}$$

Hence, the MLE $\hat{\theta}_\varepsilon = \theta_0 + \hat{u}_\varepsilon$, where $\hat{u}_\varepsilon$ is one of the solutions of the equation

$$\varepsilon \int_0^T \dot{S}_t(\theta_0 + u, X_t) dW_t - \int_0^T \dot{S}_t(\theta_0 + u, X_t)[S_t(\theta_0 + u, X_t) - b_t(\theta_0, X_t)] dt = 0.$$

Considering this equation as an implicit function $u = u(\varepsilon)$, we solve the problem as was done above. Of course, this is a particular case of the above considered problem.

Bayes estimators. Let us consider the properties of the BE $\tilde{\theta}_\varepsilon$ for the same (misspecified) situation. We suppose, that the observed process is (2.26) but the conditional expectation

$$\tilde{\theta}_\varepsilon = \int_\Theta y\pi(y)p(y|X_t, 0 \le t \le T)dy$$

is constructed with the a posterior density

$$p(y|X_t, 0 \le t \le T) = \frac{d\mathbf{P}_y^{(\varepsilon)}}{d\mathbf{P}_{\theta_0}^{(\varepsilon)}}(X) \Big/ \int_\Theta \pi(z) \frac{d\mathbf{P}_z^{(\varepsilon)}}{d\mathbf{P}_{\theta_0}^{(\varepsilon)}}(X) dz$$

based on the model (2.27). As before, we assume that $\Theta = (\alpha, \beta)$, $|\alpha| + |\beta| < \infty$ and the drifts are integral-type functionals. The prior density $\pi(y), y \in \Theta$ is positive and has a bounded derivative.

Changing the variable $y = \theta^* + \varepsilon u$ and denoting

$$Z_\varepsilon(u) = \frac{d\mathbf{P}_{\theta^*+\varepsilon u}^{(\varepsilon)}}{d\mathbf{P}_{\theta^*}^{(\varepsilon)}},$$

we can write

$$\tilde{\theta}_\varepsilon = \theta^* + \varepsilon \int_{(\alpha-\theta^*)\varepsilon^{-1}}^{(\beta-\theta^*)\varepsilon^{-1}} u\pi(\theta^* + \varepsilon u) Z_\varepsilon(u) du \Big/ \int_{(\alpha-\theta^*)\varepsilon^{-1}}^{(\beta-\theta^*)\varepsilon^{-1}} \pi(\theta^* + \varepsilon v) Z_\varepsilon(v) dv.$$

The limit for the $Z_\varepsilon(u)$ process is now

$$\begin{aligned} Z(u) &= \exp\left\{ u \int_0^T \dot{S}_t(\theta^*, x) dW_t - u \int_0^T [S_t(\theta^*, x) - b_t(\mu, x)] \times \right. \\ &\quad \times \left(\int_0^t \dot{K}'(\theta^*, t, s, x_s) x_s^{(1)} ds + \dot{V}'(\theta^*, t, x_t) x_t^{(1)} \right) dt - \\ &\quad - u \int_0^T \dot{S}_t(\theta^*, x) \left(\int_0^t K'(\theta^*, t, s, x_s) x_s^{(1)} ds + V'(\theta^*, t, x_t) x_t^{(1)} \right) dt - \\ &\quad \left. - \frac{1}{2} u^2 \ddot{G}(\theta^*, \mu, x) \right\} \equiv \exp\left\{ u\tilde{\xi} - \frac{1}{2} u^2 \ddot{G}(\theta^*, \mu, x) \right\}, \end{aligned}$$

where $\tilde{\xi}$ is a Gaussian zero-mean variable.

The direct calculation provides

$$\int_{-\infty}^{\infty} uZ(u) du \Big/ \int_{-\infty}^{\infty} Z(u) du = \ddot{G}(\theta^*, \mu, x)^{-1} \tilde{\xi} = \xi,$$

where ξ is a random variable from Theorem 2.14.

Theorem 2.17. *Let the conditions I, II be fulfilled and let $\pi(\cdot)$ be as described above, then the BE $\tilde{\theta}_\varepsilon$ is consistent and asymptotically normal:*

$$\mathcal{L}\{\varepsilon^{-1}(\tilde{\theta}_\varepsilon - \theta^*)\} \Longrightarrow \mathcal{L}\{\xi\}. \tag{2.63}$$

Proof is not given in detail because we shall use the similar results in the next section, where we introduce a set $\mathcal{A}$, such that for $\omega \in \mathcal{A}$ and some "slowly increasing" function $\lambda_\varepsilon \to \infty$, we have an estimate (see (2.72))

$$\sup_{|u|<\lambda_\varepsilon} |\ln Z_\varepsilon(u) - \ln Z(u)| < \varepsilon^\delta.$$

The circumstances in §2.7 are slightly different but the necessary modification could be done to establish this inequality and the convergence $\mathbf{P}_b^{(\varepsilon)}\{\mathcal{A}^c\} \to 0$, as $\varepsilon \to 0$. We put $L_\varepsilon(u) = \ln Z_\varepsilon(u)$ and $L(u) = \ln Z(u)$ and write

$$\mathbf{P}_b^{(\varepsilon)}\left\{\int_{-\lambda_\varepsilon}^{\lambda_\varepsilon} u^2|Z_\varepsilon(u) - Z(u)|du > \Delta\right\} \leq$$
$$\leq \mathbf{P}_b^{(\varepsilon)}\left\{\int_{-\lambda_\varepsilon}^{\lambda_\varepsilon} u^2 Z(u)|1 - \exp\{L_\varepsilon(u) - L(u)\}|du > \Delta, \mathcal{A}\right\} +$$
$$+\mathbf{P}_b^{(\varepsilon)}\{\mathcal{A}^c\} \leq \mathbf{P}_b^{(\varepsilon)}\{2\varepsilon^\delta \int_{-\infty}^{\infty} u^2 Z(u)du > \Delta, \mathcal{A}\} + \mathbf{P}_b^{(\varepsilon)}\{\mathcal{A}^c\} \longrightarrow 0$$

for any $\Delta > 0$. The integral

$$\int_{\lambda_\varepsilon}^{\infty} u^2 Z(u)du \to 0$$

because $\lambda_\varepsilon \to \infty$.

We have to show that

$$\int_{\lambda_\varepsilon}^{(\beta-\theta^*)\varepsilon^{-1}} u^2 Z_\varepsilon(u)du \to 0, \quad \int_{(\alpha-\theta^*)\varepsilon^{-1}}^{-\lambda_\varepsilon} u^2 Z_\varepsilon(u)du \to 0. \tag{2.64}$$

Then, for $m = 1$ and $m = 0$, we shall have the convergence

$$\left|\int_{(\alpha-\theta^*)\varepsilon^{-1}}^{(\beta-\theta^*)\varepsilon^{-1}} u^m \pi(\theta^* + \varepsilon u)Z_\varepsilon(u)du - \pi(\theta^*)\int_{-\infty}^{\infty} u^m Z(u)du\right| \longrightarrow 0$$

for $\omega \in \mathcal{A}$ and, as a consequence, (2.63).

We have

$$\mathbf{P}_b^{(\varepsilon)}\left\{\int_{\lambda_\varepsilon}^{(\beta-\theta^*)\varepsilon^{-1}} u^2 Z_\varepsilon(u)du > \Delta\right\} \leq$$
$$\leq \mathbf{P}_b^{(\varepsilon)}\left\{\frac{1}{3}(\beta-\theta^*)^3\varepsilon^{-3}\sup_{u>\lambda_\varepsilon} Z_\varepsilon(u) > \Delta\right\} =$$
$$= \mathbf{P}_b^{(\varepsilon)}\left\{\sup_{u>\lambda_\varepsilon}\left[\int_0^T [S(\theta^*+u, X_t) - S(\theta^*, X_t)]dW_t - \frac{1}{2\varepsilon}\hat{G}(\theta^*+u, \mu, X)\right] >\right.$$
$$\left. > \varepsilon\ln\left(\frac{3\Delta\varepsilon^3}{(\beta-\theta^*)^3}\right)\right\}$$

and we can use the derivation (2.54) which provides (2.64) and, hence, (2.63).

Example 2.5. Let the drift $b_t(X)$ in the observations

$$dX_t = b_t(X)dt + \varepsilon dW_t, \quad X_0 = x_0, \quad 0 \le t \le T$$

be unknown to the observer and in the family $\{S_t(\theta, X), \theta \in (\alpha, \beta)\}$ of (known) functions $S_t(\theta, X) = g_t(X) + \theta f_t(X)$ he seeks an element $S_t(\theta^*, X)$ which is closest to $b_t(x)$ in $L_2[0, T]$ along the trajectory of the nonperturbated real system

$$\frac{dx_t}{dt} = b_t(x), \quad x_0.$$

Then the "MLE" $\hat{\theta}_\varepsilon$ has the representation

$$\hat{\theta}_\varepsilon = \zeta_\varepsilon \chi_{\{\zeta_\varepsilon \in (\alpha,\beta)\}} + \alpha \chi_{\{\zeta_\varepsilon \le \alpha\}} + \beta \chi_{\{\zeta_\varepsilon \ge \beta\}},$$

where

$$\zeta_\varepsilon = \frac{\int_0^T f_t(X)[dX_t - g_t(X)dt]}{\int_0^T f_t(X)^2 dt} = \frac{\int_0^T f_t(X)[b_t(X) - g_t(X)]dt}{\int_0^T f_t(X)^2 dt} + $$
$$+\varepsilon \frac{\int_0^T f_t(X)dW_t}{\int_0^T f_t(X)^2 dt} \longrightarrow \theta^*,$$

and

$$\varepsilon^{-1}(\zeta - \theta^*) \to \xi$$

with

$$\theta^* = \left(\int_0^T f_t(x)^2 dt\right)^{-1} \int_0^T f_t(x)[b_t(x) - g_t(x)]dt$$

and

$$\mathcal{L}\{\xi\} = \mathcal{N}\left(0, \left(\int_0^T f_t(x)^2 dt\right)^{-1}\right)$$

under the "natural" conditions (smoothness, $\theta^* \in (\alpha, \beta)$, etc.).

2.7 Nonconsistent estimation

In the preceding paragraphs, the possibility of the consistent estimation of the parameter θ by the observations

$$dX_t = S_t(\theta, X)dt + \varepsilon dW_t, \quad X_0 = x_0, \quad 0 \le t \le T, \tag{2.65}$$

besides the smoothness of $S(\cdot)$ with respect to θ and X, was provided by the condition of "separability (or identifiability) of the models":

(I) *The function*

$$G(\theta, \theta_0; x) = \int_0^T [S_t(\theta, x) - S_t(\theta_0, x)]^2 dt, \quad \theta \in \Theta,$$

where $x = \{x_t, 0 \le t \le T\}$ *is a solution of the equation*

$$\frac{dx_t}{dt} = S_t(\theta_0, x), \quad x_0, \quad 0 \le t \le T \tag{2.66}$$

has a unique minimum on θ *at point* $\theta = \theta_0$.

This is a necessary condition for the consistence of any estimator. Indeed, suppose, that the equation

$$G(\theta, \theta_0, x) = 0$$

has a second root $\theta^* \ne \theta_0$. Then, for almost all $t \in [0, T]$, we have $S_t(\theta_0, x) = S_t(\theta^*, x)$ and the solutions of (2.66) and

$$\frac{dx_t}{dt} = S_t(\theta^*, x), \quad x_0, \quad 0 \le t \le T \tag{2.67}$$

are nondistinguishable, even "without noise" ($\varepsilon = 0$). It is not good to have such a parametric family, but sometimes the real systems are described by models with $G(\theta^*, \theta_0, x) = 0$, so it is interesting to know the asymptotic behavior of the MLE and BE in situations when consistent estimation is not possible. We begin with the case of

Entirely coincidence. MLE. We suppose that Θ is an open convex subset of $\mathbf{R}^d$.

I. *There are* k *points* $\theta_1, \ldots, \theta_k \in \Theta$ *such that with* $\mathbf{P}_{\theta_1}^{(\varepsilon)}$ *probability 1*

$$S_t(\theta_1, X) \equiv S_t(\theta_j, X), \quad j = 2, \ldots, k, \quad 0 \le t \le T$$

and there exists a subdivision $\Theta_1, \ldots, \Theta_k$ *of* $\Theta = \sum_{j=1}^k \Theta_j$ *such that for every* $j \in \{1, \ldots, k\}$, *the value* $\theta_j \in \Theta_j$ *with a nonempty vicinity* (θ_j *does not belong to the border of* Θ_j) *and for any* $\nu > 0$

$$\min_{j=1,\ldots,k} \inf_{|\theta - \theta_j| > \nu} \int_0^T [S_t(\theta, x) - S_t(\theta_j, x)]^2 dt > 0.$$

Of course, the value ν is such that the set

$$\{\theta : \ |\theta - \theta_j| > \nu\} \subset \Theta_j$$

and is not empty.

The condition of regularity here is

II. *The functions* $S_t(\theta, X), 0 \le t \le T, \theta \in \Theta$ *satisfy the conditions* $\mathcal{L}$ *and conditions III–V of §2.2 are fulfilled. We impose the condition of nondegeneracy of the matrix* $I(\theta)$ *for the points* $\theta_1, \ldots, \theta_k$ *only.*

REMARK 2.10 The last restriction is concerned with the following equality

$$0=\int_0^T [S_t(\theta_2,x)-S_t(\theta_1,x)]^2dt = \int_0^T (\theta_2-\theta_1,\dot S_t(\tilde\theta_t,x))^2dt =$$
$$= \left(\int_0^T \dot S_t(\tilde\theta_t,x)\dot S_t(\tilde\theta_t,x)^T dt(\theta_2-\theta_1),(\theta_2-\theta_1)\right).$$

We introduce the following notations:

$$\xi(\theta)=I(\theta)^{-\frac12}\int_0^T \dot S_t(\theta,x)dW_t,\qquad \zeta(\theta)=\int_0^T \dot S_t(\theta,x)dW_t,$$
$$\xi_i=\xi(\theta_i),\qquad \zeta_i=\zeta(\theta_i),\quad i=1,\ldots,k,$$
$$H_i=\{\omega:\ |\xi_i|>\max_{j\neq i}|\xi_j|\},\qquad p_i=\mathbf{P}(H_i),$$

$$\hat\theta=\sum_{i=1}^k \theta_i\chi_{\{H_i\}}.$$

So $\hat\theta$ is a discreet random variable with $\mathbf{P}(\hat\theta=\theta_i)=p_i$.

We also impose the condition

III. *For all* $i\neq j, i,j=1,\ldots,k$ *the probabilities*

$$\mathbf{P}\{|\xi_i|=|\xi_j|\}=0.$$

Under this condition, we have $\sum_{j=1}^k p_i=1$.

Theorem 2.18. *Let conditions I–III be fulfilled, then*

$$\mathbf{P}_{\theta_1}-\lim_{\varepsilon\to0}\hat\theta_\varepsilon=\hat\theta.$$

Proof. We only consider the case $k=2$. The general case is a direct generalization of this one.

Introduce the normed likelihood ratios

$$Z_\varepsilon^{(1)}(u)=\frac{d\mathbf{P}^{(\varepsilon)}_{\theta_1+\varepsilon u}}{d\mathbf{P}^{(\varepsilon)}_{\theta_1}}(X),\quad u\in U_\varepsilon=\{u:\ \theta_1+\varepsilon u\in\Theta_1\},\tag{2.68}$$

$$Z_\varepsilon^{(2)}(v)=\frac{d\mathbf{P}^{(\varepsilon)}_{\theta_2+\varepsilon v}}{d\mathbf{P}^{(\varepsilon)}_{\theta_1}}(X),\quad v\in V_\varepsilon=\{v:\ \theta_2+\varepsilon v\in\Theta_2\},\tag{2.69}$$

and the set

$$H_\varepsilon=\{\omega:\ \sup_{u\in U_\varepsilon}Z_\varepsilon^{(1)}(u)>\sup_{v\in V_\varepsilon}Z_\varepsilon^{(2)}(v)\}.\tag{2.70}$$

Also denote $\hat\theta_\varepsilon^{(1)}$ and $\hat\theta_\varepsilon^{(2)}$ as the maximum likelihood estimators defined on Θ_1 and Θ_2, respectively , i.e.,

$$\hat\theta_\varepsilon^{(1)}=arg\sup_{\theta\in\Theta_1}\frac{d\mathbf{P}^{(\varepsilon)}_\theta}{d\mathbf{P}^{(\varepsilon)}_{\theta_0}}(X)$$

where θ_0 is a fixed value. We put

$$\hat{\theta}_\varepsilon = \hat{\theta}_\varepsilon^{(1)}\chi_{\{H_\varepsilon\}} + \hat{\theta}_\varepsilon^{(2)}\chi_{\{H_\varepsilon^c\}}$$

(supposing for simplicity that $\hat{\theta}_\varepsilon = \hat{\theta}_\varepsilon^{(2)}$ if the maximums coincide).

The conditions of theorem 2.2 are fulfilled In each domain Θ_1 and Θ_2 , so $\theta_\varepsilon^{(1)} \to \theta_1$ and $\theta_\varepsilon^{(2)} \to \theta_2$. Moreover, the random two-dimensional field $(Z_\varepsilon^{(1)}(u), Z_\varepsilon^{(2)}(v), u \in U_\varepsilon, v \in V_\varepsilon)$ defined outside U_ε and V_ε decreasing to zero ([53], theorem A19) converges weakly to the random field $(Z^{(1)}(u), Z^{(2)}(v), u \in \mathbf{R}^d, v \in \mathbf{R}^d)$ with

$$\begin{aligned}
Z^{(1)}(u) &= \exp\left\{(u,\zeta_1) - \frac{1}{2}(I(\theta_1)u,u)\right\}, \\
Z^{(2)}(u) &= \exp\left\{(v,\zeta_2) - \frac{1}{2}(I(\theta_2)v,v)\right\}.
\end{aligned}$$

Hence, we have a convergence of distributions of all continuous functionals and

$$\begin{aligned}
\mathbf{P}_{\theta_1}^{(\varepsilon)}\{H_\varepsilon\} &\longrightarrow \mathbf{P}\{\sup_{u\in\mathbf{R}^d} Z^{(1)}(u) > \sup_{v\in\mathbf{R}^d} Z^{(2)}(v)\} = \\
&= \mathbf{P}\left\{\exp\{\frac{1}{2}|\xi_1|^2\} > \exp\{\frac{1}{2}|\xi_2|^2\right\} = \mathbf{P}\{|\xi_1| > |\xi_2|\} = \mathbf{P}(H_1).
\end{aligned}$$

So

$$\begin{aligned}
&\mathbf{P}_{\theta_1}^{(\varepsilon)}(|\hat{\theta}_\varepsilon - \hat{\theta}| > \delta) = \\
&= \mathbf{P}_{\theta_1}^{(\varepsilon)}(|\hat{\theta}_\varepsilon^{(1)} - \hat{\theta}| > \delta, H_\varepsilon) + \mathbf{P}_{\theta_1}^{(\varepsilon)}(|\hat{\theta}_\varepsilon^{(2)} - \hat{\theta}| > \delta, H_\varepsilon^c) = \\
&= \mathbf{P}_{\theta_1}^{(\varepsilon)}(|\hat{\theta}_\varepsilon^{(1)} - \theta_1| > \delta, H_\varepsilon, H_1) + \mathbf{P}_{\theta_1}^{(\varepsilon)}(|\hat{\theta}_\varepsilon^{(1)} - \theta_2| > \delta, H_\varepsilon, H_2) + \\
&\quad + \mathbf{P}_{\theta_1}^{(\varepsilon)}(|\hat{\theta}_\varepsilon^{(2)} - \theta_2| > \delta, H_\varepsilon^c, H_2) + \mathbf{P}_{\theta_1}^{(\varepsilon)}(|\hat{\theta}_\varepsilon^{(2)} - \theta_1| > \delta, H_\varepsilon^c, H_1) \leq \\
&\leq \mathbf{P}_{\theta_1}^{(\varepsilon)}(|\hat{\theta}_\varepsilon^{(1)} - \theta_1| > \delta) + \mathbf{P}_{\theta_1}^{(\varepsilon)}(H_\varepsilon \bigcap H_2) + \\
&\quad + \mathbf{P}_{\theta_2}^{(\varepsilon)}(|\hat{\theta}_\varepsilon^{(2)} - \theta_2| > \delta) + \mathbf{P}_{\theta_2}^{(\varepsilon)}(H_\varepsilon^c \bigcap H_1) \longrightarrow 0.
\end{aligned}$$

If we have k points $\theta_1, \ldots, \theta_k$, then similar arguments (and similar notations) allow us to prove the convergencies

$$\begin{aligned}
\mathbf{P}_{\theta_1}^{(\varepsilon)}\{\hat{\theta}_\varepsilon = \hat{\theta}_\varepsilon^{(i)}\} &\longrightarrow \mathbf{P}\{\sup_{u\in\mathbf{R}^d} Z^{(i)}(u) > \max_{j\neq i} \sup_{v\in\mathbf{R}^d} Z^{(j)}(v)\} = \\
&= \mathbf{P}\{|\xi_i| > \max_{j\neq i} |\xi_j|\} = p_i.
\end{aligned}$$

Theorem 2.19. *Let conditions I–III be fulfilled, then*

$$\mathbf{P}_{\theta_1} - \lim_{\varepsilon\to 0} \varepsilon^{-1}(\hat{\theta}_\varepsilon - \hat{\theta}) = \hat{\xi}$$

with

$$\hat{\xi} = \sum_{i=1}^k \hat{\xi}_i \chi_{\{H_i\}}, \quad \hat{\xi}_i = I(\theta_i)^{-1}\zeta(\theta_i).$$

Proof. We can write the representation

$$\begin{aligned}\varepsilon^{-1}(\hat\theta_\varepsilon-\hat\theta) &= \sum_{i=1}^{k}\varepsilon^{-1}(\hat\theta_\varepsilon^{(i)}-\theta_i)\chi_{\{H_i\}}+\\ &+\sum_{i=1}^{k}\varepsilon^{-1}\hat\theta_\varepsilon^{(i)}(\chi_{\{H_\varepsilon^{(i)}\}}-\chi_{\{H_i\}}),\end{aligned}$$

where

$$H_\varepsilon^{(i)}=\left\{\omega\ \sup_{u\in U_{\varepsilon,i}} Z_\varepsilon^{(i)}(u)>\max_{j\neq i}\sup_{u\in U_{\varepsilon,i}} Z_\varepsilon^{(j)}(u)\right\}.$$

Applying Theorem 2.2 , as was mentioned above, we establish the convergence

$$(\varepsilon^{-1}(\hat\theta_\varepsilon^{(1)}-\theta_1),\ldots,\varepsilon^{-1}(\hat\theta_\varepsilon^{(k)}-\theta_k))\Longrightarrow(\hat\xi_1,\ldots,\hat\xi_k).$$

Also note that for any $\delta>0$

$$\begin{aligned}\mathbf{P}_{\theta_1}^{(\varepsilon)}\left\{|\varepsilon^{-1}(\chi_{\{H_\varepsilon^{(i)}\}}-\chi_{\{H_i\}})|>\delta\right\} &= \mathbf{P}_{\theta_1}^{(\varepsilon)}\left\{(H_\varepsilon^{(i)}\bigcap H_i^C)\bigcup(H_i\bigcap H_\varepsilon^{(i)C})\right\}\leq\\ &\leq \mathbf{P}_{\theta_1}^{(\varepsilon)}\left\{H_\varepsilon^{(i)}\bigcap H_i^C\right\}+\mathbf{P}_{\theta_1}^{(\varepsilon)}\left\{H_i\bigcap H_\varepsilon^{(i)C}\right\}\end{aligned}$$

and from the joint weak convergence of $(Z_\varepsilon^{(i)}(\cdot),Z_\varepsilon^{(j)}(\cdot))$ to $(Z^{(i)}(\cdot),Z^{(j)}(\cdot))$ we have

$$\begin{aligned}&\mathbf{P}_{\theta_1}^{(\varepsilon)}\left\{H_\varepsilon^{(i)}\bigcap H_i^C\right\}=\\ &=\mathbf{P}_{\theta_1}^{(\varepsilon)}\left\{\sup_{u\in U_{\varepsilon,i}} Z_\varepsilon^{(i)}(u)>\max_{j\neq i}\sup_{u\in U_{\varepsilon,i}} Z_\varepsilon^{(j)}(u),\sup_u Z^{(i)}(u)<\max_{j\neq i}\sup_u Z^{(j)}(u)\right\}\longrightarrow\\ &\longrightarrow\mathbf{P}\left\{\sup_u Z^{(i)}(u)>\max_{j\neq i}\sup_u Z^{(j)}(u),\sup_u Z^{(i)}(u)<\max_{j\neq i}\sup_u Z^{(j)}(u)\right\}=0.\end{aligned}$$

REMARK 2.11 It is interesting to note that if $k=2$ and G_1,G_2 is a subdivision of Ω, i.e. $\Omega=G_1\cup G_2$, $G_1\cap G_2=\emptyset$ and we would like to minimize $\hat\xi^2$, $\hat\xi=\xi_1\,\chi_{\{G_1\}}+\xi_2\,\chi_{\{G_2\}}$ w.r.t. G_1, then

$$\inf_{G_1}\hat\xi^2=\xi_2^2\ +\ \inf_{G_1}(\xi_1^2-\xi_2^2)\chi_{\{G_1\}}$$

and we have to choose

$$G_1=\{\omega\ :\ |\xi_2|>|\xi_1|\}$$

which is almost opposite to the "choice" of MLE.

Bayes estimator. Suppose now that θ is a random variable with a priori density $\pi(y),y\in\Theta$ and the loss function is quadratic. Then the Bayes estimate is a conditional mathematical expectation

$$\tilde\theta_\varepsilon=\mathbf{E}(\theta|X_t,0\leq t\leq T)=\int_\Theta y\pi(y)p(y;X_t,0\leq t\leq T)dt$$

with

$$p(y;X_t,0\leq t\leq T)=\frac{d\mathbf{P}_y^{(\varepsilon)}}{d\mathbf{P}_{\theta_0}^{(\varepsilon)}}\left(\int_\Theta\pi(z)\frac{d\mathbf{P}_z^{(\varepsilon)}}{d\mathbf{P}_{\theta_0}^{(\varepsilon)}}(X)dz\right)^{-1},$$

where θ_0 is a fixed value of $\theta, \theta_0 \in \Theta$.

Denote

$$q_j = Q\pi(\theta_j)\det(I(\theta_j)^{-\frac{1}{2}})\exp\left\{\frac{1}{2}|\xi_j|^2\right\},$$

where Q is defined by the norming $\sum_{j=1}^k q_j = 1$. Introduce a random variable

$$\tilde{\theta} = \sum_{j=1}^k \theta_j q_j.$$

Theorem 2.20. *Suppose, that the function $\pi(\cdot)$ is continuous positive and bounded and conditions I, II are fulfilled, then*

$$\mathbf{P}_{\theta_1} - \lim_{\varepsilon\to 0}\tilde{\theta}_\varepsilon = \tilde{\theta}.$$

Proof. First consider the case $k = 2$. Changing the variables in the integrals, we can write

$$\varepsilon^{-1}\int_\Theta \pi(z)\frac{d\mathbf{P}_z^{(\varepsilon)}}{d\mathbf{P}_{\theta_1}^{(\varepsilon)}}(X)dz =$$

$$= \varepsilon^{-1}\int_{\Theta_1}\pi(z)\frac{d\mathbf{P}_z^{(\varepsilon)}}{d\mathbf{P}_{\theta_1}^{(\varepsilon)}}(X)dz + \varepsilon^{-1}\int_{\Theta_2}\pi(z)\frac{d\mathbf{P}_z^{(\varepsilon)}}{d\mathbf{P}_{\theta_1}^{(\varepsilon)}}(X)dz =$$

$$= \int_{\varepsilon^{-1}(\Theta_1-\theta_1)}\pi(\theta_1+\varepsilon u)\frac{d\mathbf{P}_{\theta_1+\varepsilon u}^{(\varepsilon)}}{d\mathbf{P}_{\theta_1}^{(\varepsilon)}}(X)du + \int_{\varepsilon^{-1}(\Theta_2-\theta_2)}\pi(\theta_2+\varepsilon v)\frac{d\mathbf{P}_{\theta_2+\varepsilon v}^{(\varepsilon)}}{d\mathbf{P}_{\theta_2}^{(\varepsilon)}}(X)dv \equiv$$

$$\equiv I_\varepsilon^{(1)} + I_\varepsilon^{(2)},$$

where we have used the identity $\mathbf{P}_{\theta_1}^{(\varepsilon)} = \mathbf{P}_{\theta_2}^{(\varepsilon)}$.

From the demonstration of theorem 10.2 in [53] follow the convergences

$$\int_{\varepsilon^{-1}(\Theta_1-\theta_1)}\pi(\theta_1+\varepsilon u)Z_\varepsilon^{(1)}(u)du \longrightarrow$$

$$\longrightarrow \pi(\theta_1)\int_{\mathbf{R}^d}\exp\left\{(u,\zeta(\theta_1)) - \frac{1}{2}(I(\theta_1)u,u)\right\} =$$

$$= \pi(\theta_1)\det(I(\theta_1)^{-\frac{1}{2}})(2\pi)^{-\frac{d}{2}}\exp\left\{\frac{1}{2}|\xi_1|^2\right\},$$

$$\varepsilon^{-1}\int_\Theta y\pi(y)\frac{d\mathbf{P}_y^{(\varepsilon)}}{d\mathbf{P}_{\theta_1}^{(\varepsilon)}}(X)dy =$$

$$= \int_{\varepsilon^{-1}(\Theta_1-\theta_1)}(\theta_1+\varepsilon u)\pi(\theta_1+\varepsilon u)Z_\varepsilon^{(1)}(u)du +$$

$$+ \int_{\varepsilon^{-1}(\Theta_2-\theta_2)}(\theta_2+\varepsilon v)\pi(\theta_2+\varepsilon v)Z_\varepsilon^{(2)}(v)dv \equiv I_\varepsilon^{(3)} + I_\varepsilon^{(4)} \longrightarrow$$

$$\longrightarrow \theta_1\pi(\theta_1)\det(I(\theta_1)^{-\frac{1}{2}})(2\pi)^{-\frac{d}{2}}\exp\left\{\frac{1}{2}|\xi_1|^2\right\} +$$

$$+\theta_2\pi(\theta_2)\det(I(\theta_2)^{-\frac{1}{2}})\exp\left\{\frac{1}{2}|\xi_2|^2\right\}.$$

A slight modification of that demonstration allows us to prove the joint convergence of these integrals and this gives a demonstration of Theorem 2.20 because

$$\tilde{\theta}_\varepsilon = (I_\varepsilon^{(1)} + I_\varepsilon^{(2)})^{-1}(I_\varepsilon^{(3)} + I_\varepsilon^{(4)}) \longrightarrow \tilde{\theta}.$$

As in Theorem 2.18, we also have convergence of $Z_\varepsilon^{(j)}(\cdot)$ in probability to $Z^{(j)}(\cdot)$ because these processes are defined on the same probability space.

REMARK 2.12 Note that in the case of coincidence of the values $\pi(\theta_1) = \pi(\theta_j), j = 2,\ldots,k$ and $\dot{S}_t(\theta_1, x) \equiv \dot{S}_t(\theta_j, x), j = 2,\ldots,k$, the BE converges to the arithmetic mean

$$\tilde{\theta} = \frac{\theta_1 + \ldots + \theta_k}{k}$$

and, except for the case $\theta_1 = \tilde{\theta}$(center of gravity), it does not converge to the true value.

REMARK 2.13 It is also interesting to note that in this problem the normed likelihood ratio

$$Z_\varepsilon(u) = \frac{d\mathbf{P}^{(\varepsilon)}_{\theta_1+\varepsilon u}}{d\mathbf{P}^{(\varepsilon)}_{\theta_1}}(X), \quad u \in U_\varepsilon = \{u : \theta_1 + \varepsilon u \in \Theta\}$$

has no limit process but in a certain sense is close to the "running" process

$$\begin{aligned} Z(u) &= \exp\left\{(u,\zeta_1) - \frac{1}{2}(I(\theta_1)u,u)\right\} \\ &\quad + \sum_{j=2}^{k} \exp\left\{((u-u_\varepsilon^{(j)},\zeta_j) - \frac{1}{2}(I(\theta_j)(u-u_\varepsilon^{(j)},(u-u_\varepsilon^{(j)}))\right\}, \end{aligned}$$

where $u_\varepsilon^{(j)} = \varepsilon^{-1}(\theta_1 - \theta_j)$. So we have a family of k "domes", one of which stands and all the others run (as $\varepsilon \to 0$) to infinity in different directions. The furthest "dome" has the fastest speed (as galaxies in astronomy).

Partial coincidence. Consider now the case when the functions $S_t(\theta_1, x_t)$ and $S(\theta_2, x_t)$ coincide along the trajectory $x_t, 0 \le t \le T$, which is a solution of

$$\frac{dx_t}{dt} = S(\theta_1, x_t), \quad x_0, \quad 0 \le t \le T,$$

and may differ outside it. So, if we denote $x_m = \min_{0\le t\le T} x_t$ and $x_M = \max_{0\le t\le T} x_t$, then the main condition is

IV. *We have* $S(\theta_1, x) = S(\theta_2, x)$ *for* $x \in [x_m, x_M]$. *Suppose also that the function* $S(\theta, x)$ *has two continuous derivatives w.r.t.* θ *and* $x, \theta \in (\alpha, \beta)$

$$\inf_{|\theta-\theta_i|>\nu, \theta\in\Theta_i} \int_0^T [S(\theta, x_t) - S(\theta_i, x_t)]^2 dt > 0$$

and $I(\theta_i) > 0, i = 1, 2.$

Denote

$$\zeta = I(\theta_2)^{-1}\left\{\int_0^T \dot S(\theta_2,x_t)dW_t - \int_0^T \dot S(\theta_2,x_t)g'(x_t)x_t^{(1)}dt\right\},$$

where

$$g(x) = S(\theta_2,x) - S(\theta_1,x), \quad g'(x) = \frac{dg(x)}{dx},$$

$x_t^{(1)}$ is a Gaussian process,

$$dx_t^{(1)} = S'(\theta_1,x_t)x_t^{(1)}dt + dW_t, \quad x_0^{(1)} = 0, \quad 0 \le t \le T \tag{2.71}$$

and the random function

$$L^{(2)}(v) = v\int_0^T \dot S(\theta_2,x_t)dW_t + \int_0^T g'(x_t)x_t^{(1)}dW_t - \frac{1}{2}\int_0^T [\dot S(\theta_2,x_t)v + g'(x_t)x_t^{(1)}]^2 dt.$$

We now put

$$\hat\theta = \theta_1\chi_{\{H_1\}} + \theta_2\chi_{\{H_2\}},$$

where

$$\begin{aligned} H_1 &= \{\omega :\ |\xi_1|^2 > 2L^{(2)}(\zeta)\}, \\ H_0 &= \{\omega :\ |\xi_1|^2 = 2L^{(2)}(\zeta)\}, \quad H_2 = \Omega\setminus H_1. \end{aligned}$$

Theorem 2.21. *Suppose that* $\mathbf{P}(H_0) = 0$ *and condition IV is fulfilled, then*

$$\mathbf{P}_{\theta_1} - \lim_{\varepsilon\to 0}\hat\theta_\varepsilon = \hat\theta.$$

Proof. Let $Z_\varepsilon^{(1)}(u)$ and $Z_\varepsilon^{(2)}(v)$ be the normed likelihood ratios (2.68) and (2.69). The limit processes are now

$$\begin{aligned} Z^{(1)}(u) &= \exp\left\{u\zeta_1 - \frac{1}{2}I(\theta_1)u^2\right\}, \quad u \in \mathbf{R}^1, \\ Z^{(2)}(v) &= \exp\left\{L(v)\right\}, \quad v \in \mathbf{R}^1. \end{aligned}$$

We also denote

$$L_\varepsilon^{(1)}(u) = \ln Z_\varepsilon^{(1)}(u), \quad L_\varepsilon^{(2)}(u) = \ln Z_\varepsilon^{(2)}(u), \quad L^{(1)}(u) = \ln Z^{(1)}(u).$$

As before

$$\hat\theta_\varepsilon = \hat\theta_\varepsilon^{(1)}\chi_{\{H_\varepsilon\}} + \hat\theta_\varepsilon^{(2)}\chi_{\{H_\varepsilon^c\}},$$

where

$$H_\varepsilon = \left\{\omega :\ \sup_{u\in U_\varepsilon} L_\varepsilon^{(1)}(u) > \sup_{v\in V_\varepsilon} L_\varepsilon^{(2)}(v)\right\}.$$

Introduce a function $\lambda_\varepsilon \to \infty$ as $\varepsilon \to 0$ and represent

$$\sup_{u\in U_\varepsilon} L_\varepsilon^{(1)}(u) = \max(\sup_{|u|<\lambda_\varepsilon} L_\varepsilon^{(1)}(u), \sup_{|u|\geq\lambda_\varepsilon, u\in U_\varepsilon} L_\varepsilon^{(1)}(u)).$$

So, if we denote

$$M_1 = \left\{\omega : \sup_{|u|\geq\lambda_\varepsilon, u\in U_\varepsilon} L_\varepsilon^{(1)}(u) < 0\right\},$$

then, for $\omega \in M_1$

$$\sup_{u\in U_\varepsilon} L_\varepsilon^{(1)}(u) = \sup_{|u|<\lambda_\varepsilon} L_\varepsilon^{(1)}(u).$$

To obtain an estimate on the probability $\mathbf{P}_{\theta_1}^{(\varepsilon)}\{H_\varepsilon\}$, we consider the difference in the processes

$$\begin{aligned}
&\sup_{|u|<\lambda_\varepsilon} |L_\varepsilon^{(1)}(u) - L^{(1)}(u)| \leq \\
&\leq \sup_{|u|<\lambda_\varepsilon} |\varepsilon^{-1}\int_0^T [S(\theta_1+\varepsilon u, X_t) - S(\theta_1, X_t)]dW_t - u\int_0^T \dot{S}(\theta_1, x_t)dW_t| + \\
&+\frac{1}{2}\sup_{|u|<\lambda_\varepsilon}\int_0^T |\varepsilon^{-2}[S(\theta_1+\varepsilon u, X_t) - S(\theta_1, X_t)]^2 - u^2\dot{S}(\theta_1, x_t)^2|dt \leq \\
&\leq \lambda_\varepsilon \sup_{|u|<\lambda_\varepsilon} |\int_0^T [\dot{S}(\theta_1+\varepsilon u, X_t) - \dot{S}(\theta_1, x_t)]dW_t| + \\
&+\frac{\lambda_\varepsilon^2}{2}\sup_{|u|<\lambda_\varepsilon}\int_0^T |\dot{S}(\theta_1+\varepsilon u, X_t)^2 - \dot{S}(\theta_1, x_t)^2|dt \leq \\
&\leq \lambda_\varepsilon \sup_{|u|<\lambda_\varepsilon} |\int_0^T [\dot{S}(\theta_1+\varepsilon u, X_t) - \dot{S}(\theta_1, x_t)]dW_t| + \\
&+C_1\lambda_\varepsilon^2 \sup_{|u|<\lambda_\varepsilon}\int_0^T (|\dot{S}(\theta_1+\varepsilon u, X_t) - \dot{S}(\theta_1+\varepsilon u, x_t)| + \\
&+|\dot{S}(\theta_1+\varepsilon u, x_t) - \dot{S}(\theta_1, x_t)|)dt,
\end{aligned}$$

where C_1 is a constant from the condition $|\dot{S}(\theta_1, x)| < C_1$.

Further, we introduce a "good" set

$$M_2 = \left\{\omega : \begin{array}{l}
\sup_{|u|<\lambda_\varepsilon} |\int_0^T [\dot{S}(\theta_1+\varepsilon u, X_t) - \dot{S}(\theta_1+\varepsilon u, x_t)]dW_t| < \frac{\varepsilon^\delta}{3\lambda_\varepsilon}, \\
\sup_{|u|<\lambda_\varepsilon} |\int_0^T [\dot{S}(\theta_1+\varepsilon u, x_t) - \dot{S}(\theta_1, x_t)]dW_t| < \frac{\varepsilon^\delta}{3\lambda_\varepsilon}, \\
\sup_{|u|<\lambda_\varepsilon} \int_0^T (|\dot{S}(\theta_1+\varepsilon u, X_t) - \dot{S}(\theta_1+\varepsilon u, x_t)| + \\
\quad +|\dot{S}(\theta_1+\varepsilon u, x_t) - \dot{S}(\theta_1, x_t)|)dt < \frac{\varepsilon^\delta}{3\lambda_\varepsilon^2 C_1}.
\end{array}\right.$$

As follows from the definition of this set, for $\omega \in M_1 \cap M_2$, we have

$$\sup_{u\in U_\varepsilon} |L_\varepsilon^{(1)}(u) - L^{(1)}(u)| < \varepsilon^\delta. \tag{2.72}$$

and, hence, on $M_1 \cap M_2$

$$\sup_{u\in U_\varepsilon} L_\varepsilon(u) = \sup_{|u|<\lambda_\varepsilon, u\in U_\varepsilon} L_\varepsilon(u) \le \sup_u L^{(1)}(u) + \varepsilon^\delta = \frac{1}{2}\xi_1^2 + \varepsilon^\delta.$$

In a similar way,

$$\sup_{u\in U_\varepsilon} L_\varepsilon^{(1)}(u) \ge \sup_u L^{(1)}(u) - \varepsilon^\delta = \frac{1}{2}\xi_1^2 - \varepsilon^\delta.$$

For the processes $L_\varepsilon^{(2)}(\cdot)$ and $L^{(2)}(v)$, the corresponding inequalities are

$$\begin{aligned}
&\sup_{|v|<\lambda_\varepsilon}\left|\int_0^T \varepsilon^{-1}[S(\theta_2+\varepsilon v, X_t) - S(\theta_1, X_t)]dW_t - \right.\\
&\quad\left. -v\int_0^T \dot S(\theta_2, x_t)dW_t - \int_0^T g'(x_t)x_t^{(1)}dW_t\right| +\\
&\quad +\frac{1}{2}\sup_{|v|<\lambda_\varepsilon}\left|\int_0^T (\varepsilon^{-2}[S(\theta_2+\varepsilon v, X_t) - S(\theta_2, X_t) + g(X_t)]^2 - \right.\\
&\quad\left. -[\dot S(\theta_2, x_t)v + g'(x_t)x_t^{(1)}]^2)dt\right| \le\\
&\quad \le \lambda_\varepsilon \sup_{|v|<\lambda_\varepsilon}\left|\int_0^T [\dot S(\theta_2+\varepsilon v, X_t) - \dot S(\theta_2, x_t)]dW_t\right| +\\
&\quad +\left|\int_0^T [\varepsilon^{-1}(g(X_t) - g(x_t)) - g'(x_t)x_t^{(1)}]dW_t\right| +\\
&\quad +\frac{1}{2}\sup_{|v|<\lambda_\varepsilon}\int_0^T (|\dot S(\theta_2+\varepsilon v, X_t) - \dot S(\theta_2, x_t)|v +\\
&\quad +|g'(\tilde X_t)\varepsilon^{-1}(X_t - x_t) - g'(x_t)x_t^{(1)}|) \times\\
&\quad \times|\dot S(\theta_2+\varepsilon v, X_t)v + g'(\tilde X_t)\varepsilon^{-1}(X_t - x_t) + \dot S(\theta_2, x_t)v + g'(x_t)x_t^{(1)}|dt \equiv R_\varepsilon.
\end{aligned}$$

Introduce the set

$$\begin{aligned}
M_3 &= \{\omega : \sup_{|v|\ge\lambda_\varepsilon} L_\varepsilon^{(2)}(v) < 0\},\\
M_4 &= \{\omega : R_\varepsilon < \varepsilon^\delta\},
\end{aligned}$$

then we have a similar estimate on $M_3 \cap M_4$:

$$\begin{aligned}
\sup_{v\in V_\varepsilon} L_\varepsilon^{(2)}(v) &= \sup_{|v|<\lambda_\varepsilon} L_\varepsilon^{(2)}(v) \le \sup_v L^{(2)}(v) + \varepsilon^\delta = L^{(2)}(\zeta) + \varepsilon^\delta,\\
\sup_{v\in V_\varepsilon} L_\varepsilon^{(2)}(v) &\ge \sup_v L^{(2)}(v) - \varepsilon^\delta = L^{(2)}(\zeta) - \varepsilon^\delta.
\end{aligned}$$

Hence, introducing

$$M = \bigcap_{j=1}^4 M_j,$$

we can obtain the estimates

$$\begin{aligned}
\mathbf{P}_{\theta_1}^{(\varepsilon)}\{\hat\theta_\varepsilon = \hat\theta_\varepsilon^{(1)}\} &= \mathbf{P}_{\theta_1}^{(\varepsilon)}\{H_\varepsilon \bigcap M\} + \mathbf{P}_{\theta_1}^{(\varepsilon)}\{H_\varepsilon \bigcap M^C\} \le \\
&\le \mathbf{P}_{\theta_1}^{(\varepsilon)}\{\frac{1}{2}\xi_1^2 + \varepsilon^\delta > L^{(2)}(\zeta) - \varepsilon^\delta, M\} + \mathbf{P}_{\theta_1}^{(\varepsilon)}\{M^C\} \le \\
&\le \mathbf{P}\{\frac{1}{2}\xi_1^2 > L^{(2)}(\zeta) - 2\varepsilon^\delta\} + \mathbf{P}_{\theta_1}^{(\varepsilon)}\{M^C\}, \\
\mathbf{P}_{\theta_1}^{(\varepsilon)}\{\hat\theta_\varepsilon = \hat\theta_\varepsilon^{(1)}\} &\ge \mathbf{P}_{\theta_1}^{(\varepsilon)}\{H_\varepsilon \bigcap M\} \ge \mathbf{P}_{\theta_1}^{(\varepsilon)}\{\frac{1}{2}\xi_1^2 - \varepsilon^\delta > L^{(2)}(\zeta) + \varepsilon^\delta, M\} \ge \\
&\ge \mathbf{P}_{\theta_1}^{(\varepsilon)}\{\frac{1}{2}\xi_1^2 > L^{(2)}(\zeta) + 2\varepsilon^\delta\} - \mathbf{P}_{\theta_1}^{(\varepsilon)}\{M^C\}.
\end{aligned}$$

The last probability could be estimated as follow. Fix $\delta \in (0, \frac{1}{2})$ and put $\gamma \in (0, \frac{1-\delta}{2})$, then $\varepsilon^\delta \lambda_\varepsilon^{-2}\varepsilon^{-1} \to \infty$ and we can use Lemma 1.5 in the estimation of large deviations for integrals like

$$\mathbf{P}_{\theta_1}^{(\varepsilon)}\left\{\sup_{|v|<\lambda_\varepsilon} \lambda_\varepsilon^{-1}\varepsilon^{-1}\left|\int_0^T [\dot S(\theta_2 + \varepsilon v, X_t) - \dot S(\theta_2, x_t)]dW_t\right| > \varepsilon^{-(1-\delta-2\gamma)}\right\}.$$

For $g'(\cdot)$ we have the estimate

$$\begin{aligned}
&|g'(\tilde X_t)\varepsilon^{-1}(X_t - x_t) - g'(x_t)x_t^{(1)}| \le \\
&\le |g'(\tilde X_t) - g'(x_t)||x_t^{(1)}| + |g'(x_t)||\varepsilon^{-1}(X_t - x_t) - x_t^{(1)}| \le \\
&\le C_2 \sup_{0\le t\le T}(|X_t - x_t||x_t^{(1)}|) + C_2 \sup_{0\le t\le T}|W_t|^2 \le C\varepsilon \sup_{0\le t\le T}|W_t|^2.
\end{aligned}$$

A more detailed study of all the terms in $\mathbf{P}_{\theta_1}^{(\varepsilon)}\{M^C\}$ repeating the corresponding calculations of Theorem 3.1, allows us to write the estimate

$$\mathbf{P}_{\theta_1}^{(\varepsilon)}\{M^C\} \le C_1 \exp\{-C_1\varepsilon^{-C_2}\}. \tag{2.73}$$

So

$$\begin{aligned}
\mathbf{P}_{\theta_1}^{(\varepsilon)}\{|\hat\theta_\varepsilon - \theta_1| < \delta\} &= \mathbf{P}_{\theta_1}^{(\varepsilon)}\{|\hat\theta_\varepsilon - \theta_1| < \delta, H_\varepsilon\} + \mathbf{P}_{\theta_1}^{(\varepsilon)}\{|\hat\theta_\varepsilon^{(1)} - \theta_1| < \delta, H_\varepsilon^c\} \le \\
&\le \mathbf{P}_{\theta_1}^{(\varepsilon)}\{\sup_{|u|<\delta\varepsilon^{-1}} Z_\varepsilon^{(1)}(u) > \sup_{|u|>\delta\varepsilon^{-1}} Z_\varepsilon^{(1)}(u)\} + \mathbf{P}_{\theta_1}^{(\varepsilon)}\{H_\varepsilon^c\}.
\end{aligned}$$

The behavior of the Bayes estimator is also now clear in this situation. The asymptotic of the integral in the vicinity of the point θ_2 is

$$\begin{aligned}
\int_{\Theta_2} \frac{d\mathbf{P}_y^{(\varepsilon)}}{d\mathbf{P}_{\theta_1}^{(\varepsilon)}}dy &= \int_{\varepsilon^{-1}(\Theta_2-\theta_2)} \pi(\theta_2 + \varepsilon v)Z_\varepsilon^{(2)}(v)dv \approx \varepsilon\pi(\theta_2)\int_{-\infty}^{\infty} Z^{(2)}(v)dv = \\
&= \varepsilon\pi(\theta_2)\sqrt{\frac{2\pi}{I(\theta_2)}}\exp\left\{\frac{\xi_2 - \int_0^T \dot S(\theta_2, x_t)g'(x_t)x_t^{(1)}dt}{2I(\theta_2)} + \right. \\
&\quad \left. + \int_0^T g'(x_t)x_t^{(1)}dW_t - \frac{1}{2}\int_0^T g'(x_t){x_t^{(1)}}^2 dt\right\} = \varepsilon q_2^*.
\end{aligned}$$

Of course, as before

$$\int_{\Theta_1} \frac{d\mathbf{P}_y^{(\varepsilon)}}{d\mathbf{P}_{\theta_1}^{(\varepsilon)}}(X)dy \approx \varepsilon\sqrt{\frac{2\pi}{I(\theta_1)}}\pi(\theta_1)\exp\left\{\frac{\xi_1^2}{2I(\theta_1)}\right\} = \varepsilon q_1^*.$$

Hence, the BE $\tilde{\theta}_\varepsilon$ converges to the value

$$\tilde{\theta} = \frac{\theta_1 q_1^* + \theta_2 q_2^*}{q_1^* + q_2^*}.$$

Entirely coincidence. MLE ($\rho = 1$)
Suppose that conditions I and II are fulfilled and instead of III we have:

IV.

$$\mathbf{P}\{|\zeta_1| = |\zeta_2|\} = 1.$$

Then the limit processes $Z^{(1)}(\cdot)$ and $Z^{(2)}(\cdot)$ have two identical maximums and the limit variable $\hat{\theta}$ is not well defined.

Consider this case in detail for the diffusion process

$$dX_t = S(\theta_1, X_t)dt + \varepsilon dW_t, \quad X_0 = x_0, \quad 0 \le t \le T$$

with $S(\theta_1, x) = S(\theta_2, x)$ for all $x \in R$. To study the limit behavior of $\hat{\theta}_\varepsilon$ we take into account the preceding term in the expansion of the likelihood ratio. Supposing that the function $S(\theta, x)$ has the required number of derivatives, we can write

$$\begin{aligned} L_\varepsilon^{(1)}(u) &= u\int_0^T \dot{S}(\theta_1, x_t)dW_t - \frac{u^2}{2}\int_0^T \dot{S}(\theta, x_t)^2 dt + \\ &+\varepsilon\left\{u\int_0^T \dot{S}'(\theta_1, x_t)x_t^{(1)}dW_t + u^2\int_0^T \ddot{S}(\theta_1, x_t)dW_t - \right. \\ &-u^3\int_0^T \dot{S}(\theta_1, x_t)\ddot{S}(\theta_1, x_t)dt - \\ &\left. -\,u^2\int_0^T \dot{S}(\theta_1, x_t)\dot{S}'(\theta, x_t)x_t^{(1)}dt\right\} + \varepsilon^2 R_\varepsilon(u,\varepsilon) = \\ &= u\zeta_1 - \frac{1}{2}I(\theta_1)u^2 + \varepsilon\psi_1(u) + \varepsilon^2 R_1(u,\varepsilon). \end{aligned}$$

A similar expansion for the second point is

$$L^{(2)}(v) = u\zeta_2 - \frac{1}{2}I(\theta_2)v^2 + \varepsilon\psi_2(v) + \varepsilon^2 R_2(u,\varepsilon).$$

The process $x_t^{(1)}$ is defined by (2.71).

Introducing the appropriate sets M_i as before, we obtain (on M) the representation

$$\begin{aligned} \sup_{|u|<\lambda_\varepsilon} L^{(1)}(u) &= \frac{1}{2}\xi_1^2 + \varepsilon\psi_1(\xi_1) + o(\varepsilon), \\ \sup_{|u|<\lambda_\varepsilon} L^{(2)}(u) &= \frac{1}{2}\xi_2^2 + \varepsilon\psi_2(\xi_2) + o(\varepsilon). \end{aligned}$$

Hence, from the equality $|\xi_1| = |\xi_2|$ we have the asymptotic

$$\begin{aligned} \mathbf{P}_{\theta_1}^{(\varepsilon)}\{\hat\theta_\varepsilon = \hat\theta_\varepsilon^{(1)}\} &\sim \mathbf{P}_{\theta_1}^{(\varepsilon)}\{\sup_{|u|<\lambda_\varepsilon} L^{(1)}(u) > \sup_{|v|<\lambda_\varepsilon} L^{(2)}(v)\} \sim \\ &\sim \mathbf{P}_{\theta_1}^{(\varepsilon)}\left\{\frac{1}{2}\xi_1^2 + \varepsilon\psi_1(\xi_1) > \frac{1}{2}\xi_2^2 + \varepsilon\psi_2(\xi_2)\right\} \longrightarrow \\ &\longrightarrow \mathbf{P}\{\psi_1(\xi) > \psi_2(\xi)\} = p_1, \quad \xi = \xi_1. \end{aligned}$$

So, if we suppose that $\mathbf{P}\{\psi_1(\xi) = \psi_2(\xi)\} = 0$, then the limit random variable $\hat\theta$ has the distribution

$$\mathbf{P}\{\hat\theta = \theta_1\} = p_1 = 1 - \mathbf{P}\{\hat\theta = \theta_2\}.$$

If $\mathbf{P}\{\psi_1(\xi) = \psi_2(\xi)\} = 1$, then we should consider the next (previous) term of expansion ($\sim \varepsilon^2$) and reduce the problem to a comparison of these terms, etc.

The behavior of the MLE in the case when $S(\theta_1 + h, x) = S(\theta_2 + h, x)$ for all $x \in R$ and all $|h| < \Delta$, with some $\Delta > 0$ (all derivatives coincide), depends on the definition of the MLE when the likelihood ratio has coinciding maximums. If, for example, we choose the minimal value, then $\hat\theta_\varepsilon \to \min(\theta_1, \theta_2)$.

We now return to the problem with k points. Suppose, that

$$\mathbf{P}\{|\xi_1| = |\xi_2|\} = 1, \quad \mathbf{P}\{|\xi_i| - |\xi_j|\} - 0, \quad i \neq j, \quad i - 2, \ldots, k, \quad j = 3, \ldots, k.$$

Then it can be shown that the limit random variable has the distribution

$$\mathbf{P}\{\hat\theta = \theta_1\} = q\mathbf{P}\{\psi_1(\xi) > \psi_2(\xi)\} = q - \mathbf{P}\{\hat\theta = \theta_2\},$$

$$\mathbf{P}\{\hat\theta = \theta_i\} = \mathbf{P}\{|\xi_i| > \max_{j\neq i, j\neq 1} |\xi_j|\}, \quad i = 2, \ldots, k,$$

$$q = 1 - \sum_{i=3}^{k} \mathbf{P}(\hat\theta = \theta_i).$$

2.8 Boundary of the parametric set

In the preceding parts of this work, it has always been supposed that the true value θ is an interior point of an open set Θ. Below, we consider the situation when the set Θ is closed and the true value belongs to the boundary of this set. So, let $\Theta = [\alpha, \beta]$ and the true value $\theta = \alpha$. The observed diffusion-type process is

$$dX_t = S_t(\alpha, X)dt + \varepsilon dW_t, \quad X_0 = x_0, \quad 0 \le t \le T. \tag{2.74}$$

First we consider the MLE

$$\hat\theta_\varepsilon = \arg\sup_{\theta\in\Theta} \frac{d\mathbf{P}_\theta^{(\varepsilon)}}{d\mathbf{P}_\theta^{(\varepsilon)}}(X),$$

where θ_0 is some fixed value. We prove that under regularity conditions, the estimator $\hat\theta_\varepsilon$ is consistent and asymptotically "half-normal". The limit function $x_t, 0 \le t \le T$ is a solution of the equation

$$\frac{dx_t}{dt} = S_t(\alpha, x), \quad x_0, \quad 0 \le t \le T$$

and we denote

$$I(\alpha+) = \int_0^T \dot{S}_t(\alpha+, x)^2 dt,$$

where $\dot{S}_t(\alpha+, x)$ is the right derivative of the function $S_t(\theta, x)$ w.r.t. θ at point $\theta = \alpha$.

Theorem 2.22. *Suppose that the measurable function $S_t(\theta, X)$ satisfies the condition $\mathcal{L}$ and has two continuous bounded derivatives w.r.t. θ in the vicinity of point $\theta = \alpha$, the derivative $\dot{S}_t(\alpha+, X)$ satisfies the condition $\mathcal{L}_1$, the quantity $I(\alpha+) > 0$ and*

$$\inf_{|\theta-\alpha|>\nu} \int_0^T [S_t(\alpha, x) - S_t(\theta, x)]^2 dt > 0$$

for any $\nu > 0$. Then the MLE $\hat{\theta}_\varepsilon$ is

(a) consistent:

$$\mathbf{P}_\alpha - \lim_{\varepsilon \to 0} \hat{\theta}_\varepsilon = \alpha;$$

(b) asymptotically "half-normal":

$$\lim_{\varepsilon \to 0} \mathbf{P}_\alpha^{(\varepsilon)} \{ I(\alpha+)^{\frac{1}{2}} \varepsilon^{-1} (\hat{\theta}_\varepsilon - \alpha) < x \} = \begin{cases} \Phi(x), & x > 0 \\ \frac{1}{2}, & x = 0 \\ 0, & x < 0 \end{cases}$$

(c) the moments also converge

$$\lim_{\varepsilon \to 0} \mathbf{E}_\alpha | I(\alpha+)^{\frac{1}{2}} \varepsilon^{-1} (\hat{\theta}_\varepsilon - \alpha)|^p = \frac{2^{\frac{p}{2}-1}}{\sqrt{\pi}} \Gamma\left(\frac{p-1}{2}\right).$$

Here $\Phi(x)$ is the distribution function of $\mathcal{N}(0,1)$.

Proof. Introduce the notations

$$Z(u) = \exp\left\{ u\Delta(\alpha, x) - \frac{1}{2} u^2 I(\alpha+) \right\},$$

where $u \geq 0$ and

$$\Delta(\alpha, x) = \int_0^T \dot{S}_t(\alpha+, x) dW_t,$$

$$\mathcal{L}\{\Delta(\alpha, x)\} = \mathcal{N}(0, I(\alpha+)).$$

The maximum of this process is

$$\max_{u \geq 0} Z(u) = Z(\hat{u}),$$

where

$$\hat{u} = \begin{cases} \Delta(\alpha, x), & \text{if } \Delta(\alpha, x) > 0 \\ 0, & \text{if } \Delta(\alpha, x) \leq 0. \end{cases}$$

Note that $\mathbf{P}(\Delta(\alpha,x)>0)=\frac{1}{2}$.

For the process

$$Z_\varepsilon(u)=\frac{d\mathbf{P}^{(\varepsilon)}_{\alpha+\varepsilon u}}{d\mathbf{P}^{(\varepsilon)}_{\alpha}}(X),\quad u\geq 0 \tag{2.75}$$

repeating the corresponding calculations in the proof of Lemmas 2.2 and 2.3, we obtain the inequalities

$$\mathbf{E}_\alpha|Z_\varepsilon^{\frac{1}{2}}(u_2)-Z_\varepsilon^{\frac{1}{2}}(u_1)|^2\leq C|u_2-u_1|^2 \tag{2.76}$$

$$\mathbf{E}_\alpha Z_\varepsilon^{\frac{1}{2}}(u)\leq C\exp\{-Cg(u)\}, \tag{2.77}$$

where $g(\cdot)\in\mathbf{G}$.

Moreover, according to the Lemma 2.1 the process $Z_\varepsilon(\cdot)$ admits the representation

$$Z_\varepsilon(u_\varepsilon)=\exp\{u\Delta(\alpha,x)-\frac{1}{2}u^2I(\alpha+)+\psi_\varepsilon(u_\varepsilon,u,X)\}, \tag{2.78}$$

where $u_\varepsilon\to u$ and

$$\mathbf{P}_\alpha-\lim_{\varepsilon\to 0}\psi_\varepsilon(u_\varepsilon,u,X)=0.$$

So all the (nonuniform on $\theta\in\mathbf{K}$) conditions of Theorem 1.4 are fulfilled and, hence, the estimator $\hat{\theta}_\varepsilon$ is consistent, the random variable $\varepsilon^{-1}(\hat{\theta}_\varepsilon-\alpha)$ converges in distribution to $\hat{u}$, the unique point of the maximum of the limiting process $Z(u), u\geq 0$, and for any $p>0$

$$\begin{aligned}\lim_{\varepsilon\to 0}\mathbf{E}_\alpha|I(\alpha+)^{\frac{1}{2}}\varepsilon^{-1}(\hat{\theta}_\varepsilon-\alpha)|^p=\\ =\mathbf{E}|I(\alpha+)^{-\frac{1}{2}}\Delta(\alpha,x)\chi_{\{\Delta(\alpha,x)>0\}}|^p=\\ =\frac{1}{\sqrt{2\pi}}\int_0^\infty x^pe^{-\frac{x^2}{2}}dx=\frac{2^{\frac{p}{2}-1}}{\sqrt{\pi}}\Gamma\left(\frac{p-1}{2}\right).\end{aligned}$$

Now consider the behavior of the Bayes estimator $\tilde{\theta}_\varepsilon$ constructed by the observations (2.74). We suppose that θ is a random variable with a positive continuous density $\pi(\cdot)$ and the loss function is quadratic, then the Bayes estimator is a conditional mathematical expectation $\tilde{\theta}_\varepsilon=\mathbf{E}(\theta|X_t,0\leq t\leq T)$ which admits the representation

$$\tilde{\theta}_\varepsilon=\left(\int_\alpha^\beta\pi(z)\frac{d\mathbf{P}^{(\varepsilon)}_z}{d\mathbf{P}^{(\varepsilon)}_{\theta_0}}(X)dz\right)^{-1}\int_\alpha^\beta y\pi(y)\frac{d\mathbf{P}^{(\varepsilon)}_z}{d\mathbf{P}^{(\varepsilon)}_{\theta_0}}(X)dy,$$

where, as before, θ_0 is a fixed value.

We assume that the true value $\theta=\alpha$. The probability of this event is zero but it is also zero for all other fixed values of θ. After a change of measures and a change of variables $y=\alpha+\varepsilon u$, we obtain the representation

$$\tilde{\theta}_\varepsilon=\alpha+\varepsilon\left(\int_0^{(\beta-\alpha)\varepsilon^{-1}}\pi(\alpha+\varepsilon v)Z_\varepsilon(v)dv\right)^{-1}\int_0^{(\beta-\alpha)\varepsilon^{-1}}u\pi(\alpha+\varepsilon u)Z_\varepsilon(u)du,$$

where $Z_\varepsilon(\cdot)$ is defined in (2.75).

Theorem 2.23. *If the conditions of Theorem 2.9 are fulfilled, then the Bayes estimator $\tilde{\theta}_\varepsilon$ is consistent*

$$\mathbf{P}_\alpha - \lim_{\varepsilon\to 0} \tilde{\theta}_\varepsilon = \alpha$$

and the normed difference converges in distribution,

$$\mathcal{L}_\alpha\{\varepsilon^{-1}(\tilde{\theta}_\varepsilon - \alpha)\} \Longrightarrow \{\zeta\},$$

where the random variable

$$\zeta = \Delta(\alpha,x)I(\alpha+)^{-1} + I(\alpha+)^{-\frac{1}{2}}\Phi(-\Delta(\alpha,x)I(\alpha+)^{-\frac{1}{2}})^{-1}\exp\left\{-\frac{\Delta(\alpha,x)^2}{2I(\alpha+)}\right\}. \quad (2.79)$$

Proof. The properties of the likelihood ratio $Z_\varepsilon(u)$ established in (2.76), (2.77) and the "half-LAN" of the family of measures $\{\mathbf{P}_\theta^{(\varepsilon)}, \theta \in \Theta\}$ at point $\theta = \alpha$, allow us to use Theorem 1.5, which provides the convergence

$$\varepsilon^{-1}(\tilde{\theta}_\varepsilon - \alpha) \Longrightarrow \left(\int_{-\infty}^{\infty} Z(u)du\right)^{-1} \int_{-\infty}^{\infty} uZ(u)du$$

and the direct computation of these two integrals gives the desired representation of ζ. Note that the first term of this representation corresponds to the usual asymptotic normality of the estimator and the second is due to the boundary.

Now let us consider another situation, when

The true value θ_0 does not belong to the prescribed parametric set Θ. Such a problem can arise if the statistician has chosen a wrong set Θ. We suppose that the observed process is

$$dX_t = S_t(\theta_0, X)dt + \varepsilon dW_t, \quad X_0 = x_0, \quad 0 \le t \le T$$

and the MLE $\hat{\theta}_\varepsilon$ is constructed on the set $\Theta = [\alpha, \beta]$, when the true value $\theta_0 < \alpha$.

Denote, as usual, x_t as the solution of the limit equation

$$\frac{dx_t}{dt} = S_t(\theta_0, x), \quad x_0, \quad 0 \le t \le T$$

and

$$G(\theta, \theta_0, x) = \int_0^T [S_t(\theta, x) - S_t(\theta_0, x)]^2 dt.$$

It is natural to suppose that

$$\inf_{\theta\in[\alpha,\beta]} G(\theta, \theta_0, x) = G(\alpha, \theta_0, x) \quad (2.80)$$

and we shall assume that this minimum is unique. Of course, sometimes this condition is not satisfied and the point of minimum θ_* can belong to the interval (α, β), then we can denote $b_t(x) = S_t(\theta_*, x)$ and consider this problem as a particular case of the misspecified model of §2.6.

It is also natural to suppose that

$$\dot G(\alpha+,\theta_0,x) \equiv \int_0^T \dot S_t(\alpha+,x)[S_t(\alpha,x)-S_t(\theta_0,x)]dt > 0 \qquad (2.81)$$

because we have the minimum at the left of Θ. Moreover, we suppose that

$$\ddot G(\alpha+,\alpha;x) > 0, \qquad (2.82)$$

which corresponds to the convexity of the function $G(\cdot)$ at point α.

Theorem 2.24. *Let the function $S_t(\theta,X)$ have three continuous bounded derivatives w.r.t. θ and satisfies the conditions (2.80)- (2.82), then there exist such ε_0, the set $\mathcal{A}$, and the random variable ζ, that for $\varepsilon<\varepsilon_0$ the MLE $\hat\theta_\varepsilon$ admits the representation*

$$\hat\theta_\varepsilon = \alpha\chi_{\{\mathcal{A}\}} + \zeta\chi_{\{\mathcal{A}^c\}} \qquad (2.83)$$

and

$$\mathbf{P}_{\theta_0}^{(\varepsilon)}\{\mathcal{A}^c\} \le C_1\exp\{-c_1\varepsilon^{-\gamma_1}\}, \qquad (2.84)$$

$$\mathbf{P}_{\theta_0}^{(\varepsilon)}\{|\zeta|>\varepsilon^\delta\} \le C_2\exp\{-c_2\varepsilon^{-\gamma_2}\} \qquad (2.85)$$

with some positive constants $C_i, c_i, \gamma_i, i=1,2$ and $\delta\in(0,\frac12)$.

Proof. The logarithm of the likelihood ratio

$$L(\alpha+u,\alpha;X) = \ln\frac{d\mathbf{P}_{\alpha+u}^{(\varepsilon)}}{d\mathbf{P}_{\alpha}^{(\varepsilon)}}(X), \quad u\ge 0$$

with $\mathbf{P}_{\theta_0}^{(\varepsilon)}$ probability 1 has the form

$$\begin{aligned} L(\alpha+u,\alpha;X) &= \frac{1}{\varepsilon}\int_0^T [S_t(\alpha+u,X)-S_t(\alpha,X)]dW_t - \\ &\quad -\frac{1}{2\varepsilon^2}\int_0^T (S_t(\alpha+u,X)^2 - S_t(\alpha,X)^2 - \\ &\quad -2[S_t(\alpha+u,X)-S_t(\alpha,X)]S_t(\theta_0,X))dt = \\ &= \frac{1}{\varepsilon}\int_0^T [S_t(\alpha+u,X)-S_t(\alpha,X)]dW_t - \\ &\quad -\frac{1}{2\varepsilon^2}[G(\alpha+u,\theta_0;X)-G(\alpha,\theta_0;X)]. \end{aligned}$$

Put $\nu=\varepsilon^\delta, \delta\in(0,\frac12)$ and introduce a "good set"

$$\mathcal{A}_1 = \{\omega:\ \sup_{u\ge\nu} L(\alpha+u,\alpha;X) < 0\}$$

localizing the problem. Then we have $\hat\theta_\varepsilon\in[\alpha,\alpha+\nu)$ for $\omega\in\mathcal{A}_1$.

We need the convexity of the function $L(\cdot)$ in the vicinity of the point α and so we introduce a second "good set"

$$\mathcal{A}_2 = \{\omega:\ \sup_{0\le u\le\nu} \ddot L(\alpha+u,\alpha;X) < 0\}.$$

Moreover, we introduce a set of decreasing $L(\cdot)$ in the vicinity of α:

$$\mathcal{A}_3 = \{\omega : \ \dot{L}(\alpha+,\alpha;X) < 0\}.$$

Now denote $\mathcal{A} = \bigcap_{j=1}^3 \mathcal{A}_i$, then for $\omega \in \mathcal{A}$ we have the equality

$$\hat{\theta}_\varepsilon = \alpha.$$

Hence

$$\mathbf{P}_{\theta_0}^{(\varepsilon)}\{|\hat{\theta}_\varepsilon - \alpha| > 0\} = \mathbf{P}_{\theta_0}^{(\varepsilon)}(\mathcal{A}^c) \le \sum_{j=1}^3 \mathbf{P}_{\theta_0}^{(\varepsilon)}(\mathcal{A}_j^c).$$

To prove (2.84) and (2.85) we proceed as in §2.6. First, we establish

$$|G(\alpha+u,\theta_0,X) - G(\alpha+u,\theta_0,x)| \le C\varepsilon \sup_{0\le t\le T}|W_t|.$$

Then, using the smoothness of $S(\cdot)$ on θ and (2.81), we derive

$$G(\alpha+u,\theta_0,x) - G(\alpha,\theta_0,x) \ge \kappa u$$

with some positive κ and, repeating the corresponding calculation of §2.6, obtain the first inequality

$$\mathbf{P}_{\theta_0}^{(\varepsilon)}\{\mathcal{A}_1^c\} \le C\exp(-c\varepsilon^{-\gamma_1}).$$

For the second probability

$$\mathbf{P}_{\theta_0}^{(\varepsilon)}\{\mathcal{A}_2^c\} = \mathbf{P}_{\theta_0}^{(\varepsilon)}\left\{\sup_{0<u<\alpha}[\int_0^T \ddot{S}_t(\alpha+u,X)dW_t - \frac{1}{2\varepsilon}\ddot{G}(\alpha+u,\theta_0;X)] > 0\right\} \le$$

$$\le \mathbf{P}_{\theta_0}^{(\varepsilon)}\left\{\sup_{0<u<\alpha}[\int_0^T \ddot{S}_t(\alpha+u,X_t)dW_t - \frac{1}{4\varepsilon}\ddot{G}(\alpha+u,\theta_0x)] > 0\right\} +$$

$$+\mathbf{P}_{\theta_0}^{(\varepsilon)}\left\{\sup_{0<u<\nu}|\ddot{G}(\alpha+u,\theta_0,X) - \ddot{G}(\alpha+u,\theta_0,x)| > \ddot{G}(\alpha+u,\theta_0,X)\right\}. \quad (2.86)$$

The following inequalities could be derived using standard arguments

$$\sup_{0<u<\nu}|\ddot{G}(\alpha+u,\theta_0,x) - \ddot{G}(\alpha,\theta_0,x)| \le C\nu; \quad (2.87)$$

$$\sup_{0<u<\nu}|\ddot{G}(\alpha+u,\theta_0,X) - \ddot{G}(\alpha+u,\theta_0,x)| \le C\varepsilon \sup_{0\le t\le T}|W_t|; \quad (2.88)$$

$$\inf_{0<u<\nu}\ddot{G}(\alpha+u,\theta_0,x) = \inf_{0\le u\le\nu}[\ddot{G}(\alpha,\theta_0,x) + uG^{(3)}(\tilde{\alpha},\theta_0,x)] \ge \ddot{G} - \nu C,$$

where $\ddot{G} = \ddot{G}(\alpha+,\theta_0,x)$ and the constant C is derived from the condition $|S^{(3)}| < C$.

Denote

$$\varepsilon_1 = \left(\frac{\ddot{G}}{2C}\right)^{\frac{1}{\delta}}.$$

Then, for $\varepsilon < \varepsilon_1$

$$\inf_{0<u<\nu}\ddot{G}(\alpha+u,\theta_0;x) \ge \frac{1}{2}\ddot{G}.$$

For the last probability in (2.86), using (2.87) and (2.88) we obtain an upper estimate

$$\mathbf{P}\left\{\sup_{0\leq t\leq T}|W_t|>\varepsilon^{-1}\frac{\ddot{G}}{2C}\right\}\leq 2\exp\left\{-\frac{\ddot{G}^2}{8c^2}\varepsilon^{-2}\right\}.$$

Then

$$\mathbf{P}_{\theta_0}^{(\varepsilon)}\left\{\sup_{0<u<\nu}[\int_0^T\ddot{S}_t(\alpha+u,X_t)dW_t-\frac{1}{4\varepsilon}\ddot{G}(\alpha+u,\theta_0,x)]>0\right\}\leq$$

$$\leq\mathbf{P}_{\theta_0}^{(\varepsilon)}\left\{\sup_{0<u<\nu}\int_0^T\ddot{S}_t(\alpha+u,X_t)dW_t>\frac{\ddot{G}}{8\varepsilon}\right\}\leq C\exp\{-C\varepsilon^{-1}\}.$$

We have used Lemma 1.5 in this last step here.

The probability $\mathbf{P}_{\theta_0}^{(\varepsilon)}(\mathcal{A}_3^C)$ could be estimated by using the same standard arguments.

From the representation (2.83) it follows that the MLE is equal to α with positive probability. Remember also that this is not an asymptotic result because the representation (2.83) and the inequalities (2.84),(2.85) are valid for all $\varepsilon<\varepsilon_1$.

The properties of Bayes estimate could also be described.

Chapter 3

Expansions

The consistency and asymptotic normality of the MLE $\hat{\theta}_\varepsilon$ (in the regular case) allow us to write a formal representation

$$\hat{\theta}_\varepsilon = \theta + \varepsilon\xi(1 + o(1)),$$

where ξ is Gaussian, and to understand this equality as an expansion of MLE by the powers of ε. So it is natural to consider the other terms also in the asymptotic expansion

$$\hat{\theta}_\varepsilon = \theta + \sum_{j=1}^{k} \psi_j \varepsilon^j + o(\varepsilon^k),$$

where the variables $\psi_j, j = 1, \ldots, k$ do not depend on ε and the rest are small in a certain sense.

Under the regularity conditions we find the representation of the type

$$\hat{\theta}_\varepsilon = \theta + \left\{ \sum_{j=1}^{k} \psi_j \varepsilon^j + \eta\varepsilon^{k+\frac{1}{2}} \right\} \chi_{\{\mathcal{A}\}} + \zeta\chi_{\{\mathcal{A}^C\}},$$

where $|\eta| < 1$ and $\mathbf{P}(\mathcal{A}^C)$ is small. We also discuss some generalizations of this problem. Particularly, we consider the similar expansion in the case of ergodic observations as $T \to \infty$. The last section is devoted to the problem of expansion of the distribution function and the moments by the powers of ε.

3.1 Expansion of the MLE

Let us consider the problem of parameter estimation by the observations of the diffusion process

$$dX_t = S(\theta, X_t)dt + \varepsilon dW_t, \; X_0 = x_0, \; 0 \le t \le T, \tag{3.1}$$

where θ is one-dimensional, $\theta \in (\alpha, \beta)$, and $S(\cdot, \cdot)$ is a known smooth function. For simplicity of exposition, we assume that the observed process is homogeneous (does not depend on t) and the trend $S(\theta, X_t)$ does not depend on the past. In Section 3.2, we consider some generalizations of this scheme.

Let x_t be a solution of the ordinary differential equation

$$\frac{dx_t}{dt} = S(\theta, x_t), \quad x_0, \quad 0 \le t \le T, \tag{3.2}$$

let the Fisher information

$$I(\theta) = \int_0^T \dot{S}(\theta, x_t)^2 dt$$

and the function

$$G(\tilde{\theta}, \theta) = \int_0^T [S(\tilde{\theta}, x_t) - S(\theta, x_t)]^2 dt,$$

where x_t corresponds to (3.2).

As before, a point over the function corresponds to differentiation w.r.t. θ and a prime w.r.t. x. Moreover, $S_\theta^{(j)}(\theta, x)$ and $S_x^{(i)}(\theta, x)$ will denote the j-th derivative of $S(\theta, x)$ w.r.t. θ and the i-th derivative of $S(\theta, x)$ w.r.t. x. So $S_{\theta,x}^{(j+i)}(\theta, x)$ is the $j+i$-th mixed derivative. For example, $S''(\theta, x) = S_x^{(2)}(\theta, x), \dot{S}'(\theta, x) = S_{\theta,x}^{(1+1)}(\theta, x)$, etc. We denote the complement of the set $\mathcal{A}$ by $\mathcal{A}^c$ and $\mathbf{K}$ is arbitrary compact in $\Theta = (\alpha, \beta)$.

The following theorem presents an *asymptotic expansion of the MLE.*

Theorem 3.1. *Let the function $S(\theta, x)$ have $k+3$ continuous bounded derivatives w.r.t. θ and x, $\inf_{\theta\in\Theta} I(\theta) > 0$ and, for all $\nu > 0$,*

$$\inf_{\theta\in\mathbf{K}} \inf_{|\tilde{\theta}-\theta|>\nu} G(\tilde{\theta}, \theta) > 0, \tag{3.3}$$

then there are random variables $\psi_j, j = 1, \ldots, k, \eta_1, \zeta$, a set $\mathcal{A}$ and the number ε_0, such that for all $\varepsilon < \varepsilon_0$, the MLE $\hat{\theta}_\varepsilon$ admits the representation

$$\hat{\theta}_\varepsilon = \theta + \left\{ \sum_{j=1}^k \psi_j \varepsilon^j + \eta_1 \varepsilon^{k+\frac{1}{2}} \right\} \chi_{\{\mathcal{A}\}} + \zeta \chi_{\{\mathcal{A}^c\}}, \tag{3.4}$$

where $|\eta_1| < 1$, the random variables $\psi_1, \ldots, \psi_k$ do not depend on ε and, with the positive constants $C_i, c_i, \gamma_i, i = 1, 2, \delta \in (0, \frac{1}{2})$, we have

$$\sup_{\theta\in\mathbf{K}} \mathbf{P}_\theta^{(\varepsilon)}\{\mathcal{A}^c\} \le C_1 \exp\{-c_1 \varepsilon^{-\gamma_1}\}, \tag{3.5}$$

$$\sup_{\theta\in\mathbf{K}} \mathbf{P}_\theta^{(\varepsilon)}\{|\zeta| > \varepsilon^\delta\} \le C_2 \exp\{-c_2 \varepsilon^{-\gamma_2}\}. \tag{3.6}$$

Proof. The method applied here is similar to that used in the Section 2.6.

Let $\delta \in (0, \frac{1}{2}), \nu_\varepsilon = \varepsilon^\delta$ and $\mathcal{A}_1$ is the first "good set"

$$\mathcal{A}_1 = \{\omega : \sup_{|u|>\nu_\varepsilon} L(\theta + u, \theta, X) < 0\},$$

where

$$L(\theta + u, \theta, X) = \ln \frac{d\mathbf{P}_{\theta+u}^{(\varepsilon)}}{d\mathbf{P}_\theta^{(\varepsilon)}}(X), \quad u \in (\alpha - \theta, \beta - \theta),$$

and $\mathbf{P}_\theta^{(\varepsilon)}, \theta \in \Theta$, are the measures induced in $(\mathcal{C}_T, \mathcal{B}_T)$ by the process (3.1). Note that all the measures of this family are equivalent and with $\mathbf{P}_\theta^{(\varepsilon)}$ probability 1,

$$\begin{aligned} L(\theta_1, \theta; X) &= \frac{1}{\varepsilon}\int_0^T [S(\theta_1, X_t) - S(\theta, X_t)]dW_t - \\ &- \frac{1}{2\varepsilon^2}\int_0^T [S(\theta_1, X_t) - S(\theta, X_t)]^2 dt. \end{aligned}$$

For $\omega \in \mathcal{A}_1$ we have $\hat{\theta}_\varepsilon \in (\theta - \nu_\varepsilon, \theta + \nu_\varepsilon)$ because $L(\theta, \theta, X) = 0$. We consider the problem with a fixed set $\mathbf{K}$ and, for such values of ε, that

$$\alpha < \min_{\theta \in \mathbf{K}} \theta - \nu_\varepsilon, \quad \beta > \max_{\theta \in \mathbf{K}} \theta + \nu_\varepsilon.$$

Under the previously made assumptions, the process $L(\theta + u, \theta_1; X), u \in (\alpha - \theta, \beta - \theta)$ is differentiable w.r.t. u with $\mathbf{P}_\theta^{(\varepsilon)}$ probability 1 (Lemma 1.1) and the MLE $\hat{\theta}_\varepsilon = \theta + \hat{u}_\varepsilon$, where $\hat{u}_\varepsilon$ is one of the solutions of the *maximum likelihood equation*

$$\begin{aligned} &\varepsilon \int_0^T \dot{S}(\theta + u, X_t)dW_t - \int_0^T \dot{S}(\theta + u, X_t)[S(\theta + u, X_t) - \\ &-S(\theta, X_t)]dt = 0, \quad u \in (-\nu_\varepsilon, \nu_\varepsilon). \end{aligned} \tag{3.7}$$

The condition of the boundedness of the derivatives allows us to write

$$\begin{aligned} |\dot{S}(\theta, x)| < C_1, &\qquad |\ddot{S}(\theta, x)| < C_2, \\ |S(\theta, x) - S(\theta, y)| &\leq L_0|x - y|, \\ |\dot{S}(\theta, x) - \dot{S}(\theta, y)| &\leq L_1|x - y| \end{aligned} \tag{3.8}$$

and these constants take part in the definition of

$$\varepsilon_1 = (2 + 6C_1C_2T + 4C_1L_1Te^{L_0T})^{-\frac{1}{\delta}} I(\theta)^{\frac{1}{\delta}}.$$

We now introduce the second "good set"

$$\mathcal{A}_2 = \left\{ \begin{aligned} &\sup\nolimits_{0 \leq t \leq T} |W_t| < \varepsilon^{-1+\delta}, \\ \omega :\ &\sup\nolimits_{|u| < \nu_\varepsilon} \textstyle\int_0^T \ddot{S}(\theta + u, X_t)dW_t < \varepsilon^{-1+\delta}, \\ &\textstyle\int_0^T \dot{S}(\theta, X_t)dW_t < \frac{1}{2}\varepsilon^{-1+\delta} I(\theta). \end{aligned} \right.$$

Lemma 3.1. *Equation (3.7) for $\omega \in \mathcal{A}_1 \cap \mathcal{A}_2$ and $\varepsilon < \varepsilon_1$, has a unique solution $\hat{u}_\varepsilon$.*

Proof. First we rewrite this equation in the form

$$F(u, \varepsilon) = 0, \quad u \in (-\nu_\varepsilon, \nu_\varepsilon). \tag{3.9}$$

Then

$$\begin{aligned} \frac{\partial F(u, \varepsilon)}{\partial u} &= \varepsilon \int_0^T \ddot{S}(\theta + u, X_t)dW_t - \int_0^T \ddot{S}(\theta + u, X_t) \times \\ &\times [S(\theta + u, X_t) - S(\theta, X_t)]dt - \int_0^T \dot{S}(\theta + u, X_t)^2 dt. \end{aligned}$$

In our assumptions

$$\left|\int_0^T \ddot{S}(\theta+u,X_t)[S(\theta+u,X_t)-S(\theta,X_t)]dt\right| \le C_1C_2\nu_\varepsilon T.$$

Hence, for $\omega \in \mathcal{A}_1 \cap \mathcal{A}_2$,

$$\begin{aligned}
&\int_0^T |\dot{S}(z,X_t)^2 - \dot{S}(\theta,x_t)^2|dt = \\
&\quad = \int_0^T |\dot{S}(z,X_t) - \dot{S}(\theta,x_t)||\dot{S}(z,X_t) + \dot{S}(\theta,x_t)|dt \le \\
&\quad \le 2C_1\int_0^T |\dot{S}(z,X_t) - \dot{S}(z,x_t)|dt + 2C_1\int_0^T |\dot{S}(z,x_t) - \dot{S}(\theta,x_t)|dt \le \\
&\quad \le 2C_1L_1T \sup_{0\le t\le T} |X_t - x_t| + 2C_1C_2T\nu_\varepsilon.
\end{aligned}$$

By Lipschitz condition (3.8) and Lemma 1.13

$$\sup_{0\le t\le T} |X_t - x_t| \le \varepsilon e^{L_0T} \sup_{0\le t\le T} |W_t|.$$

So on the set $\mathcal{A}_2$

$$\int_0^T |\dot{S}(z,X_t)^2 - \dot{S}(\theta,x_t)^2|dt \le (2C_1L_1Te^{L_0T} + 2C_1C_2T)\varepsilon^\delta.$$

Hence

$$\begin{aligned}
\sup_{|u|<\nu_\varepsilon} \frac{\partial F(u,\varepsilon)}{\partial u} &\le \varepsilon^\delta + C_1C_2T\varepsilon^\delta - I(\theta) + (2C_1L_1Te^{L_0T} + 2C_1C_2T)\varepsilon^\delta = \\
&= (1 + 3C_1C_2T + 2C_1L_1Te^{L_0T})\varepsilon^\delta - I(\theta) < -\frac{1}{2}I(\theta)
\end{aligned} \tag{3.10}$$

for $\varepsilon < \varepsilon_1$. On $\mathcal{A}_1 \cap \mathcal{A}_2$, we have

$$\begin{aligned}
F(-\nu_\varepsilon,\varepsilon) &= \varepsilon\int_0^T \dot{S}(\theta,X_t)dW_t + \int_0^{-\nu_\varepsilon} \frac{\partial F}{\partial u}du = \\
&= \varepsilon\int_0^T \dot{S}(\theta,X_t)dW_t - \int_{-\nu_\varepsilon}^0 \frac{\partial F}{\partial u}du \ge \\
&\ge \varepsilon\int_0^T \dot{S}(\theta,X_t)dW_t + \frac{1}{2}\varepsilon^\delta I(\theta) > 0.
\end{aligned} \tag{3.11}$$

Similar inequalities provide

$$F(\nu_\varepsilon,\varepsilon) < 0. \tag{3.12}$$

From (3.10) – (3.12) follows the existence and uniqueness of the solution of (3.9).

Lemma 3.2. *Let the function $S(\theta,x)$ have $k+1$ continuous bounded derivatives w.r.t. x, then the stochastic process (3.1) has, with probability 1, k continuous* *derivatives on ε and the processes $X_t^{(j)} = \frac{\partial^j}{\partial \varepsilon^j} X_t, j = 1, \ldots, k$ satisfy the system of equations*

$$
\begin{aligned}
dX_t^{(1)} &= S_x^{(1)}(\theta, X_t) X_t^{(1)} dt + dW_t \\
\frac{dX_t^{(2)}}{dt} &= S_x^{(2)}(\theta, X_t)(X_t^{(1)})^2 + S_x^{(1)}(\theta, X_t) X_t^{(2)} \\
&\cdots \quad \cdots \\
\frac{dX_t^{(k)}}{dt} &= \sum_I C_I S_x^{(p_I)}(\theta, X_t) \prod_{j=1}^{k} (X_t^{(j)})^{i_j}
\end{aligned}
\tag{3.13}
$$

with the initial values $X_0^{(j)} = 0, j = 1, \ldots, k$. In the last equation, the summation is realized by all non-negative integer-valued sets of indices $I = \{i_1, \ldots, i_k\}$, such that

$$\sum_{j=1}^{k} j\, i_j = k$$

and C_I are some integer valued positive constants that do not depend on $S(\cdot,\cdot), \theta, x$. The parameter p_I takes the values $1, \ldots, k$.

Proof (also to be found in [39]). Let us consider the difference

$$
\begin{aligned}
Z(t) &= h^{-1}[X_t(\varepsilon + h) - X_t(\varepsilon)] - X_t^{(1)} = \\
&= \int_0^t \{[S(\theta, X_s(\varepsilon + h) - S(\theta, X_s(\varepsilon))]h^{-1} - S_x^{(1)}(\theta, X_s(\varepsilon)) X_s^{(1)}\} ds.
\end{aligned}
$$

By the theorem of averages, we have

$$
\begin{aligned}
|Z(t)| &= |\int_0^t \{S_x^{(1)}(\theta, \tilde{X}_s)[X_s(\varepsilon + h) - X_s(\varepsilon)]h^{-1} - S_x^{(1)}(\theta, X_s) X_s^{(1)}\} ds| \leq \\
&\leq \int_0^t |S_x^{(1)}(\theta, \tilde{X}_s) - S_x^{(1)}(\theta, X_s)||X_s^{(1)}| ds + \int_0^t |S_x^{(1)}(\theta, \tilde{X}_s)||Z(s)| ds.
\end{aligned}
$$

Using Lemma 1.11, we derive

$$\lim_{h \to 0} \sup_{\theta \in \mathbf{K}} \sup_{0 \leq t \leq T} |X_t(\varepsilon + h) - X_t(\varepsilon)| = 0.$$

So $\tilde{X}_s \to X_s$ uniformly on s and $\theta \in \mathbf{K}$ as $h \to 0$, and using this lemma once more, we obtain

$$\lim_{h \to 0} \sup_{\theta \in \mathbf{K}} \sup_{0 \leq t \leq T} |Z(t)| = 0.$$

So the derivative $X_t^{(1)}$ satisfies the first equation of the system (3.13). For the second, we have

$$
\begin{aligned}
&\left|[X_t^{(1)}(\varepsilon + h) - X_t^{(1)}(\varepsilon)]h^{-1} - X_t^{(2)}\right| \leq \\
&\leq \int_0^t |S'(\theta, X_s(\varepsilon + h)) - S'(\theta, X_s(\varepsilon))| h^{-1} |X_s^{(1)}(\varepsilon)| ds + \\
&\quad + \int_0^t |S'(\theta, X_s(\varepsilon))||[X_s^{(1)}(\varepsilon + h) - X_s^{(1)}(\varepsilon)]h^{-1} - X_s^{(2)}| ds.
\end{aligned}
$$

So as $h \longrightarrow 0$

$$\frac{dX_t^{(2)}}{dt} = S''(\theta, X_t)(X_t^{(1)})^2 + S'(\theta, X_t)X_t^{(2)}.$$

Hence, using induction, we obtain (3.13).

Lemma 3.3. *Let the functions $S(\theta, x)$ and $f(x)$ have $k+1$ continuous bounded derivatives on x, then the stochastic integral*

$$\eta(\varepsilon) = \int_0^T f(X_t(\varepsilon))dW_t$$

has, with probability 1, k continuous derivatives on ε and

$$\frac{\partial^k \eta(\varepsilon)}{\partial \varepsilon^k} = \sum_I C_I \int_0^T f^{(p_I)}(X_t) \prod_{j=1}^k (X_t^{(j)})^{i_j} dW_t,$$

where the constants C_I, p_I and the summation are the same as in Lemma 3.2.

Proof is an evident consequence of Lemmas 1.2 and 3.2.

Lemma 3.4. *Let the function $S(\theta, x)$ have $k+3$ continuous bounded derivatives w.r.t. θ and x, and $I(\theta) > 0$, then for $\varepsilon < \varepsilon_1$ and $\omega \in \mathcal{A}_1 \cap \mathcal{A}_2$,*

$$\hat{\theta} = \theta + \sum_{j=1}^k \psi_j \varepsilon^j + \varepsilon^{k+1} \psi_{k+1}, \tag{3.14}$$

where $\psi_j, j = 1, \ldots, k$, some random variables which do not depend on ε.

Proof. The equation

$$F(u, \varepsilon) = 0, \quad u \in (-\nu_\varepsilon, \nu_\varepsilon), \quad \varepsilon \in (0, \varepsilon_1)$$

defines an implicit function $u = u(\varepsilon)$.

By the rule of differentiating implicit functions we can calculate the first $k+1$ derivatives of u w.r.t. ε. The general form of the j-th derivative is

$$\frac{\partial^j u(\varepsilon)}{\partial \varepsilon^j} = \left(\frac{\partial F}{\partial u}\right)^{1-2j} \sum_I H_I \frac{\partial^{l_1} F(u, \varepsilon)}{\partial u^{i_1} \partial \varepsilon^{l_1 - i_1}} \cdots \frac{\partial^{l_{2j-1}} F(u, \varepsilon)}{\partial u^{i_{2j-1}} \partial \varepsilon^{l_{2j-1} - i_{2j-1}}}, \tag{3.15}$$

where H_I are constants that do not depend on $F(\cdot)$ and the summation is over all non-negative integer-valued collections of indices $I = \{l_r, i_s\}$ satisfying the equalities

$$l_1 + \ldots + l_{2j-1} = 3j - 2,$$

$$i_1 + \ldots + i_{2j-1} = 2j - 2.$$

Of course, in (3.15) we have all derivatives of the function $F(u, \varepsilon)$ up to order $j+1$. For the higher derivatives the coefficients $H_I = 0$.

The validity of (3.15) could be verified by direct computation of the first derivatives

$$\begin{aligned} u'_\varepsilon &= -F'_\varepsilon(F'_u)^{-1}, \\ u''_{\varepsilon\varepsilon} &= -[(F''_{\varepsilon\varepsilon}(F'_u)^2 - 2F''_{u\varepsilon}F'_uF'_\varepsilon + F''_{uu}(F'_\varepsilon)^2](F'_u)^{-3} \end{aligned}$$

and then applying induction.

The partial derivative of $F(u,\varepsilon)$ w.r.t. u is $(Z = \theta + u)$

$$\begin{aligned} \frac{\partial^r F(u,\varepsilon)}{\partial u^r} &= \varepsilon \int_0^T S_\theta^{(r+1)}(Z,X_t)dW_t - \\ &- \int_0^T S_\theta^{(r+1)}(Z,X_t)[S(Z,X_t) - S(\theta,X_t)]dt + \\ &+ \sum_{i=0}^{r-1} G_i \int_0^T S_\theta^{(i+1)}(Z,X_t)S_\theta^{(r-i)}(Z,X_t)dt \end{aligned}$$

with some constants G_i.

So the general form is

$$\begin{aligned} \frac{\partial^{r+s} F(u,\varepsilon)}{\partial u^r \partial \varepsilon^s} &= \sum_I h_I \int_0^T S_{\theta,x}^{(r+1,m_I)}(Z,X_t) \prod_{j=1}^s (X_t^{(j)})^{i_j} dW_t + \\ &+ \sum_J \sum_{i=0}^{r-1} \int_0^T S_{\theta,x}^{(i+1,l_J)}(Z,X_t) \prod_{j=1}^s (X_t^{(j)})^{l_j} S_{\theta,x}^{(r-i,n_J)}(Z,X_t)dt + \\ &+ \sum_J f_J \int_0^T S_{\theta,x}^{(r+1,q_J)}(Z,X_t)[S_x^{(v_J)}(Z,X_t) - \\ &- S_x^{(v_J)}(\theta,X_t)] \prod_{j=1}^s (X_t^{(j)})^{q_j} dt. \end{aligned} \tag{3.16}$$

Summing is taking over the sets of indices satisfying the equalities

$$\sum_{j=1}^s j\, i_j = s, \quad \sum_{j=1}^s j\, l_j = s, \quad \sum_{j=1}^s j\, q_j = s. \tag{3.17}$$

Joining the different terms in (3.16), we can write

$$\begin{aligned} \frac{\partial^{r+s} F(u,\varepsilon)}{\partial u^r \partial \varepsilon^s} &= \sum_I \int_0^T R^{(I)}(Z,X_t) \prod_{j=1}^s (X_t^{(j)})^{i_j} dW_t + \\ &+ \sum_j \int_0^T Q^{(J)}(Z,X_t) \prod_{j=1}^s (X_t^{(j)})^{i_j} dt, \end{aligned} \tag{3.18}$$

where $R^{(I)}(\cdot)$ and $Q^{(J)}(\cdot)$ are the bounded smooth functions of their arguments.

Equation (3.9) has a unique solution for all $\varepsilon < \varepsilon_1$ on $\mathcal{A}_1 \cap \mathcal{A}_2$ and the value $u = 0$ solves this equation as $\varepsilon = 0$, so $u(0) = 0$ and $\hat\theta_\varepsilon|_{\varepsilon=0} = \theta$ (on $\mathcal{A}_1 \cap \mathcal{A}_2$). The Taylor formula for $u(\varepsilon)$ provides the equality

$$u(\varepsilon) = \sum_{j=1}^k \frac{u^{(j)}(0)}{j!}\varepsilon^j + \frac{u^{(k+1}(\tilde\varepsilon)}{(k+1)!}\varepsilon^{k+1}.$$

Let $\psi_j = (j!)^{-1}u^{(j)}(0)$. Note that, at $\varepsilon = 0$, the system (3.13) is reduced to

$$\begin{aligned} dx_t^{(1)} &= S_x^{(1)}(\theta, x_t)x_t^{(1)}dt + dW_t \\ \frac{dx_t^{(2)}}{dt} &= S_x^{(2)}(\theta, x_t)(x_t^{(1)})^2 + S_x^{(1)}(\theta, x_t)x_t^{(2)} \\ &\quad \dots \quad \dots \\ \frac{dx_t^{(k)}}{dt} &= \sum_I C_I S_x^{(P_I)}(\theta, x_t)\prod_{j=1}^{k}(x_t^{(j)})^{i_j} \end{aligned} \tag{3.19}$$

with the same initial values.

The solution of the first equation gives

$$x_t^{(1)} = \int_0^t \exp\left\{\int_s^t S_x^{(1)}(\theta, x_v)dv\right\} dW_s,$$

so $x_t^{(1)}$ is a Gaussian process with zero mean. The second equation has the solution

$$x_t^{(2)} = \int_0^t \exp\left\{\int_s^t S_x^{(1)}(\theta, x_v)dv\right\} S_x^{(2)}(\theta, x_s)(x_s^{(1)})^2 ds,$$

which is a bilinear functional of the Wiener process. Continuing this process, we obtain a general representation of the derivative

$$x_t^{(k)} = \int_0^t \exp\left\{\int_s^t S_x^{(1)}(\theta, x_v)dv\right\} \sum_I C_I S_x^{(P_I)}(\theta, x_s)\prod_{j=1}^{k-1}(x_s^{(j)})^{i_j} ds.$$

We now introduce the last "good set"

$$\mathcal{A}_3 = \left\{\omega : \sup_{|u|<\nu_\varepsilon} |u^{(k+1)}(\varepsilon_0, u)| < (k+1)!\varepsilon^{-\frac{1}{2}}\right\},$$

then, on $\mathcal{A} = \mathcal{A}_1 \cap \mathcal{A}_2 \cap \mathcal{A}_3$, we have the expansion

$$\hat{\theta}_\varepsilon = \theta + \sum_{j=1}^{k} \psi_j \varepsilon^j + \eta_1 \varepsilon^{k+\frac{1}{2}},$$

where $|\eta_1| < 1$, coinciding with (3.4) (on $\mathcal{A}$). So it only remains to prove the inequalities (3.5) and (3.6). We recall that this scheme is similar to the one considered in the proof of Theorem 2.14.

The function $G(\tilde{\theta}, \theta)$ is continuous and, moreover, derivable on θ and $\tilde{\theta}$, because $x_t = x_t(\theta)$ is also sufficiently smooth on θ so by the conditions of Theorem 3.1, we can write

$$\lim_{u\to 0} \frac{G(\theta + u, \theta)}{u^2} = I(\theta) > 0$$

and this convergence is uniform on $\theta \in \mathbf{K}$. On the other side, by (3.3) we have

$$\inf_{\theta\in\mathbf{K}} \inf_{|u|>h} G(\theta + u, \theta) > 0$$

for any $h > 0$ such that $\theta + u \in \Theta$. These two inequalities allow us to write

$$\inf_{\theta \in \mathbf{K}} G(\theta + u, \theta) \geq \kappa u^2,$$

where $\kappa = \kappa(\mathbf{K})$ is a positive constant. Below we denote $\Delta S_\varepsilon(u) = S(\theta + u, X_t) - S(\theta, X_t)$ and $\Delta S_0(u) = S(\theta + u, x_t) - S(\theta, x_t)$. For the probability of the complement of $\mathcal{A}_1$ we have

$$\begin{aligned}
\mathbf{P}_\theta^{(\varepsilon)}\{\mathcal{A}_1^c\} &= \mathbf{P}_\theta^{(\varepsilon)}\left\{\sup_{|u|>\nu_\varepsilon}[\int_0^T \Delta S_\varepsilon(u)dW_t - \frac{1}{2\varepsilon}\|\Delta S_\varepsilon(u)\|^2] \geq 0\right\} \leq \\
&\leq \mathbf{P}_\theta^{(\varepsilon)}\left\{\sup_{|u|>\nu_\varepsilon}[\int_0^T \Delta S_\varepsilon(u)dW_t - \frac{1}{4\varepsilon}\|\Delta S_0(u)\|^2] \geq 0\right\} + \\
&\quad + \mathbf{P}_\theta^{(\varepsilon)}\left\{\sup_{|u|>\nu_\varepsilon}\frac{1}{2\varepsilon}\int_0^T |\Delta S_0(u)^2 - \Delta S_\varepsilon(u)^2|dt - \frac{1}{4\varepsilon}\|\Delta S_0(u)\|^2 \geq 0\right\} \leq \\
&\leq \mathbf{P}_\theta^{(\varepsilon)}\left\{\sup_{|u|>\nu_\varepsilon}\int_0^T \Delta S_\varepsilon(u)dW_t \geq \inf_{|u|>\nu_\varepsilon}\frac{1}{4\varepsilon}\|\Delta S_0(u)\|^2\right\} + \\
&\quad + \mathbf{P}_\theta^{(\varepsilon)}\left\{\sup_{|u|>\nu_\varepsilon}\frac{1}{2\varepsilon}\int_0^T |\Delta S_0(u) - \Delta S_\varepsilon(u)||\Delta S_0(u) + \Delta S_\varepsilon(u)|dt \geq \right. \\
&\quad \left. \geq \inf_{|u|>\nu_\varepsilon}\frac{1}{4\varepsilon}\|\Delta S_0(u)\|^2\right\} \leq \\
&\leq \mathbf{P}_\theta^{(\varepsilon)}\left\{\sup_{|u|>\nu_\varepsilon}\int_0^T \Delta S_\varepsilon(u)dW_t \geq \frac{\kappa}{4\varepsilon}\nu_\varepsilon^2\right\} + \\
&\quad + \mathbf{P}_\theta^{(\varepsilon)}\left\{\sup_{|u|>\nu_\varepsilon}\frac{1}{\varepsilon}\int_0^T |\Delta S_0(u) - \Delta S_\varepsilon(u)|dt \geq \frac{\kappa\nu_\varepsilon^2}{8C_0\varepsilon}\right\}.
\end{aligned}$$

Consider separately on $u \in (\nu_\varepsilon, \beta - \theta)$ and $u \in (\alpha - \theta, -\nu_\varepsilon)$

$$\begin{aligned}
&\mathbf{P}_\theta^{(\varepsilon)}\left\{\sup_{\nu_\varepsilon<u<\beta-\theta}\int_0^T \Delta S_\varepsilon(u)dW_t \geq \frac{\kappa}{4}\varepsilon^{-1+2\delta}\right\} \leq \\
&\leq \mathbf{P}_\theta^{(\varepsilon)}\left\{\sup_{\nu_\varepsilon<u<\beta-\theta}\int_0^T [\Delta S_\varepsilon(u) - \Delta S_\varepsilon(\nu_\varepsilon)]dW_t \geq \frac{\kappa}{8}\varepsilon^{-1+2\delta}\right\} + \\
&+ \mathbf{P}_\theta^{(\varepsilon)}\left\{\int_0^T \Delta S_\varepsilon(\nu_\varepsilon)dW_t \geq \frac{\kappa}{8}\varepsilon^{-1+2\delta}\right\} \leq C\varepsilon^{1-2\delta}\exp\left\{-\kappa_1\frac{\varepsilon^{-1+2\delta}}{(\beta - \theta - \nu_\varepsilon)}\right\} + \\
&+ C\exp\{-\kappa_2\varepsilon^{-1+2\delta}\} < C\exp\{-\kappa_0\varepsilon^{-1+2\delta}\},
\end{aligned}$$

where we have used Lemmas 1.4 and 1.5, and consider $\varepsilon < \tilde{\varepsilon}$ with $\tilde{\varepsilon}$ being properly defined.

Analogously

$$\mathbf{P}_\theta^{(\varepsilon)}\left\{\sup_{\alpha-\theta<u<-\nu_\varepsilon}\int_0^T \Delta S_\varepsilon(u)dW_t \geq \frac{\kappa}{4}\varepsilon^{-1+2\delta}\right\} < C\exp\{-\kappa_1\varepsilon^{-1+2\delta}\}.$$

Then

$$\frac{1}{\varepsilon}\int_0^T |\Delta S_0(u) - \Delta S_\varepsilon(u)|dt \leq C\sup_{0\leq t\leq T}|W_t|.$$

Hence

$$\mathbf{P}_\theta^{(\varepsilon)}\left\{\sup_{|u|>\nu_\varepsilon}\frac{1}{\varepsilon}\int_0^T|\Delta S_0(u)-\Delta S_\varepsilon(u)|dt\geq\frac{\kappa}{8C_0}\varepsilon^{-1+2\delta}\right\}\leq$$
$$\leq\mathbf{P}\left\{\sup_{0\leq t\leq T}|W_t|>\frac{\kappa}{8C_0CT}\varepsilon^{-1+2\delta}\right\}<$$
$$<4\mathbf{P}\left\{W_T>\frac{\kappa}{8C_0CT}\varepsilon^{-1+2\delta}\right\}\leq 2\exp\left\{-\frac{\kappa^2\varepsilon^{-2+4\delta}}{128C_0^2C^2T^3}\right\}.$$

Joining all the estimates, we obtain

$$\sup_{\theta\in\mathbf{K}}\mathbf{P}_\theta^{(\varepsilon)}\{\mathcal{A}_1^c\}\leq C\exp\{-\kappa_0\varepsilon^{-1+2\delta}\}\tag{3.20}$$

with some C and $\kappa_0>0$, because all the inequalities are uniform on $\theta\in\mathbf{K}$.

Further, for the complement of $\mathcal{A}_2$, we have

$$\begin{aligned}\mathbf{P}_\theta^{(\varepsilon)}\{\mathcal{A}_2^c\}\ &\leq\ \mathbf{P}\left\{\sup_{0\leq t\leq T}|W_t|>\varepsilon^{-1+\delta}\right\}+\\
&+\mathbf{P}_\theta^{(\varepsilon)}\left\{\sup_{|u|<\nu_\varepsilon}\int_0^T\dddot{S}(\theta+u,X_t)dW_t>\varepsilon^{-1+\delta}\right\}+\\
&+\mathbf{P}_\theta^{(\varepsilon)}\left\{\int_0^T\dot{S}(\theta,X_t)dW_t>\frac{1}{2}\varepsilon^{-1+\delta}I(\theta)\right\}\leq 4\mathbf{P}\{W_T>\varepsilon^{-1+\delta}\}+\\
&+\mathbf{P}_\theta^{(\varepsilon)}\left\{\sup_{-\nu_\varepsilon<u<\nu_\varepsilon}\int_0^T[\ddot{S}(\theta+u,X_t)-\ddot{S}(\theta-\nu_\varepsilon,X_t)]dW_t>\frac{1}{2}\varepsilon^{-1+\delta}\right\}+\\
&+\mathbf{P}_\theta^{(\varepsilon)}\left\{\int_0^T\ddot{S}(\theta-\nu_\varepsilon,X_t)dW_t>\frac{1}{2}\varepsilon^{-1+\delta}\right\}+\\
&+\mathbf{P}_\theta^{(\varepsilon)}\left\{\int_0^T\dot{S}(\theta,X_t)dW_t>\frac{1}{2}\varepsilon^{-1+\delta}I(\theta)\right\}\leq\\
&\leq\ 2\exp\{-\frac{1}{2T}\varepsilon^{-2+2\delta}\}+C\varepsilon\exp\{-\kappa\varepsilon^{-1+\delta}\}+C\exp\{-\kappa\varepsilon^{-1+\delta}\}<\\
&<\ C\exp\{-\kappa\varepsilon^{-1+\delta}\},\end{aligned}$$

where we have used Lemmas 1.4 and 1.5 with $p=1$.

So we have obtained the second estimate

$$\sup_{\theta\in\mathbf{K}}\mathbf{P}_\theta^{(\varepsilon)}\{\mathcal{A}_2^c\}\leq C\exp\{-\kappa\varepsilon^{-1+\delta}\}\tag{3.21}$$

because the right sides of the inequalities do not depend on $\theta\in\mathbf{K}$.

According to (3.18), the random variable ψ_{k+1} contains the sum of the products of integrals and the estimate of $\mathbf{P}_\theta^{(\varepsilon)}\{\mathcal{A}_3^c\}$ could be reduced to estimates of the probabilities

$$\mathbf{P}_\theta^{(\varepsilon)}\left\{\sup_{|u|<\nu_\varepsilon}\int_0^T R(\theta+u,X_t)\prod_{j=1}^{k+1}(X_t^{(j)})^{i_j}dW_t>C_1\varepsilon^{-r_1}\right\},\tag{3.22}$$

$$\mathbf{P}_\theta^{(\varepsilon)}\left\{\sup_{|u|<\nu_\varepsilon}\int_0^T Q(\theta+u,X_t)\prod_{j=1}^{k+1}(X_t^{(j)})^{i_j}dt > C_2\varepsilon^{-r_2}\right\}, \tag{3.23}$$

where $r_1 > 0$ and $r_2 > 0$ are some numbers.

Lemma 3.5. *Under the conditions of Lemma 3.2, the solutions $X_t^{(j)}, 0 \le t \le T$ of the system (3.13) with probability 1 satisfy the inequalities*

$$\sup_{\theta\in\mathbf{K}}\sup_{0\le t\le T}|X_t^{(j)}| \le M_j\eta^j,\ j=1,\dots,k, \tag{3.24}$$

where $\eta = \sup_{0\le t\le T}|W_t|$ and $M_j, j=1,\dots,k$ are some positive constants.

Proof. From the first equation of the system (3.13) and Lemma 1.11, we derive

$$|X_t^{(1)}| \le \int_0^t |S_x^{(1)}(\theta,X_s)||X_s^{(1)}|ds + \sup_{0\le t\le T}|W_t|.$$

So

$$\sup_{0\le t\le T}|X_t^{(1)}| \le e^{C_1T}\eta.$$

Further

$$\begin{aligned}|X_t^{(2)}| &\le \int_0^t |S_x^{(2)}(\theta,X_s)|(X_s^{(1)})^2ds + \int_0^t |S_x^{(1)}(\theta,X_s)||X_s^{(2)}|ds \le \\ &\le C_2Te^{2C_1T}\eta^2 + C_1\int_0^t |X_s^{(2)}|ds.\end{aligned}$$

So once more applying Lemma 1.11, we obtain

$$\sup_{0\le t\le T}|X_t^{(2)}| \le M_2\eta^2.$$

The general case is treated by mathematical induction.
The functions $R(u,X_t)$ have a bounded derivative w.r.t. u, so

$$\int_0^T |\dot R(u,X_t)|^2\Big(\prod_{j=1}^{k+1}(X_t^{(j)})^{i_j}\Big)^2dt \le$$

$$\le C^2T\prod_{j=1}^{k+1}(M_j\eta^j)^{2i_j} \le M\eta^{2\sum ji_j} = M\eta^{2k+2},$$

where the indices of summation satisfy (3.17).
Hence, the condition (1.16) of Lemma 1.5 is fulfilled with $p = 2k+2$ and

$$\sup_{\theta\in\mathbf{K}}\mathbf{P}_\theta^{(\varepsilon)}\left\{\sup_{-\nu_\varepsilon<u<\nu_\varepsilon}\int_0^T[R(\theta+u,X_t)-R(\theta-\nu_\varepsilon,X_t)]\times\right.$$

$$\left.\times\prod_{j=1}^{k+1}(X_t^{(j)})^{i_j}dW_t > C_1\varepsilon^{-r_1}\right\} \le C\varepsilon^{-\frac{r_1}{2k+2}}\exp\left\{-\kappa\varepsilon^{\frac{r_1}{2k+2}}\right\}. \tag{3.25}$$

The estimates (3.24) allow us to derive an analogous exponential inequality for the probability (3.23) also.

Joining the inequalities (3.20), (3.21), and (3.25) and its analogs for (3.23), we obtain (3.5).

Note also that the inequality (3.6) has already been derived because

$$\begin{aligned} \mathbf{P}_\theta^{(\varepsilon)}\{|\zeta| > \varepsilon^\delta\} &\leq \mathbf{P}_\theta^{(\varepsilon)}\{\sup_{|u|<\nu_\varepsilon} L(\theta+u,\theta;X) < \sup_{|u|>\nu_\varepsilon} L(\theta+u,\theta;X)\} \leq \\ &\leq \mathbf{P}_\theta^{(\varepsilon)}\{\sup_{|u|>\nu_\varepsilon} L(\theta+u,\theta;X) > 0\} = \mathbf{P}_\theta^{(\varepsilon)}\{\mathcal{A}_1^c\} \end{aligned}$$

and Theorem 3.1 is proved.

REMARK 3.1 The Gaussian term of the expansion (3.4) is

$$\psi_1 = I(\theta)^{-1}\int_0^T \dot S(\theta,x_t)dW_t.$$

Note also that the random variable $\psi_1\chi_{\{\mathcal{A}\}}$ is not Gaussian because $\mathcal{A}$ contains the conditions of the same Wiener process. Direct computation of the second term of the sum in (3.4) allows us to write (on $\mathcal{A}$) the expansion

$$\begin{aligned} \hat\theta_\varepsilon &= \theta + \varepsilon I(\theta)^{-1}\int_0^T \dot S(\theta,x_t)dW_t + \varepsilon^2 I(\theta)^{-3}\times \\ &\times\left\{I(\theta)\int_0^T \dot S(\theta,x_t)dW_t\int_0^T \ddot S(\theta,x_t)dW_t - \right.\\ &-\frac{3}{2}\left(\int_0^T \dot S(\theta,x_t)dW_t\right)^2\int_0^T \dot S(\theta,x_t)\ddot S(\theta,x_t)dt - \\ &-2I(\theta)\int_0^T \dot S'(\theta,x_t)\dot S(\theta,x_t)x_t^{(1)}dt\int_0^T \dot S(\theta,x_t)dW_t + \\ &\left.+I(\theta)^2\int_0^T \dot S'(\theta,x_t)x_t^{(1)}dW_t\right\} + o(\varepsilon^2). \end{aligned} \tag{3.26}$$

It is possible, of course, to write the third and even other terms.

3.2 Possible generalizations

1. Nonhomogeneous process

Let the observed process be

$$dX_t = S(\theta,t,X_t)dt + \varepsilon\sigma(t,X_t)dW_t,\quad X_0 = x_0,\quad 0\leq t\leq T \tag{3.27}$$

and $\theta\in(\alpha,\beta)=\Theta, |\alpha|+|\beta|<\infty$. The expansion of the MLE $\hat\theta_\varepsilon$ by the powers of ε could be done using the same method as above and the only difference between these two problems lies in the estimates of the probabilities (3.5), (3.6).

Let us introduce the notations: $f_t(\theta,x) = S(\theta,t,x)\sigma(t,x)^{-1}, \dot f_t(\theta,x) = \frac{\partial}{\partial\theta} f_t(\theta,x)$,

$$\begin{aligned}\frac{dx_t}{dt} &= S(\theta,t,x_t), \quad x_0, \quad 0 \le t \le T,\\ I(\theta) &= \int_0^T \dot f_t(\theta,x_t)^2 dt,\\ G(\tilde\theta,\theta) &= \int_0^T [f_t(\tilde\theta,x_t) - f_t(\theta,x_t)]^2 dt.\end{aligned}$$

The expansion of the MLE for this model is given by

Theorem 3.2. *Let the functions $S(\theta,t,x), \sigma(t,x)$ have $k+3$ continuous bounded derivatives w.r.t. θ and x, the function $\sigma(t,x) \in [c_0, C_0]$ with some positive constants c_0, C_0 and, for any compact $\mathbf{K} \subset \Theta$ and $h > 0$,*

$$\inf_{\theta\in\mathbf{K}} I(\theta) > 0, \quad \inf_{\theta\in\mathbf{K}} \inf_{|\tilde\theta-\theta|>h,\tilde\theta\in\Theta} G(\tilde\theta,\theta) > 0.$$

Then there exist random variables $\tilde\psi_j, j = 1,\dots,k, \tilde\eta_1, \tilde\zeta$ and set $\tilde A$ such that

$$\hat\theta_\varepsilon = \theta + \left\{\sum_{j=1}^k \tilde\psi_j\varepsilon^j + \tilde\eta_1\varepsilon^{k+\frac12}\right\}\chi_{\{\tilde A\}} + \tilde\zeta\chi_{\{\tilde A^c\}}, \tag{3.28}$$

where the random variables $\tilde\psi_1,\dots,\tilde\psi_k$ do not depend on ε and, for any positive number n,

$$\sup_{\theta\in\mathbf{K}} \mathbf{P}_\theta^{(\varepsilon)}\{\tilde A^c\} \le c_n\varepsilon^n, \tag{3.29}$$

$$\sup_{\theta\in\mathbf{K}} \mathbf{P}_\theta^{(\varepsilon)}\{|\tilde\zeta| > \varepsilon^\delta\} \le C\exp\{-c\varepsilon^{-\gamma}\} \tag{3.30}$$

with $\delta \in (0,\frac12), C > 0, c > 0, \gamma > 0, c_n > 0$.

Proof of this theorem is similar to the proof of Theorem 3.1 because the maximum likelihood equation is now

$$\begin{aligned}\varepsilon^2\dot L(\theta+u,\theta,x) &= \varepsilon\int_0^T \dot f_t(\theta+u,X_t)dW_t -\\ &\quad -\int_0^T \dot f_t(\theta+u,X_t)[f_t(\theta+u,X_t) - f_t(\theta,X_t)]dt = 0\end{aligned}$$

and all formal calculations of the coefficients $\tilde\psi_j$ of the expansion (3.28) are similar to the calculation of ψ_j in (3.4). System (3.13) is now replaced by

$$\begin{aligned}dX_t^{(1)} &= S_x^{(1)}(\theta,t,X_t)X_t^{(1)}dt + [\sigma(t,X_t) + \varepsilon\sigma'(t,X_t)X_t^{(1)}]dW_t\\ dX_t^{(2)} &= [S_x^{(2)}(\theta,t,X_t)(X_t^{(1)})^2 + S_x^{(1)}(\theta,t,X_t)X_t^{(2)}]dt+\\ &\quad + [2\sigma'(t,X_t)X_t^{(1)} + \varepsilon\sigma''(t,X_t)(X_t^{(1)})^2 + \varepsilon\sigma'(t,X_t)X_t^{(2)}]dW_t\end{aligned} \tag{3.31}$$

and so on. The k-th equation of this system is of the form

$$\begin{aligned} dX_t^{(k)} &= \sum_I h_i S_x^{(I)}(\theta,t,X_t)\prod_{j=1}^k (X_t^{(j)})^{i_j}dt + \\ &+ [\varepsilon \sum_I h_I \sigma_x^{(|I|)}(t,X_t)\prod_{j=1}^k (X_t^{(j)})^{i_j} + \sum_p c_p\sigma_x^{(|p|)}(t,X_t)\prod_{j=1}^{k-1}(X_t^{(j)})^{p_j}]dW_t \end{aligned}$$

with some constants h_i and c_p.

From Equation (3.27) and the boundedness of the derivative $S'(\theta,t,x)$, we have the inequalities

$$\begin{aligned} |X_t - x_t| &\leq \int_0^t |S(\theta,s,X_s) - S(\theta,s,x_s)|ds + \varepsilon\left|\int_0^t \sigma(s,X_s)dW_s\right| \leq \\ &\leq L\int_0^t |X_s - x_s|ds + \varepsilon \sup_{0\leq t\leq T}\left|\int_0^t \sigma(s,X_s)dW_s\right| \end{aligned}$$

and, by Lemma 1.11 with probability 1,

$$\sup_{0\leq t\leq T}|X_t - x_t| \leq \varepsilon e^{LT}\sup_{0\leq t\leq T}\left|\int_0^t \sigma(s,X_s)dW_s\right|. \tag{3.32}$$

The probability of the complement of the "good" set

$$\tilde{\mathcal{A}}_1 = \{\omega : \sup_{|u|>\nu_\varepsilon} L(\theta+u,\theta;X) < 0\}$$

is calculated by repeating the corresponding calculations of Theorem 3.1 and the only difference is in deriving the estimate

$$\begin{aligned} &\mathbf{P}_\theta^{(\varepsilon)}\left\{\sup_{|u|>\nu_\varepsilon}\frac{1}{\varepsilon}\int_0^T |\Delta f_0(u) - \Delta f_\varepsilon(u)|dt > \frac{\kappa}{8C_0}\varepsilon^{-1+2\delta}\right\} \leq \\ &\leq \mathbf{P}_\theta^{(\varepsilon)}\left\{\sup_{0\leq t\leq T}\left|\int_0^t \sigma(s,X_s)dW_s\right| > \kappa\varepsilon^{-1+2\delta}\right\} \leq \\ &\leq 2\exp\left\{-\frac{\kappa^2}{2C}\varepsilon^{-2+4\delta}\right\}, \end{aligned}$$

where we have used the estimate (1.20).

Similar arguments allow us to obtain the necessary estimate for $\mathbf{P}_\theta^{(\varepsilon)}\{\tilde{\mathcal{A}}_2^c\}$, but they are not sufficient to derive the last estimates, because the solution of the first equation of system (3.31) is

$$X_t^{(1)} = V_t\left\{\int_0^t V_s^{-1}\sigma(t,X_s)dW_s - \varepsilon\int_0^t V_s^{-1}\sigma(s,X_s)\sigma'(s,X_s)ds\right\},$$

where

$$V_t = \exp\left\{\int_0^t [S'(\theta,s,X_s) - \frac{1}{2}\varepsilon^2\sigma'(s,X_s)^2]ds + \varepsilon\int_0^t \sigma'(s,X_s)dW_s\right\}$$

and condition (1.16) of Lemma 1.5 is not fulfilled for any $p > 0$. Nevertheless, at least we can have a polynomial rate. To explain this, we first have to consider linear model

$$dy_t = [\alpha(t) + \beta(t)y_t]dt + [\gamma(t) + \delta(t)y_t]dW_t, \quad y_0 = 0, \quad 0 \le t \le T$$

and, using the inequality

$$(a+b)^{2m} \le 2^{2m-1}a^{2m} + 2^{2m-1}b^{2m}$$

for some integer $m > 0$, we can write

$$\begin{aligned}
\mathbf{E}y_t^{2m} &\le 2^{2m-1}\mathbf{E}\left(\int_0^t [\alpha(s) + \beta(s)y_s]ds\right)^{2m} + \\
&\quad +2^{2m-1}\mathbf{E}\left(\int_0^t (\gamma(s) + \delta(s)y_s)dW_s\right)^{2m} \le \\
&\le 2^{2m-1}t^{2m-1}\int_0^t \mathbf{E}(\alpha(s) + \beta(s)y_s)^{2m}ds + \\
&\quad +2^{2m-1}[m(2m-1)]^m t^{m-1}\int_0^t \mathbf{E}(\gamma(s) + \delta(s)y_s)^{2m}ds \le \\
&\le 2^{4m-2}T^{2m-1}\left(\int_0^T \mathbf{E}\alpha(s)^{2m}ds + \int_0^t \mathbf{E}(\beta(s)y_s)^{2m}ds\right) + \\
&\quad +2^{4m-1}[m(2m-1)]^m T^{m-1}\left(\int_0^T \mathbf{E}\gamma(s)^{2m}ds + \int_0^t \mathbf{E}(\delta(s)y_s)^{2m}ds\right).
\end{aligned}$$

Suppose now that the random functions $\beta(t)$ and $\delta(t)$ are bounded and the integrals

$$\int_0^T \mathbf{E}(\alpha(s))^{2m}ds + \int_0^T \mathbf{E}(\gamma(s))^{2m}ds < C$$

with some $C > 0$, then we have

$$\mathbf{E}y_t^{2m} \le C_1(m,T) + C_2(m,T)\int_0^t \mathbf{E}(y_s)^{2m}ds$$

and, by Lemma 1.11, we obtain

$$\sup_{0\le t\le T} \mathbf{E}y_t^{2m} < C_1(m,T)\exp\{C_2(m,T)T\}. \tag{3.33}$$

System (3.31) consists of equations linear with respect to the higher derivative, so, by (3.33), all the moments of the solutions are bounded:

$$\sup_{\theta\in\Theta}\sup_{0\le t\le T} \mathbf{E}_\theta(X_t^{(j)})^{2m} < C_j \tag{3.34}$$

for all $j = 1, \ldots, k$ and all integer $m > 0$.

Consider now the stochastic integral

$$\eta(\theta) = \int_0^T f_t(\theta)dW_t,$$

where the random function $f_t(\theta) \in \mathcal{P}$ is derivable with probability 1 w.r.t. θ and

$$\sup_{\theta\in\Theta} \mathbf{E} \int_0^T (\dot{f}_t(\theta))^{2m} dt < \infty.$$

Then we have the estimates

$$\mathbf{P}_\theta^{(\varepsilon)} \left\{ \int_0^T [f(\theta+h) - f(\theta)] dW_t > Cg(h) \right\} \le$$

$$\le C_m C^{-2m} \left(\frac{h}{g(h)} \right)^{2m} \sup_\theta \int_0^T \mathbf{E}_\theta \dot{f}_t(\theta)^{2m} dt \le$$

$$\le C(m,T) \left(\frac{h}{Cg(h)} \right)^{2m}$$

Let $g(h) = \sqrt{h}$, then by Lemma 1.3, with

$$\begin{aligned} H(C) &= \sum_{n=1}^{\infty} 2^{n-1} q(C, 2^{-n}(\beta-\alpha)) = \\ &= \frac{C_m}{2} C^{-2m} \sum_{n=1}^{\infty} 2^{-n(m-1)} (\beta-\alpha)^m = C_m \frac{C^{-2m}(\beta-\alpha)^m}{2^m - 2}, \end{aligned}$$

we can write

$$\mathbf{P}_\theta^{(\varepsilon)} \left\{ \sup_{\alpha<\theta<\beta} \int_0^T [f_t(\theta) - f_t(\alpha)] dW_t > N \right\} < \frac{C}{N^{2m}}. \tag{3.35}$$

This estimate of large deviations and the property (3.34) allow us to obtain the estimate (3.30).

REMARK 3.2 The case of the diffusion-type process

$$dX_t = [V(\theta,t,X_t) + \int_0^t K(\theta,t,s,X_s) ds] dt + \varepsilon dW_t, \quad X_0 = x_0, \quad 0 \le t \le T$$

can be treated in a similar way and does not introduce any great difficulties.

2. Ergodic process

Let the observed process be

$$dX_t = S(\theta, X_t) dt + \sigma(X_t) dW_t, \quad X_0 = x_0, \quad 0 \le t \le T$$

and we are interested in the behavior of MLE $\hat{\theta}_T$ as $T \to \infty$. The approach proposed in section 3.1 could be applied to this scheme also.

The maximum likelihood equation is now

$$\int_0^T \dot{f}(\theta+u, X_t) dW_t - \int_0^T \dot{f}(\theta+u, X_t)[f(\theta+u, X_t) - f(\theta, X_t)] dt = 0,$$

where $f(\theta,x) = S(\theta,x)\sigma(x)^{-1}$. After dividing this equation by T and denoting $\varepsilon = \frac{1}{\sqrt{T}}$, we can write

$$\begin{aligned} &\varepsilon \frac{1}{\sqrt{T}} \int_0^T \dot{f}(\theta+u, X_t) dW_t - \\ &\qquad - \frac{1}{T} \int_0^T \dot{f}(\theta+u, X_t)[f(\theta+u, X_t) - f(\theta, X_t)] dt = 0. \end{aligned} \tag{3.36}$$

Now we can forget that ε depends on T and consider ε as an independent variable, then Equation (3.36) defines an implicit function $u = u(\varepsilon)$ and we can obtain the first terms of the expansion as

$$\hat{\theta}_T = \theta + \{\psi_1 T^{-\frac{1}{2}} + \eta T^{-\frac{3}{4}}\}\chi_{\{\mathcal{A}\}} + \zeta\chi_{\{\mathcal{A}^c\}},$$

where

$$\psi_1 = \frac{1}{\sqrt{T}}\int_0^T \dot{f}(\theta, X_t)dW_t\left(\frac{1}{T}\int_0^T \dot{f}(\theta, X_t)^2 dt\right)^{-1}.$$

If the process $X_t, t \geq 0$ has ergodic properties [49], then

$$\frac{1}{T}\int_0^T \dot{f}(\theta, X_t)^2 dt \longrightarrow \int_{-\infty}^{\infty}\left(\frac{\dot{S}(\theta, x)}{\sigma(x)}\right)^2 d\mu_\theta(x) = I(\theta),$$

where $\mu_\theta(\cdot)$ is a stationary distribution and

$$\mathcal{L}_\theta\{\psi_1\} \Longrightarrow \mathcal{N}(0, I(\theta)^{-1}).$$

The expansion of MLE $\hat{\theta}_T$ is given by the following theorem.

Theorem 3.3. *Let the function $f(\theta, x)$ have $k+3$ continuous bounded derivatives w.r.t. θ and the derivative $|\dot{f}(\theta, x)| \geq \kappa > 0$, then there exist the random variables $\psi_j = \psi_j(T), j = 1, \ldots, k+1, \zeta$ and a set $\mathcal{B}$ such that*

$$\hat{\theta}_T = \theta + \left\{\sum_{j=1}^{k}\psi_j T^{-\frac{j}{2}} + \psi_{k+1}T^{-\frac{(k+1/2)}{2}}\right\}\chi_{\{\mathcal{B}\}} + \zeta\chi_{\{\mathcal{B}^c\}},$$

where the random variables $\psi_j, j = 1, \ldots, k$ are bounded in probability, $|\psi_{k+1}| < 1$ and

$$\mathbf{P}_\theta^{(T)}\{\mathcal{B}^c\} \leq C_1\exp\{-\kappa_1 T^{\gamma_1}\},$$
$$\mathbf{P}_\theta^{(T)}\{|\eta| > T^{-\delta}\} \leq C_2\exp\{-\kappa_2 T^{\gamma_2}\},$$

where $\delta \in (0, \frac{1}{2}), C_i, \kappa_i$ and γ_i are some positive constants.

The proof of this theorem is even easier, than that of Theorem 3.1 because, in Equation (3.36), we do not consider the dependence X_t on ε.

3. Misspecified model

This method has already been used in the proof of Theorem 2.13. To obtain the other terms of the expansion, we have to require the additional smoothness of the coefficients.

Let the observed process be

$$dX_t = b(\mu, X_t)dt + \varepsilon dW_t, \quad X_0 = x_0, \quad 0 \leq t \leq T, \tag{3.37}$$

where $\mu \in \mathcal{M}$, but the statistician can suppose that the process $X_t, 0 \leq t \leq T$ corresponds to the model

$$dX_t = S(\theta, X_t)dt + \varepsilon dW_t, \quad X_0 = x_0, \quad 0 \leq t \leq T$$

and try to estimate θ by observations of the process (3.37). Denote

$$G(\theta,\mu)=\int_0^T[S(\theta,x_t)-b(\mu,x_t)]^2dt,$$

where

$$\frac{dx_t}{dt}=b(\mu,x_t),\quad x_0.$$

In [74] was proved the following theorem.

Theorem 3.4. *Let the functions $S(\theta,x)$ and $b(\mu,x)$ have $k+3$ continuous bounded derivatives with respect to x and θ, the function $G(\theta,\mu)$ has a unique minimum at point $\theta^*=\theta^*(\mu)\in(\alpha,\beta)$ for any $\mu\in\mathcal{M}$, and*

$$\inf_{\mu\in\mathcal{M}}\ddot{G}(\theta^*,\mu)>0,$$

then the MLE $\hat{\theta}_\varepsilon$ admits the representation

$$\hat{\theta}_\varepsilon=\theta^*+\left\{\sum_{j=1}^k\psi_j\varepsilon^j+\psi_{k+1}\varepsilon^{k+\frac{1}{2}}\right\}\chi_{\{\mathcal{A}\}}+\zeta\chi_{\{\mathcal{A}^c\}},$$

where the random variables $\psi_1,\ldots,\psi_k$ do not depend on ε, $|\psi_{k+1}|<1$ and for $\mathcal{A}$ and ζ, estimates of the type (3.29), (3.30) are valid.

4. Partially observed process

This model will be studied in details in Chapter 6. Here we only outline the proof.

Let $\{X_t, 0\le t\le T\}$ be an observed component of the two-dimensional process

$$dX_t=[f_t(\theta)+h_t(\theta)Y_t]dt+\varepsilon dW_t,\quad X_0=0,$$

$$dY_t=[a_t(\theta)+b_t(\theta)Y_t]dt+\varepsilon\sigma_t(\theta)dV_t,\quad Y_0=y_0,$$

where $f_t(\cdot),h_t(\cdot),a_t(\cdot),b_t(\cdot)$ and $\sigma_t(\cdot)$ are known functions, W_t and V_t are independent Wiener processes, and the value of parameter $\theta\in(\alpha,\beta)$ have to be estimated. Accordingly innovation theorem

$$dX_t=[f_t(\theta)+h_t(\theta)m_t(\theta)]dt+\varepsilon d\bar{W}_t,\quad X_0=0,\tag{3.38}$$

where $m_t(\theta)=\mathbf{E}_\theta(Y_t|X_s,0\le s\le t)$ satisfies the equation

$$\begin{aligned}dm_t(\theta)\ &=\ [a_t(\theta)+b_t(\theta)m_t(\theta)]dt+\\ &\quad+\frac{\gamma_t(\theta)h_t(\theta)}{\varepsilon^2}\left[dX_t-[f_t(\theta)+h_t(\theta)m_t(\theta)]dt\right],\qquad m_0=y_0\end{aligned}$$

with a mean square error $\gamma_t(\theta)=\mathbf{E}_\theta(Y_t-m_t(\theta))^2$ taken from the Riccati-type equation

$$\frac{\partial\gamma_t(\theta)}{\partial t}=2b_t(\theta)\gamma_t(\theta)-\left(\frac{\gamma_t(\theta)h_t(\theta)}{\varepsilon}\right)^2+\varepsilon^2\sigma_t(\theta),\quad\gamma_0(\theta)=0$$

and $\bar{W}_t$ is an innovation process. The process (3.38) can be written as

$$dX_t = S_t(\theta, X)dt + \varepsilon d\bar{W}_t, \quad X_0 = 0$$

with the corresponding (integral-type) nonanticipative trend coefficient $S_t(\theta, X)$. This model is close to the one considered in section 2.2, but the Wiener process $\bar{W}_t$ now depends on θ and ε.

The maximum likelihood equation is

$$\varepsilon \int_0^T \dot{S}_t(\theta+u, X)d\bar{W}_t - \int_0^T \dot{S}_t(\theta+u, X)[S_t(\theta+u, X) - S_t(\theta, X)]dt = 0,$$

where $\dot{S}_t(\theta, X) = \dot{f}_t(\theta) + \dot{h}_t(\theta)m_t(\theta) + h_t(\theta)\dot{m}_t(\theta)$ and the stochastic process $\dot{m}_t(\theta)$ is obtained by formally differentiating the equation for $m_t(\theta)$ by θ.

Also introduce the notations

$$y_t(\theta) = y_0 \exp\left\{\int_0^t b_s(\theta)ds\right\} + \int_0^t \exp\left\{\int_s^t b_v(\theta)dv\right\} a_s(\theta)ds,$$

$$\dot{y}_t(\theta) = \frac{\partial}{\partial\theta} y_t(\theta),$$

$$I(\theta) = \int_0^T [\dot{f}_t(\theta) + \dot{h}_t(\theta)y_t(\theta) + h_t(\theta)\dot{y}_t(\theta)]^2 dt,$$

$$\begin{aligned}\frac{\partial y_t(\theta_1, \theta)}{\partial t} &= a_t(\theta_1) + [b_t(\theta_1) - Z_t(\theta_1)h_t(\theta_1)^2]y_t(\theta_1, \theta) + \\ &+ Z_t(\theta_1)h_t(\theta_1)[f_t(\theta) - f_t(\theta_1) + h_t(\theta)y_t(\theta)], \quad y_0(\theta_1, \theta) = y_0,\end{aligned}$$

$$G(\theta_1, \theta) = \int_0^T [f_t(\theta_1) - f_t(\theta) + h_t(\theta_1)y_t(\theta_1, \theta) - h_t(\theta)y_t(\theta)]^2 dt.$$

The Wiener process in the maximum likelihood equation (3.38) now depends on the value of ε and $d\bar{W}_t = \varepsilon^{-1}[dX_t - S_t(\theta, X)dt]$, so we can write this equation as

$$\varepsilon \int_0^T \dot{S}_t(\theta+u, X)dW_t - \int_0^T \dot{S}_t(\theta+u, X)[S_t(\theta+u, X) - f_t(\theta) - h_t(\theta)Y_t(\theta)]dt = 0$$

and, considering this equality as an implicit function $u = u(\varepsilon)$, we can obtain the desired expansion.

Theorem 3.5. *Let the functions $f_t(\cdot), h_t(\cdot), b_t(\cdot)$ and $\sigma_t(\cdot)$ be bounded and have $k+3$ continuous bounded derivatives w.r.t. θ, the quantity $I(\theta) > 0$ and, for any $\nu > 0$ and $\theta \in \Theta$,*

$$\inf_{|\theta_1 - \theta| > h} G(\theta_1, \theta) > 0,$$

then the MLE $\hat{\theta}_\varepsilon$ admits the representation

$$\hat{\theta}_\varepsilon = \theta + \left\{\sum_{j=1}^k \psi_j \varepsilon^j + \psi_{k+1}\varepsilon^{k+\frac{1}{2}}\right\} \chi_{\{A\}} + \zeta\chi_{\{A^c\}},$$

where the random variables $\psi_1, \ldots, \psi_k$ *do not depend on* ε, $|\psi_{k+1}| < 1$ *and we have the estimates:*

$$\mathbf{P}_\theta^{(\varepsilon)}\{\mathcal{A}^c\} \leq C_1 \exp\{-\kappa_1 \varepsilon^{-\gamma_1}\},$$

$$\mathbf{P}_\theta^{(\varepsilon)}\{|\zeta| > \varepsilon^\delta\} < C_2 \exp\{-\kappa_2 \varepsilon^{-\gamma_2}\}$$

with the positive constants $C_i, \kappa_i, \gamma_i, i = 1, 2$.

Proof is not presented here because it is a repetition of the proof of Theorem 3.1. The only difference lies in the calculation of the coefficients $\psi_1, \ldots, \psi_{k+1}$ according to the equations of the optimal filtration.

The first (normal) term of this expansion is

$$\psi_1 = I(\theta)^{-1} \int_0^T [\dot{f}_t(\theta) + \dot{h}_t(\theta) y_t(\theta) + h_t(\theta) \dot{y}_t(\theta)] dW_t,$$

so the MLE $\hat{\theta}_\varepsilon$ as $\varepsilon \to 0$ is asymptotically normal

$$\mathcal{L}_\theta\left\{\varepsilon^{-1}(\hat{\theta}_\varepsilon - \theta)\right\} \Longrightarrow \mathcal{N}(0, I(\theta)^{-1}). \tag{3.39}$$

We return to this problem in Chapter 6 and we shall prove (3.39) under less restrictive conditions.

3.3 Expansion of the distribution function

Below, we obtain *asymptotic expansions of the distribution function* and of the moments of the random variable $\hat{u}_\varepsilon = \varepsilon^{-1} I(\theta)^{1/2}(\hat{\theta}_\varepsilon - \theta)$ by the powers of ε in the case of observations

$$dX_t = S(\theta, X_t)\, dt + \varepsilon\, dW_t,\ X_0 = x_0,\ 0 \leq t \leq T, \tag{3.40}$$

of homogeneous diffusion processes. For simplicity of notation, we also suppose that the function $S(\theta, x_t) > 0$, for all $t \in [0, T]$. Remember, that if $S(\theta, x_\tau) = 0$ for some $\tau \in [0, T]$, then $x_t = x_\tau$ and $S(\theta, x_t) = 0$ for all $t \in [\tau, T]$.

The representation

$$\hat{\theta}_\varepsilon = \theta + \left\{\sum_{j=1}^k \psi_j \varepsilon^j + \eta \varepsilon^{k+\frac{1}{2}}\right\} \chi_{\{\mathcal{A}\}} + \zeta \chi_{\{\mathcal{A}^c\}}$$

of the MLE $\hat{\theta}_\varepsilon$ with estimates (3.5) and (3.6), allow us to write, for the distribution function $F_\varepsilon(x)$ of the normed difference $\hat{u}_\varepsilon$, the two-side inequalities (here $\psi_j^* = I(\theta)^{\frac{1}{2}} \psi_j$)

$$\begin{aligned} F_\varepsilon(x) &= \mathbf{P}_\theta^{(\varepsilon)}\{\hat{u}_\varepsilon < x, \mathcal{A}\} + \mathbf{P}_\theta^{(\varepsilon)}\{\hat{u}_\varepsilon < x, \mathcal{A}^c\} \leq \\ &\leq \mathbf{P}_\theta^{(\varepsilon)}\left\{\sum_{j=1}^k \psi_j^* \varepsilon^{j-1} < x + \varepsilon^{k-\frac{1}{2}}\right\} + \mathbf{P}_\theta^{(\varepsilon)}\{\mathcal{A}^c\} \end{aligned}$$

and

$$F_\varepsilon(x) \geq \mathbf{P}_\theta^{(\varepsilon)}\{\hat u_\varepsilon < x, \mathcal{A}\} \geq \mathbf{P}_\theta^{(\varepsilon)}\left\{\sum_{j=1}^k \psi_j^* \varepsilon^{j-1} < x - \varepsilon^{k-\frac12}\right\} - \mathbf{P}_\theta^{(\varepsilon)}\{\mathcal{A}^c\}.$$

So, if

$$\mathbf{P}_\theta^{(\varepsilon)}\left\{\sum_{j=1}^k \psi_j^* \varepsilon^{j-1} < x \pm \varepsilon^{k-\frac12}\right\} = \mathbf{P}_\theta^{(\varepsilon)}\left\{\sum_{j=1}^k \psi_j^* \varepsilon^{j-1} < x\right\} + o(\varepsilon^k),$$

then the problem of the expansion of $F_\varepsilon(x)$ by the powers of ε, is reduced to the problem of expansion

$$\mathbf{P}_\theta^{(\varepsilon)}\left\{\sum_{j=1}^k \psi_j^* \varepsilon^{j-1} < x\right\},$$

where the random variables $\psi_1^*, \ldots, \psi_k^*$ do not depend on ε and $\psi_1^* \sim \mathcal{N}(0,1)$.

First, we consider the post-Gaussian term of expansion of the distribution function. So, according to (3.26), on the set $\mathcal{A}$ we have the equality

$$\hat u_\varepsilon = \psi_1^* + \varepsilon\,\psi_2^* + \eta_1\,\varepsilon^{3/2} \tag{3.41}$$

and we expand the probability

$$G_\varepsilon(x) = \mathbf{P}_\theta\{\psi_1^* + \varepsilon\,\psi_2^* < x\}$$

We need several additional notations :

$$\begin{aligned}
h_t(\theta, x) &= \int_0^t \dot S'(\theta, x_s)\,\dot S(\theta, x_s)\,S(\theta, x_s)\,ds,\\
\sigma_1^2 &= I(\theta), \qquad \sigma_2^2 = \int_0^T \ddot S(\theta, x_t)^2\,dt,\\
\sigma_3^2 &= \int_0^T [h_T(\theta, x) - h_t(\theta, x)]^2\,S(\theta, x_t)^{-2}\,dt,\\
\sigma_4^2 &= \int_0^T \dot S'(\theta, x_t)^2 \int_0^t \frac{dv}{S(\theta, x_v)^2}\,S(\theta, x_t)^2\,dt,\\
\xi_1 &= \sigma_1^{-1}\int_0^T \dot S(\theta, x_t)\,dW_t\ (= \psi_1^*),\ \xi_2 = \sigma_2^{-1}\int_0^T \ddot S(\theta, x_t)\,dW_t\\
\xi_3 &= \sigma_3^{-1}\int_0^T \dot S'(\theta, x_t)\,\dot S(\theta, x_t)\,x_t^{(1)}\,dt,\\
\xi_4 &= \sigma_4^{-1}\int_0^T \dot S'(\theta, x_t)\,x_t^{(1)}\,dW_t,\\
\rho_{ij} &= \mathbf{E}\xi_i\,\xi_j, \qquad i,j = 1,\ldots,4, \qquad a_1 = -\frac{3\rho_{12}\,\sigma_2}{2\sigma_1^2},\\
a_2 &= \sigma_2\,\sigma_1^{-2}, \qquad a_3 = -2\,\sigma_3\,\sigma_1^{-2}, \qquad a_4 = \sigma_4\sigma_1^{-1}, \qquad R = \sum_{i=1}^3 a_i\,\rho_{1i},\\
Q &= \sigma_1^{-3}\int_0^T \dot S'(\theta, x_t)\,\dot S(\theta, x_t)\,S(\theta, x_t)\int_0^t \frac{\dot S(\theta, x_v)}{S(\theta, x_v)}dv\,dt.
\end{aligned}$$

Above, we make an agreement that if $\sigma_j = 0$, then we put $\xi_j = 0$. Note that if $\sigma_j \neq 0$, then

$$\mathcal{L}\{\xi_j\} = \mathcal{N}(0,1), \qquad j = 1,2,3 \tag{3.42}$$

and $\mathbf{E}\xi_4^2 = 1$. The last equality can be proved by direct calculation. To explain the equality (3.42) for $j = 3$, we recall that the process $x_t^{(1)}$, as a solution of the first equation of system (3.19), has the representation

$$\begin{aligned} x_t^{(1)} &= \int_0^t \exp\left\{\int_s^t S'(\theta, x_v)\, dv\right\} dW_s \\ &= S(\theta, x_t) \int_0^t \frac{dW_v}{S(\theta, x_v)}, \end{aligned} \tag{3.43}$$

because

$$\int_s^t S'(\theta, x_v)\, dv = \int_s^t \frac{S'(\theta, x_v)\, dx_v}{S(\theta, x_v)} = \ln \frac{S(\theta, x_t)}{S(\theta, x_s)}.$$

So, for the random variable ξ_3 integrating by parts, we can write

$$\begin{aligned} &\int_0^T \dot{S}'(\theta, x_t)\, \dot{S}(\theta, x_t)\, x_t^{(1)}\, dt = \\ &= \int_0^T \dot{S}'(\theta, x_t)\, \dot{S}(\theta, x_t)\, S(\theta, x_t) \int_0^t \frac{dW_v}{S(\theta, x_v)}\, dl = \\ &= \int_0^T \left(\int_0^t \frac{dW_v}{S(\theta, x_v)}\right) dh_t(\theta, x) = h_T(\theta, x) \int_0^T \frac{dW_t}{S(\theta, x_t)} - \\ &- \int_0^T h_t(\theta, x) S(\theta, x_t)^{-1}\, dW_t = \int_0^T [h_T(\theta, x) - h_t(\theta, x)]\, S(\theta, x_t)^{-1}\, dW_t. \end{aligned}$$

Hence, from this representation follows (3.42) for $j = 3$.

The equality (3.41) could be rewritten in new notations as

$$\hat{u}_\varepsilon = \xi_1 + \varepsilon \sum_{j=1}^{3} a_j\, \xi_1\, \xi_j + \varepsilon\, a_4\, \xi_4 + \eta_1\, \varepsilon^{3/2},$$

where (modifying the definition of the set $\mathcal{A}_3$) we can have $|\eta_1| < 1$. Below, we denote by $n(y)$ the density of the normal $\mathcal{N}(0,1)$ law.

Theorem 3.6. *Let the conditions of Theorem 3.1 be fulfilled for $k = 1$, then*

$$\begin{aligned} &|\mathbf{P}_\theta^{(\varepsilon)}\{\varepsilon^{-1} I(\theta)^{1/2}(\hat{\theta}_\varepsilon - \theta) < x\} - \int_{-\infty}^{x} n(y)\, dy + \varepsilon R\, n(x) - \\ &-\varepsilon (R + Q)(1 - x^2)\, n(x)| \le C_* \varepsilon^{3/2} \end{aligned} \tag{3.44}$$

with some positive constant C_*.

Proof. As was mentioned above, we can write two inequalities:

$$\begin{aligned} &\mathbf{P}_\theta^{(\varepsilon)}\{\varepsilon^{-1} I(\theta)^{1/2}(\hat{\theta}_\varepsilon - \theta) < x\} = \\ &= \mathbf{P}_\theta^{(\varepsilon)}\{\hat{u}_\varepsilon < x, \mathcal{A}\} + \mathbf{P}_\theta^{(\varepsilon)}\{\hat{u}_\varepsilon < x, \mathcal{A}^c\} \le \\ &\le \mathbf{P}_\theta^{(\varepsilon)}\{\xi_1 + \varepsilon \psi_2^* + \varepsilon^{3/2} \eta_1 < x\} + \mathbf{P}_\theta^{(\varepsilon)}\{\mathcal{A}^c\} \le \\ &\le \mathbf{P}_\theta^{(\varepsilon)}\{\xi_1 + \varepsilon \psi_2^* < x + \varepsilon^{3/2}\} + C_1 \exp\{-c_1 \varepsilon^{-\gamma_1}\} \end{aligned} \tag{3.45}$$

And analogously

$$\mathbf{P}_\theta^{(\varepsilon)}\{\hat{u}_\varepsilon < x\} \geq \mathbf{P}_\theta^{(\varepsilon)}\{\hat{u}_\varepsilon < x, \mathcal{A}^c\} \geq \mathbf{P}_\theta^{(\varepsilon)}\{\xi_1 + \varepsilon\psi_2^* < x - \varepsilon^{3/2}\} - \\ - \mathbf{P}_\theta^{(\varepsilon)}\{\mathcal{A}^c\}. \tag{3.46}$$

In our notations

$$\psi_2^* = \sum_{j=1}^{3} a_j\, \xi_1\, \xi_j \;+\; a_4\, \xi_4.$$

Denote also $\psi = \psi_2^*$, so

$$G_\varepsilon(x) = \mathbf{P}_\theta^{(\varepsilon)}\{\xi_1 + \varepsilon\, \psi < x\}$$

and develop this function by the powers of ε. First, we develop the characteristic function and then estimate the remainder of the inverse Fourier transform.

Taylor's formula allows us to write

$$e^{i\lambda} = 1 + i\lambda - \lambda^2 \int_0^1 (1-s)\, e^{is\lambda}\, ds$$

So

$$\begin{aligned}\Phi_\varepsilon(\lambda) &= \mathbf{E}_\theta \exp\{i\lambda\xi_1 + i\lambda\varepsilon\psi\} = \mathbf{E}e^{i\lambda\xi_1} + i\lambda\varepsilon\mathbf{E}_\theta\psi\, e^{i\lambda\xi_1} - \\ &\quad - \lambda^2\varepsilon^2 \int_0^1 (1-s)\mathbf{E}_\theta\psi^2 \exp\{i\lambda\xi_1 + i\lambda s\varepsilon\psi\}ds = \\ &= e^{-\frac{\lambda^2}{2}} \;+\; i\lambda\varepsilon\mathbf{E}\psi e^{i\lambda\xi_1} + r_\varepsilon(\lambda) \end{aligned} \tag{3.47}$$

Direct calculation provides the following equalities

$$\begin{aligned}\mathbf{E}\xi_1^2\, e^{i\lambda\xi_1} &= -\frac{\partial^2}{\partial\lambda^2}\mathbf{E}e^{i\lambda\xi_1} = -\frac{\partial^2}{\partial\lambda^2}e^{-\lambda^2/2} = (1-\lambda^2)\, e^{-\lambda^2/2} \\ \mathbf{E}\xi_1\, \xi_j\, e^{i\lambda\xi_1} &= \rho_{1j}(1-\lambda^2)e^{-\lambda^2/2},\end{aligned}$$

because ξ_1, ξ_2 and ξ_3 are Gaussian random variables. To calculate $\mathbf{E}\psi e^{i\lambda\xi_1}$, we introduce the notations

$$\begin{aligned} I_t &= \int_0^t \dot{S}(\theta, x_s)^2\, ds, \\ V_t &= \exp\left\{\frac{i\lambda}{\sigma_1}\int_0^t \dot{S}(\theta, x_s)\, dW_s + \frac{\lambda^2}{2\sigma_1^2}\, I_t\right\}. \end{aligned}$$

By the Itô formula, the process V_t has the representation

$$V_T = 1 \;+\; \frac{i\lambda}{\sigma_1}\int_0^T V_t\, \dot{S}(\theta, x_t)dW_t.$$

Hence

$$\begin{aligned}
\mathbf{E}\xi_4\, e^{i\lambda\xi_1} &= \sigma_4^{-1}e^{-\lambda^2/2}\mathbf{E}\left\{\int_0^T \dot S'(\theta,x_t)\, x_t^{(1)}\, dW_t\, \exp\left[\frac{i\lambda}{\sigma_1}\int_0^T \dot S(\theta,x_t)\, dW_t + \frac{\lambda^2}{2\sigma_1^2}I_T\right]\right\} = \\
&= \sigma_4^{-1}e^{-\lambda^2/2}\mathbf{E}\left\{\int_0^T \dot S'(\theta,x_t)\, x_t^{(1)}\, dW_t \left[1+\frac{i\lambda}{\sigma_1}\int_0^T V_t\, \dot S(\theta,x_t)\, dW_t\right]\right\} = \\
&= \frac{i\lambda}{\sigma_1\sigma_4}\, e^{-\lambda^2/2}\int_0^T \dot S'(\theta,x_t)\, \dot S(\theta,x_t)\mathbf{E}(x_t^{(1)}V_t)\, dt = \\
&= \frac{i\lambda}{\sigma_1\sigma_4}\, e^{-\lambda^2/2}\int_0^T \dot S'(\theta,x_t)\, \dot S(\theta,x_t)\, S(\theta,x_t)\exp\left\{\frac{\lambda^2 I_t}{2\sigma_1^2}\right\}\times \\
&\quad \times\mathbf{E}\left\{\int_0^t \frac{dW_s}{S(\theta,x_s)}\, \exp\left\{\frac{i\lambda}{\sigma_1}\int_0^t \dot S(\theta,x_s)\, dW_s\right\}\right\}\, dt.
\end{aligned}$$

Both stochastic integrals are Gaussian random variables, so

$$\begin{aligned}
&\mathbf{E}\left\{\int_0^t \frac{dW_s}{S(\theta,x_s)}\, \exp\left\{\frac{i\lambda}{\sigma_1}\int_0^t \dot S(\theta,x_s)\, dW_s\right\}\right\} = \\
&= \frac{i\lambda}{\sigma_1}\int_0^t \frac{\dot S(\theta,x_s)}{S(\theta,x_s)}\, ds\, \exp\left\{-\frac{\lambda^2 I_t}{2\sigma_1^2}\right\}.
\end{aligned}$$

Substituting this equality, we obtain

$$\begin{aligned}
\mathbf{E}\xi_4\, e^{i\lambda\xi_1} &= \frac{(i\lambda)^2}{\sigma_1^2\sigma_4}\, e^{-\lambda^2/2}\int_0^T \dot S(\theta,x_t)\dot S(\theta,x_t)S(\theta,x_t)\left(\int_0^t \frac{\dot S(\theta,x_s)}{S(\theta,x_s)}ds\right)dt = \\
&= \frac{(i\lambda)^2}{\sigma_4}\, e^{-\lambda^2/2}\, Q\, \sigma_1
\end{aligned}$$

and returning to expression (3.47) , we can write

$$\begin{aligned}
\mathbf{E}\psi\, e^{i\lambda\xi_1} &= \sum_{j=1}^3 a_j\rho_{1j}(1-\lambda^2)\, e^{-\lambda^2/2} - Q\lambda^2\, e^{-\lambda^2/2} = \\
&= (1-\lambda^2)e^{-\lambda^2/2}\, R - \lambda^2\, e^{-\lambda^2/2}Q.
\end{aligned}$$

Denote by $H_k(x)$ the k-th Hermitian polynomial and recall , that

$$H_1(x) = x, \qquad H_2(x) = x^2-1, \qquad H_3(x) = x^3-3x$$

The Fourier transforms of $H_k(x)\, n(x)$ is $(i\lambda)^k\exp(-\lambda^2/2)$ and the inverse Fourier transform of $i\lambda\varepsilon\mathbf{E}\psi e^{i\lambda\xi_1}$ is

$$1/(2\pi)\int_{-\infty}^{\infty} e^{-i\lambda x}i\lambda\varepsilon\mathbf{E}(\psi e^{i\lambda\xi_1})d\lambda = \varepsilon RH_1(x)n(x) + \varepsilon(R+Q)H_3(x)n(x)$$

So the distribution function

$$\begin{aligned}
G_\varepsilon(x) &= \int_{-\infty}^x n(y)\, dy - \varepsilon R\, n(x) + \varepsilon(R+Q)(1-x^2)\, n(x) + \\
&\quad +1/(2\pi)\int_{-\infty}^{\infty} ie^{-i\lambda x}\lambda^{-1}r_\varepsilon(\lambda)d\lambda.
\end{aligned}$$

It only remains to estimate the remainder. To do this, we need two auxiliary results. The first one is due to Esseen [34]:

Lemma 3.6. *Let M, N be positive constants, $F(x)$ nondecreasing function, $G(x)$ a function of bounded variation, $F(\infty) = G(\infty)$, $F(-\infty) = G(-\infty)$, the derivative $|G'(x)| \leq M$ and $\Phi(\lambda)$, $\Psi(\lambda)$ are their Fourier-Stiltjes transforms. Then, for any $m > 1$, there exists $C(m)$ such that*

$$|F(x) - G(x)| \leq \frac{m}{2\pi} \int_{-N}^{N} \left| \frac{\Phi(\lambda) - \Psi(\lambda)}{\lambda} \right| d\lambda + C(m)\frac{M}{N}. \tag{3.48}$$

The second result is concerned with the two variables

$$\xi = \int_0^T f(t)\, dW_t, \qquad \zeta = \int_0^T \int_0^t h(t,s) dW_s\, dW_t,$$

where $f(t)$ and $h(t,s)$ are two deterministic measurable functions, such that

$$\int_0^T f(t)^2 dt = 1, \qquad \int_0^T h(t,s)^2 ds < C \tag{3.49}$$

for all $t \in [0,T]$ and some $C > 0$.

Lemma 3.7. *Let conditions (3.49) be fulfilled, then*

$$|\mathbf{E}\zeta^2\, e^{i\lambda\xi + i\varepsilon\lambda\zeta}| \leq (C_1 + C_2\lambda^{-1}\varepsilon^{-4} + C_3\lambda^{-2}\varepsilon^{-10})\, e^{-c\varepsilon^{-2}} + C_4\lambda^{-2} \tag{3.50}$$

with some constants $C_j, j = 1, \ldots, 4$ depending on $f(\cdot)$ and $h(\cdot,\cdot)$.

Proof. Let us introduce the Gaussian process

$$y_t = W_t - \xi \int_0^t f(s)\, ds \tag{3.51}$$

(the author is grateful to A.V. Skorohod for the idea of this representation of the Wiener process), and denote by $\mathcal{F}_T^y$ the sigma-algebra induced by this process, then $y_t, 0 \leq t \leq T$ and ξ are independent because

$$\mathbf{E}(y_t\xi) = 0.$$

The random variable ζ admits the representation

$$\begin{aligned}
\zeta &= \xi^2 \int_0^T \int_0^t h(t,s) f(s) ds f(t) dt + \xi \left[\int_0^T \int_0^t h(t,s) dy_s f(t) dt + \right. \\
&\quad \left. + \int_0^T \int_0^t h(t,s) f(s) ds dy_t \right] + \int_0^T \int_0^t h(t,s) dy_s dy_t = \\
&= A\xi^2 + B(y)\xi + C(y).
\end{aligned}$$

Denote also

$$q(x) = x + \varepsilon B(y) x + \varepsilon A\, x^2$$

and introduce the sets

$$\mathcal{B}_1=\left\{\omega:|A\xi|<\frac{1}{8\varepsilon}\right\},\qquad \mathcal{B}_2=\left\{\omega:|B(y)|<\frac{1}{4\varepsilon}\right\},\qquad \mathcal{B}=\mathcal{B}_1\bigcap\mathcal{B}_2. \tag{3.52}$$

Then we can write

$$\begin{aligned}\mathbf{E}\zeta^2e^{i\lambda(\xi+\varepsilon\zeta)} &= \mathbf{E}(A\xi^2+B(y)\xi+C(y))^2\exp\{i\lambda(q(\xi)+\varepsilon C(y))\}= \\ &= \sum_{l=0}^{4}\mathbf{E}b_l(y)e^{i\varepsilon\lambda C(y)}\mathbf{E}\left(\xi^le^{i\lambda q(\xi)}|\mathcal{F}_T^y\right)\end{aligned}$$

with the obvious notations for $b_l(y)$. Therefore

$$\begin{aligned}&|\mathbf{E}\zeta^2e^{i\lambda(\xi+\varepsilon\zeta)}|\le|\mathbf{E}\zeta^2\;\chi_{\{\mathcal{B}\}}e^{i\lambda(\xi+\varepsilon\zeta)}|+\mathbf{E}\zeta^2\chi_{\{\mathcal{B}^c\}}\le \\ &\le\sum_{l=0}^{4}\mathbf{E}|b_l(y)|\chi_{\{\mathcal{B}_2\}}\left|\mathbf{E}\left(\xi^l\chi_{\{\mathcal{B}_1\}}e^{i\lambda q(\xi)}|\mathcal{F}_T^y\right)\right|+\left(\mathbf{P}\{\mathcal{B}^c\}\mathbf{E}\zeta^4\right)^{1/2}\end{aligned} \tag{3.53}$$

The random variables ξ and $B(y)$ are Gaussian, so if $|A|\ne 0$, then

$$\mathbf{P}\{\mathcal{B}^c\}\le 1/2\exp\{-(128A^2\varepsilon^2)^{-1}\}+1/2\exp\{-c_0\varepsilon^{-2}\}\le\exp\{-c\varepsilon^{-2}\}$$

with some $c_0>0$ and $c=\min\{\frac{1}{128A^2},c_0\}$. If $|A|=0$, then $\mathbf{P}\{\mathcal{B}_1^c\}=0$ and we have to consider the next term with a corresponding modification of the proof. The mathematical expectation $\mathbf{E}\zeta^4$ is bounded.

Further, for $\omega\in\mathcal{B}_2$ and denoting $N_1=(8\varepsilon|A|)^{-1}$ with probability 1, we can write

$$\mathbf{E}\left(\xi^l\chi_{\{\mathcal{B}_1\}}e^{i\lambda q(\xi)}|\mathcal{F}_T^y\right)=\frac{1}{\sqrt{2\pi}}\int_{-N_1}^{N_1}x^le^{i\lambda q(x)-x^2/2}dx$$

Below, we twice integrate by parts and put $q'(x)=1+\varepsilon B(y)+2\varepsilon Ax$. Note that for $\omega\in\mathcal{B}$ we have $q'(\xi)>1/2$.

$$\begin{aligned}&\int_0^{N_1}x^le^{i\lambda q(x)-x^2/2}dx=\frac{1}{i\lambda}\frac{N_1^le^{i\lambda q(N_1)-N_1^2/2}}{q'(N_1)}+\frac{1}{i\lambda}\int_0^{N_1}\frac{P_{l+2}(x)e^{i\lambda q(x)-x^2/2}}{q'(x)^2}dx= \\ &=\frac{1}{i\lambda}\frac{N_1^le^{i\lambda q(N_1)-N_1^2/2}}{q'(N_1)}+\frac{1}{\lambda^2}\frac{P_{l+2}(N_1)e^{i\lambda q(N_1)-N_1^2/2}}{q'(N_1)^3}+ \\ &+\frac{1}{\lambda^2}\int_0^{N_1}\frac{P_{l+6}(x)e^{i\lambda q(x)-x^2/2}}{q'(x)^6}dx,\end{aligned} \tag{3.54}$$

where $P_l(x)$ is a polynomial on x of order l. The integral

$$\int_0^{N_1}\frac{|P_{l+6}(x)|e^{-x^2/2}}{q'(x)^6}dx\le 2^6\int_0^{\infty}|P_{l+6}(x)|e^{-x^2/2}dx\equiv C_0<\infty.$$

Hence

$$\left|\int_0^{N_1}x^le^{i\lambda q(x)-x^2/2}dx\right|\le\frac{2}{\lambda}N_1^l\;e^{-N_1^2/2}+\frac{1}{\lambda^2}|P_{l+2}(N_1)|\;e^{-N_1^2/2}+\frac{C_0}{\lambda^2}.$$

Remember also that all mathematical expectations $\mathbf{E}|b_l(y)|$ are bounded. So we can write

$$|\mathbf{E}\zeta^2 e^{i\lambda(\xi+\varepsilon\zeta)}| \leq (C_1 + \frac{C_2}{\lambda}\varepsilon^{-4} + \frac{C_3}{\lambda^2}\varepsilon^{-10})\, e^{-c\varepsilon^{-2}} + \frac{C_4}{\lambda^2}, \tag{3.55}$$

where the constants C_j for bounded $f(t)$ and $h(t,s)$ could be chosen do not depending on $f(\cdot)$ and $h(\cdot,\cdot)$.

Returning to the proof of Theorem 3.6, first we note that, using Itô formula, we can write ψ in the integral form like ζ, so the conditions of Lemma 3.7 are fulfilled with the bounded functions $f(\cdot)$ and $h(\cdot,\cdot)$. Then we denote by

$$G(x) = \int_{-\infty}^{x} n(y)\, dy - \varepsilon R\, n(x) \,+\, \varepsilon(R+Q)(1-x^2)\, n(x)$$

and by $g(\lambda)$ its Fourier transform. Lemmas 3.6 and 3.7 allow us to estimate the difference :

$$\begin{aligned}
|G_\varepsilon(x) \;&-\; G(x)| \leq \frac{m}{2\pi}\int_{-N}^{N}\left|\frac{\Phi_\varepsilon(\lambda)-g(\lambda)}{\lambda}\right| d\lambda \,+ C(m)\frac{M}{N} \\
&\leq \frac{m}{2\pi}\int_{-N_0}^{N_0}\left|\frac{\Phi_\varepsilon(\lambda)-g(\lambda)}{\lambda}\right| d\lambda + \frac{m}{2\pi}\int_{N_0<|\lambda|<N}\left|\frac{\Phi_\varepsilon(\lambda)-g(\lambda)}{\lambda}\right| d\lambda + C(m)\frac{M}{N} \\
&\leq \frac{m\varepsilon^2 N_0^2}{4\pi}\mathbf{E}\psi^2 + \frac{m\varepsilon^2}{4\pi}\int_{N_0}^{N}(C_1\lambda + C_2\varepsilon^{-l} + C_3\lambda^{-1}\varepsilon^{-(l+6)})d\lambda\; e^{-c\varepsilon^{-2}} + \\
&+\; \frac{m\varepsilon^2}{2\pi}C_4 \ln\frac{N}{N_0} + C(m)\frac{M}{N}
\end{aligned} \tag{3.56}$$

The proposition of the Theorem follows now from this inequality if we put $N = \varepsilon^{-3/2}$ and fix N_0 as a constant.

Now consider the general case ($k \geq 2$). Taylor's formula

$$e^{i\lambda} = \sum_{l=0}^{k}\frac{(i\lambda)^l}{l!} + \frac{(i\lambda)^{k+1}}{k!}\int_0^1 (1-s)^k\, e^{is\lambda}\, d\lambda$$

allows us to write the expansion of the characteristic function

$$\begin{aligned}
\Phi_\varepsilon(\lambda) \;&=\; \mathbf{E}\exp\left\{i\lambda\xi_1 + i\lambda\sum_{j=2}^{k}\psi_j^*\,\varepsilon^{j-1}\right\} = \mathbf{E}e^{i\lambda\xi_1}\,\{1 + i\lambda\psi_2^*\varepsilon + \\
&+\left[i\lambda\psi_3^* + \frac{(i\lambda)^2}{2}\psi_2^{*2}\right]\varepsilon^2 + \ldots + \left[i\lambda\psi_k^* + \ldots + \frac{(i\lambda)^{k-1}}{(k-1)!}\psi_2^{*(k-1)}\right]\varepsilon^{k-1} + \\
&+\left[\frac{(i\lambda)^2}{2}\sum_{r=2}^{k}\psi_r^*\psi_{k+2-r}^* + \ldots + \frac{(i\lambda)^k}{k!}\psi_2^{*k}\right]\varepsilon^k + \\
&+\; \varepsilon^{k+1}\int_0^1 g_k(s,\lambda)\exp\left\{i\lambda s\sum_{j=1}^{k}\psi_j^*\varepsilon^{j-1}\right\}\, ds\bigg\},
\end{aligned}$$

where $g_k(s,\lambda)$ is a polynomial on $(1-s)$, λ and $\psi_1^*,\ldots,\psi_k^*$. We denote by $\Psi_j(\lambda), j = 1,\ldots,k$ the coefficients of this expansion by the powers of ε (i.e., $\Psi_2(\lambda) = i\lambda\psi_3^* +$

$\frac{(i\lambda)^2}{2}\psi_2^{*2}$, etc.) and introduce the inverse of the Fourier transformation

$$\frac{d}{dx}G_j(x) = \frac{1}{2\pi}\int_{-\infty}^{\infty} e^{-i\lambda x}\mathbf{E}\Psi_j(\lambda)e^{i\lambda\xi_1}\,d\lambda$$

We now put

$$G(x) = \Phi(x) + \sum_{j=1}^{k} G_j(x)\,\varepsilon^j$$

and estimate the difference between $F_\varepsilon(x)$ and $G(x)$ in the same way as was done above. So, with the help of the Gaussian process $y_t, 0 \leq t \leq T$, we represent the sum

$$\begin{aligned} q(\xi_1) &= \sum_{j=1}^{k}\psi_j^*\varepsilon^{j-1} = \xi_1 + [a_2^{(2)}(y)\varepsilon + a_3^{(3)}(y)\varepsilon^2 + \ldots + \\ &\quad + a_k^{(k)}(y)\varepsilon^{k-1}] + \ \xi_1[a_2^{(1)}(y)\varepsilon + a_3^{(2)}(y)\varepsilon^2 + \ldots + a_k^{(k-1)}(y)\varepsilon^{k-1}] + \\ &\quad + \ldots + \xi_1^{k-1}[a_{k-1}^{(0)}\varepsilon^{k-2} + a_k^{(1)}(y)\varepsilon^{k-1}] + \xi_1^k a_k^{(0)}\varepsilon^{k-1} = \\ &= \xi_1 + \sum_{j=0}^{k}\xi_1^j A_j(y,\varepsilon), \end{aligned}$$

where we have denoted by $a_l^{(n)}(y)$ the l-tuple integral containing n integrations w.r.t. $y_t, 0 \leq t \leq T$ (n-linear functional of the process y_t) and $A_j(y,\varepsilon)$ are the corresponding coefficients. These integrals are defined as integrals on quasimartingales (see [98] for details), but we only need to know here that so-defined iterated integrals have all the polynomial moments, so in the estimates of the complement $\mathbf{P}_\theta\{\mathcal{B}^c\}$ below, we can write, for example, by Tchebyshev's inequality : for any $Q > 0$

$$\mathbf{P}\{|a_l^{(n)}(y)| > \varepsilon^{-1/2}\} \leq \varepsilon^{Q/2}\mathbf{E}|a_l^{(n)}(y)|^Q \leq C\,\varepsilon^{Q/2},$$

where the constant C could be chosen as not depending on ε. The exponential rate of convergence can also be obtained, but the polynomial estimate is sufficient for our purposes. We also define

$$q'(\xi_1) = 1 + \sum_{j=1}^{k} j\,\xi_1^{j-1}\,A_j(y,\varepsilon)$$

and introduce the sets

$$\mathcal{B}_1 = \{\omega : |\xi_1| < N_1\}, \qquad \mathcal{B}_2 = \left\{\omega : \sum_{j=1}^{k} j\,N_1^{j-1}|A_j(y,\varepsilon)| < \frac{1}{4}\right\}, \quad \mathcal{B} = \mathcal{B}_1 \cap \mathcal{B}_2,$$

where $N_1 = |8a_k^{(0)}(k-1)|^{-1}$ and we assume that $a_k^{(0)} \neq 0$.

The mathematical expectation of the remainder could be estimated with the help of similar inequalities

$$\begin{aligned} |\mathbf{E}\Psi(\lambda)\,e^{i\lambda q(\xi_1)}| &\leq |\sum_l \mathbf{E}|b_l(y,\lambda)\,\chi_{\{\mathcal{B}_2\}}|\,|\mathbf{E}(\xi_1^l\,e^{i\lambda_s q(\xi_1)}\chi_{\{\mathcal{B}_1\}}|\mathcal{F}_T^y)| + \\ &\quad + (\mathbf{P}\{\mathcal{B}^c\}\,\mathbf{E}\Psi^2)^{1/2} \end{aligned}$$

If the term $b_l(y,\varepsilon)$ contains λ of maximal power r, then we integrate (3.54) by parts r times to obtain the estimate for the last term in (3.55) of the order $C\lambda^{-r}$. Note that on the set $\mathcal{B}$, in this integral we have $q'(\xi_1) > 1/2$. So for the difference $G_\varepsilon(x) - G(x)$, where

$$G_\varepsilon(x) = \mathbf{P}_\theta\{q(\xi_1) \ < x\},$$

we shall have an estimate like (3.56)

$$|G_\varepsilon(x)-G(x)| \leq CN_0^{k+1}\varepsilon^{k+1}+\frac{m\varepsilon^{k+1}}{2\pi}\int_{N_0\leq|\lambda|\leq N}\lambda^{-1}\left|\mathbf{E}\int_0^1 g_k(s,\lambda)e^{i\lambda q(\xi_1)}ds\right|d\lambda+C(m)\frac{M}{N}$$

where we put $N = \varepsilon^{-(k+1/2)}$. Remember also that the function $G(x)$ has a bounded derivative, so

$$|G(x+\varepsilon^{k+1/2}) - G(x)| \leq M\ \varepsilon^{k+1/2}.$$

All these allow us to write the estimate

$$|\mathbf{P}_\theta^{(\varepsilon)}\{\varepsilon^{-1}I(\theta)^{1/2}(\hat\theta_\varepsilon - \theta)\} - G(x)| \leq C_{**}\ \varepsilon^{k+1/2}. \qquad (3.57)$$

Under the established assumptions, the constants in the proof above could be chosen as not depending on θ , if $\theta \in \mathbf{K}$, so this expansion is also uniform for the values of θ far from the bounds of Θ.

Moments expansion. Let us consider the expansion of the moments of the random variable

$$\hat u_\varepsilon = \varepsilon^{-1}I(\theta)^{1/2}(\hat\theta_\varepsilon - \theta)$$

by the powers of ε. We shall use the representation

$$\hat u_\varepsilon = \{\xi_1 + \psi_1^*\varepsilon + \psi_2^*\varepsilon^2 + \eta_2\varepsilon^{5/2}\}\chi_{\{A\}} + \varepsilon^{-1}\zeta\chi_{\{A\}}$$

where $|\eta_2| < 1$. So we restrict ourselves to the first three terms of expansion, i.e., we seek representations like

$$\mathbf{E}_\theta|\hat u_\varepsilon|^p = A_0 + A_1\varepsilon + A_3\varepsilon^2 + A_3(\varepsilon)\varepsilon^{5/2},$$

where $A_i, i = 1,2,3$ are constants and $A_3(\varepsilon)$ is a bounded function.

We assume that $p \geq 3$. Taylor's formula allows us to write

$$\begin{aligned}|a+\epsilon|^p &= |a|^p + p\epsilon|a|^{p-1}\mathrm{sign}a + \frac{p(p-1)}{2}\epsilon^2|a|^{p-2} + \\ &+ \frac{p(p-1)}{2}\epsilon^2\left(|a+\gamma\epsilon|^{p-2} - |a|^{p-2}\right), \qquad \gamma\in(0,1)\end{aligned} \qquad (3.58)$$

and

$$||a+\epsilon|^p - |a|^p| \leq p\,(|a| + |\epsilon|)^{p-1}. \qquad (3.59)$$

We also need an estimate on the tails of the distribution of $\hat{u}_\varepsilon$. Denote, as in Section 3.1,

$$\begin{aligned} \Delta S_\varepsilon(u) &= S(\theta+\varepsilon I(\theta)^{-1/2}u, X_t) - S(\theta, X_t), \\ \Delta S_\varepsilon(v) &= S(\theta+v, X_t) - S(\theta, X_t), \\ \Delta S_0(v) &= S(\theta+v, x_t) - S(\theta, x_t) \end{aligned}$$

and recall that

$$\inf_{|v|\geq\nu} \int_0^T [\Delta S_0(v)]^2 dt \geq \kappa\nu^2.$$

Below, we put $\nu = \varepsilon x I(\theta)^{-1/2}$ and follow with the derivation of (3.20).

$$\begin{aligned} \mathbf{P}_\theta^{(\varepsilon)}\{|\hat{u}_\varepsilon| \geq x\} &= \mathbf{P}_\theta^{(\varepsilon)}\left\{\sup_{|u|\geq x} Z_\varepsilon(u) > \sup_{|u|<x} Z_\varepsilon(u)\right\} \leq \\ &\leq \mathbf{P}_\theta^{(\varepsilon)}\left\{\sup_{|u|\geq x}\left[\int_0^T \Delta S_\varepsilon(u)\,dW_t - \frac{1}{2\varepsilon}\int_0^T \Delta S_\varepsilon(u)^2 dt\right] > 0\right\} \leq \\ &\leq \mathbf{P}_\theta^{(\varepsilon)}\left\{\sup_{|v|\geq\nu}\int_0^T \Delta S_\varepsilon(v) dW_t > \frac{1}{4\varepsilon}\inf_{|v|\geq\nu}\int_0^T \Delta S_0(v)^2 dt\right\} + \\ &+ \mathbf{P}_\theta^{(\varepsilon)}\left\{\sup_{|v|\geq\nu}\int_0^T |\Delta S_\varepsilon(v)^2 - \Delta S_0(v)^2| dt \geq \frac{1}{2}\inf_{|v|\geq\nu}\int_0^T \Delta S_0(v)^2 dt\right\} \leq \\ &\leq \mathbf{P}_\theta^{(\varepsilon)}\left\{\sup_{v}\int_0^T \Delta S_\varepsilon(v) dW_t > \frac{\kappa\nu^2}{4\varepsilon}\right\} + \mathbf{P}_\theta^{(\varepsilon)}\left\{\sup_{0\leq t\leq T}|W_t| > \frac{\kappa\nu^2}{C\varepsilon}\right\} \leq \\ &\leq C_1\exp\{-\kappa_3\varepsilon x^2\} + 4\exp\{-\kappa_4\varepsilon^2 x^4\} \leq C_* e^{-\kappa_3\varepsilon x^2} \end{aligned}$$

with some positive constants C_* and κ_3.

This estimate helps us to prove that integration over the set $\mathcal{A}^c$ gives a negligible contribution.

$$\begin{aligned} \mathbf{E}_\theta|\hat{u}_\varepsilon|^p\chi_{\{\mathcal{A}^c\}} &= \mathbf{E}_\theta|\hat{u}_\varepsilon|^p\chi_{\{\mathcal{A}^c, |\hat{u}_\varepsilon|<\varepsilon^{\delta-1}\}} + \mathbf{E}_\theta|\hat{u}_\varepsilon|^p\chi_{\{\mathcal{A}^c, |\hat{u}_\varepsilon|\geq\varepsilon^{\delta-1}\}} \leq \\ &\leq \varepsilon^{(\delta-1)p}\mathbf{P}_\theta^{(\varepsilon)}\{\mathcal{A}^c\} + \int_{\varepsilon^{\delta-1}}^\infty x^p d\mathbf{P}_\theta^{(\varepsilon)}\{|\hat{u}_\varepsilon| < x\} \leq \\ &\leq C_*\varepsilon^{(\delta-1)p}e^{-\kappa_3\varepsilon^{2\delta-1}} + p\int_{\varepsilon^{\delta-1}}^\infty x^{p-1}\mathbf{P}_\theta^{(\varepsilon)}\{|\hat{u}_\varepsilon| \geq x\}dx \leq \\ &\leq C_*\varepsilon^{(\delta-1)p}e^{-\kappa_3\varepsilon^{2\delta-1}} + C_* p\int_{\varepsilon^{\delta-1}}^\infty x^{p-1}e^{-\kappa_3\varepsilon x^2}dx = \\ &= C_*\varepsilon^{(\delta-1)p}e^{-\kappa_3\varepsilon^{2\delta-1}} + C_* p\varepsilon^{-\frac{p}{2}}\int_{\varepsilon^{\delta-1/2}}^\infty y^{p-1}e^{-\kappa_3 y^2}dy \leq \\ &\leq C_{**}\varepsilon^{-n}e^{-\kappa_3\varepsilon^{2\delta-1}} \end{aligned}$$

with a number n. Therefore, it is sufficient to study the expansion of

$$\mathbf{E}_\theta|\xi_1 + \psi_1^*\varepsilon + \psi_2^*\varepsilon^2 + \eta_2\varepsilon^{5/2}|^p\chi_{\{\mathcal{A}\}}.$$

Moreover, the contribution of the term $\eta_2\varepsilon^{5/2}$ is also negligible, because by (3.59)

$$\begin{aligned} &\mathbf{E}_\theta\left||\xi_1 + \psi_1^*\varepsilon + \psi_2^*\varepsilon^2 + \eta_2\varepsilon^{5/2}|^p\chi_{\{\mathcal{A}\}} - |\xi_1 + \psi_1^*\varepsilon + \psi_2^*\varepsilon^2|^p\chi_{\{\mathcal{A}\}}\right| \leq \\ &\leq p\varepsilon^{5/2}\mathbf{E}_\theta\left||\xi_1 + \psi_1^*\varepsilon + \psi_2^*\varepsilon^2| + \varepsilon^{5/2}\right|^{p-1} \leq C\varepsilon^{5/2}. \end{aligned}$$

So the problem reduces to the expansion of

$$\mathbf{E}_\theta|\xi_1+\psi_1^*\varepsilon+\psi_2^*\varepsilon^2|^p\chi_{\{\mathcal{A}\}},$$

but this mathematical expectation has the same first three terms of expansion as

$$\mathbf{E}_\theta|\xi_1+\psi_1^*\varepsilon+\psi_2^*\varepsilon^2|^p$$

because the contribution of the integration over $\mathcal{A}^c$ is negligible. For the last expression by Taylor's formula (3.58), we have

$$\begin{aligned}
\mathbf{E}_\theta|\xi_1+\psi_1^*\varepsilon+\psi_2^*\varepsilon^2|^p &= \mathbf{E}_\theta|\xi_1|^p+p\,\mathbf{E}_\theta\left(|\xi_1|^{p-1}\mathrm{sign}\xi_1\,(\psi_1^*\varepsilon+\psi_2^*\varepsilon^2)\right)+\\
&+\frac{p(p-1)}{2}\mathbf{E}_\theta\left(|\xi_1|^{p-2}(\psi_1^*\varepsilon+\psi_2^*\varepsilon^2)^2\right)+\\
&+\frac{p(p-1)}{2}\mathbf{E}_\theta\left(|\xi_1+\gamma(\psi_1^*\varepsilon+\psi_2^*\varepsilon^2)|^{p-2}-|\xi_1|^{p-2}\right)(\psi_1^*\varepsilon+\psi_2^*\varepsilon^2)^2=\\
&=\frac{2^{p/2}}{\sqrt{\pi}}\Gamma(\frac{p+1}{2})+p\,\mathbf{E}_\theta\left(|\xi_1|^{p-1}\psi_1^*\mathrm{sign}\xi_1\right)\varepsilon+\\
&+\left(p\,\mathbf{E}_\theta|\xi_1|^{p-1}\mathrm{sign}\xi_1\,\psi_2^*+\frac{1}{2}p(p-1)\mathbf{E}_\theta|\xi_1|^{p-2}\,\psi_1^{*2}\right)\varepsilon^2+R(\varepsilon),
\end{aligned}$$

where $\Gamma(\cdot)$ is a gamma-function and

$$|R(\varepsilon)|\le\frac{1}{2}p(p-1)(p-2)\mathbf{E}_\theta\left||\xi_1|+|\psi_1^*\varepsilon+\psi_2^*\varepsilon^2|\right|^{p-3}(\psi_1^*\varepsilon+\psi_2^*\varepsilon^2)^3\le C\,\varepsilon^3$$

because all mathematical expectations are bounded.

So finally, we can write the desired expansion as

$$\begin{aligned}
\mathbf{E}_\theta\left|\varepsilon^{-1}I(\theta)^{1/2}(\hat\theta_\varepsilon-\theta)\right|^p &= \frac{2^{p/2}}{\sqrt{\pi}}\Gamma(\frac{p+1}{2})+\\
&+\left(p\mathbf{E}_\theta|\xi_1|^{p-1}\mathrm{sign}\xi_1\,\psi_2^*+\frac{1}{2}p(p-1)\mathbf{E}_\theta|\xi_1|^{p-2}\psi_1^{*2}\right)\varepsilon^2+A(\varepsilon)\varepsilon^{5/2} \quad (3.60)
\end{aligned}$$

because

$$\mathbf{E}_\theta\left(|\xi_1|^{p-1}\psi_1^*\mathrm{sign}\xi_1\right)=0.$$

Here $A(\varepsilon)$ is a bounded function.

Chapter 4

Nonparametric Estimation

Below, we consider several problems of estimation concerning diffusion-type observations in situations when it is impossible to describe the underlying uncertainty as a parametric family. We study the asymptotic of the kernel-type estimators of trend coefficients and a "natural" estimator of the trajectory (state) of a deterministic limit system. The rate of convergence of these kernel-type estimators depends on the smoothness of the trend of the nonperturbed system and the optimality of this rate is established in an appropriate sense. For the state estimators, we introduce a low minimax bound on the risk function and then prove that the observed trajectory of the perturbed system is an asymptotically optimal estimator.

4.1 Trend estimation

We are given a probability space $\{\Omega, \mathcal{F}, \mathbf{P}\}$, a family $\hat{\mathcal{F}}$ and stochastic differential equation

$$dX_t = S_t(X)dt + \varepsilon dW_t,\ X_0 = x_0,\ 0 \leq t \leq T. \tag{4.1}$$

Suppose that the function $S(\cdot)$ is unknown and we have to estimate the function $S_t = S_t(x)$, $0 \leq t \leq T$, where x_t, $0 \leq t \leq T$ is a solution of

$$\frac{dx_t}{dt} = S_t(x),\ \ x_0,\ \ 0 \leq t \leq T. \tag{4.2}$$

As before, we suppose that $S_t(X)$, $\in \mathcal{P}_0$ and satisfies the condition $\mathcal{L}$

$$\begin{aligned} |S_t(X) - S_t(Y)| &\leq L \textstyle\int_0^t |X_s - Y_s| dK(s) + L|X_t - Y_t|, \\ |S_t(X)| &\leq L \textstyle\int_0^t (1 + |X_s|) dK(s) + L(1 + |X_t|), \end{aligned} \tag{4.3}$$

which provide the existence and uniqueness of the solutions of Equations (4.1) and (4.2). Denote $\Theta(L)$ as the class of functionals $S_t(X)$ satisfying conditions (4.3) with the same L and $K(\cdot)$. Then, repeating the proof of Lemma 1.13, we can deduce, as in (1.46), that

$$\sup_{S \in \Theta(L)} \sup_{0 \leq t \leq T} \mathbf{E}|X_t - x_t|^2 \leq c\varepsilon^2. \tag{4.4}$$

Let us also introduce a class Θ_α of (deterministic) functions $\{g_t,\ 0 \le t \le T\}$ k-times derivable with respect to t and the k-th derivative which satisfies the Hölder condition of the order $\alpha \in (0,1]$

$$|g_t^{(k)} - g_s^{(k)}| \le L_\alpha |t-s|^\alpha, \quad t,s \in [0,T] \tag{4.5}$$

with the same constant L_α. Finally, we denote $\Theta_\beta(L)$ as the class of trend coefficients $S(X) = \{S_t(X),\ 0 \le t \le T\}$ belonging to $\Theta(L)$ and such that the functions $S(x) = \{S_t(x),\ 0 \le t \le T\}$ (as $\varepsilon = 0$) belong to Θ_β, so

$$\Theta_\beta(L) = \{S(X):\ S(X) \in \Theta(L),\ S(x) \in \Theta_\beta\}.$$

Next, we introduce a *kernel* $G(u)$, $u \in R$ as a bounded function of finite support (there exist two constants A and B such that $G(u) = 0$ for $u \notin [A,B]$) and

$$\int_A^B G(u)du = 1. \tag{4.6}$$

We assume that $A < 0$ and $B > 0$.

Remind that in the statistics of i.i.d. observations $X_1, \cdots, X_n$ with unknown distribution function $F(x)$ and density $f(x)$, the empiric distribution function $\hat{F}_n(x)$ admits the representation ($x \ge 0$)

$$d\hat{F}_n(x) = f(x)\,dx + \frac{1}{\sqrt{n}}\,dW_n(x),$$

where the normed difference $W_n(x) = \sqrt{n}(\hat{F}_n(x) - F(x))$ converges to the Brownian bridge as $n \to \infty$. The kernel-type estimator of the density function is

$$\hat{f}_n(x) = \frac{1}{n\,\varphi_n} \sum_{j=1}^n G\left(\frac{X_j - x}{\varphi_n}\right) = \frac{1}{\varphi_n} \int G\left(\frac{y-x}{\varphi_n}\right) d\hat{F}_n(y).$$

We introduce a *kernel-type (nonparametric) estimator* of the function S_t analogously

$$\hat{S}_t = \frac{1}{\varphi_\varepsilon} \int_0^T G\left(\frac{\tau - t}{\varphi_\varepsilon}\right) dX_\tau, \tag{4.7}$$

where the normalizing function φ_ε will be defined below.

Theorem 4.1. *Let $S(X) \in \Theta_\alpha(L)$, the function $\varphi_\varepsilon \to 0$ and $\varepsilon^2 \varphi_\varepsilon^{-1} \to 0$, then for any $0 < c \le d < T$, the estimator $\hat{S}_t$ is uniformly consistent*

$$\lim_{\varepsilon \to 0} \sup_{S(X) \in \Theta_\alpha(L)} \sup_{c \le t \le d} \mathbf{E}_S |\hat{S}_t - S_t(x)|^2 = 0. \tag{4.8}$$

Proof. Below we use the definition of $\hat{S}_t$, the elementary inequality $(a+b+c)^2 \le$

$3a^2+3b^2+3c^2$ and the properties of stochastic integral

$$\begin{aligned}\mathbf{E}|\hat{S}_t - S_t(x)|^2 &= \mathbf{E}\left\{\frac{1}{\varphi_\varepsilon}\int_0^T G\left(\frac{\tau-t}{\varphi_\varepsilon}\right)[S_\tau(X)-S_\tau(x)]d\tau + \right.\\ &\quad + \frac{1}{\varphi_\varepsilon}\int_0^T G\left(\frac{\tau-t}{\varphi_\varepsilon}\right)S_\tau(x)d\tau - S_t(x) + \\ &\quad \left. + \frac{\varepsilon}{\varphi_\varepsilon}\int_0^T G\left(\frac{\tau-t}{\varphi_\varepsilon}\right)dW_\tau\right\}^2 \leq \\ &\leq 3\mathbf{E}\left(\frac{1}{\varphi_\varepsilon}\int_0^T G\left(\frac{\tau-t}{\varphi_\varepsilon}\right)[S_\tau(X)-S_\tau(x)]d\tau\right)^2 + \\ &\quad + 3\left(\frac{1}{\varphi_\varepsilon}\int_0^T G\left(\frac{\tau-t}{\varphi_\varepsilon}\right)S_\tau(x)d\tau - S_t(x)\right)^2 + \\ &\quad + \frac{3\varepsilon^2}{\varphi_\varepsilon^2}\int_0^T G\left(\frac{\tau-t}{\varphi_\varepsilon}\right)^2 d\tau. \end{aligned} \tag{4.9}$$

Utilizing the change of variables $u=(t-\tau)\varphi_\varepsilon^{-1}$ and denoting $\varepsilon_1 = \varepsilon' \wedge \varepsilon''$, where

$$\varepsilon' = \sup\left\{\varepsilon :\ \varphi_\varepsilon \leq -\frac{c}{A}\right\},\quad \varepsilon'' = \sup\left\{\varepsilon :\ \varphi_\varepsilon \leq \frac{T-d}{B}\right\}$$

we obtain for $\varepsilon < \varepsilon_1$

$$\frac{\varepsilon^2}{\varphi_\varepsilon^2}\int_0^T G\left(\frac{\tau-t}{\varphi_\varepsilon}\right)^2 d\tau = \frac{\varepsilon^2}{\varphi_\varepsilon}\int_{-\infty}^{\infty} G(u)^2 du = c\varepsilon^2\varphi_\varepsilon^{-1} \longrightarrow 0 \tag{4.10}$$

and

$$\begin{aligned}&\left|\frac{1}{\varphi_\varepsilon}\int_0^T G\left(\frac{\tau-t}{\varphi_\varepsilon}\right)S_\tau(x)d\tau - S_t(x)\right| = \\ &= \left|\int_{-\infty}^{\infty} G(u)[S_{t+\varphi_\varepsilon u}(x) - S_t(x)]du\right| \leq \\ &\leq \varphi_\varepsilon^\alpha L_\alpha \int |G(u)u^\alpha| du = C_\alpha \varphi_\varepsilon^\alpha \longrightarrow 0. \end{aligned} \tag{4.11}$$

Below, we use the inequalities (4.3) and (4.4)

$$\begin{aligned}&\mathbf{E}\left\{\frac{1}{\varphi_\varepsilon}\int_0^T G\left(\frac{\tau-t}{\varphi_\varepsilon}\right)[S_\tau(X)-S_\tau(x)]d\tau\right\}^2 = \\ &= \mathbf{E}\left\{\int G(u)[S_{t+\varphi_\varepsilon u}(X) - S_{t+\varphi_\varepsilon u}(x)]du\right\}^2 \leq \\ &\leq 2(B-A)\int G(u)^2\{L^2\mathbf{E}\left(\int_0^{t+\varphi_\varepsilon u}|X_s - x_s|dK(s)\right)^2 + \\ &\quad + L^2\mathbf{E}|X_{t+\varphi_\varepsilon u} - x_{t+\varphi_\varepsilon u}|^2\}du \leq c\varepsilon^2 \longrightarrow 0, \end{aligned} \tag{4.12}$$

so statement (4.8) of Theorem 4.1 is proved.

REMARK 4.1. As follows from the proof presented above, it is possible to choose the functions c_ε and d_ε such that $c_\varepsilon \to 0$ and $d_\varepsilon \to T$ and satisfy the conditions

$$\frac{c_\varepsilon}{\varphi_\varepsilon} > -A, \qquad \frac{T-d_\varepsilon}{\varphi_\varepsilon} > B,$$

even $c_\varepsilon = -A\varphi_\varepsilon$ and $d_\varepsilon = T - B\varphi_\varepsilon$, then the estimator $\hat{S}_t$ is consistent for all $t \in [0,T]$;

$$\lim_{\varepsilon\to 0} \sup_{S(X)\in\Theta_\alpha(L)} \sup_{c_\varepsilon\le t\le d_\varepsilon} \mathbf{E}_S|\hat{S}_t - S_t(x)|^2 = 0.$$

The other way to obtain an estimator which is consistent uniformly on $[0,T]$ is to introduce two one-sided kernels $G_1(u) = 0$ for $u < 0$ and $G_2(u) = 0$ for $u > 0$ and to use them in the vicinities of points $t = 0$ and $t = T$.

From here on we require the additional conditions on the kernel

$$\int_{-\infty}^{\infty} u^j G(u)du = 0, \quad j = 1, ..., k. \tag{4.13}$$

The rate of convergence of $\hat{S}_t$ is established in the following theorem.

Theorem 4.2. *Let* $S(X) \in \Theta_\beta(L)$ *and* $\varphi_\varepsilon = \varepsilon^{\frac{2}{2\beta+1}}$, *then*

$$\lim_{\varepsilon\to 0} \sup_{S(X)\in\Theta_\beta(L)} \sup_{c\le t\le d} \mathbf{E}_S|\hat{S}_t - S_t(x)|^2 \varepsilon^{-\frac{4\beta}{2\beta+1}} < \infty.$$

Proof. By Taylor's formula,

$$\begin{aligned} S_t(x) &= S_{t_0}(x) + \sum_{j=1}^{k} S_{t_0}^{(j)}(x)\frac{(t-t_0)^j}{j!} + \\ &+ \left[S_{t_0+\gamma(t-t_0)}^{(k)}(x) - S_{t_0}^{(k)}(x)\right]\frac{(t-t_0)^k}{k!}. \end{aligned}$$

Substituting this expression in (4.11) and using (4.5) and (4.13), we obtain

$$\begin{aligned} &|\int G(u)[S_{t+\varphi_\varepsilon u}(x) - S_t(x)]du| = \\ &= |\sum_{j=1}^{k} S_t^{(j)}(x) \int G(u)u^j du\, \varphi_\varepsilon^j (j!)^{-1} + \\ &+ \int G(u)u^k [S_{t+\gamma\varphi_\varepsilon u}^{(k)}(x) - S_t^{(k)}(x)]\varphi_\varepsilon^k (k!)^{-1}| \le \\ &\le L_\alpha \varphi_\varepsilon^{k+\alpha}(k!)^{-1} \int |G(u)u^{k+\alpha}|du = C_2\varphi_\varepsilon^\beta. \end{aligned} \tag{4.14}$$

Thus, for small ε, and taking into account inequalities (4.9), (4.10) and this one, we have

$$\sup_{c\le t\le d} \mathbf{E}_S|\hat{S}_t - S_t(x)|^2 < C_1\frac{\varepsilon^2}{\varphi_\varepsilon} + C_2\varphi_\varepsilon^{2\beta} + C_3\varepsilon^2 \tag{4.15}$$

with some constants C_1, C_2, C_3 which do not depend on function $S(\cdot)$. So setting

$$\varphi_\varepsilon = \varepsilon^{\frac{2}{2\beta+1}}$$

we obtain the required inequality of Theorem 4.2.

REMARK 4.2. The asymptotic normality of this estimator can be obtained as follows. Suppose that $S(X) \in \Theta(L)$, the function $S_t(x)$ has $k+1$ continuous derivatives on t and put $\varphi_\varepsilon = \varepsilon^{\frac{2}{2k+3}}$, then for any $t \in (0,T)$

$$\mathcal{L}\left\{\varepsilon^{-\frac{2k+2}{2k+3}}(\hat{S}_t - S_t(x))\right\} \Longrightarrow \mathcal{N}(m,\sigma^2)$$

where

$$\begin{aligned} m &= \frac{S_t^{(k+1)}(x)}{(k+1)!}\int G(u)\, u^{k+1}\, du \\ \sigma^2 &= \int G(u)^2\, du. \end{aligned}$$

Indeed, we can write

$$\varepsilon^{-\frac{2k+2}{2k+3}}(\hat{S}_t - S_t(x)) = \eta_\varepsilon(t) + r_1(t) + r_2(t),$$

where

$$\eta_\varepsilon(t) = \varepsilon^{-\frac{1}{2k+3}} \int_0^T G\left(\frac{\tau - t}{\varphi_\varepsilon}\right) dW_\tau \quad \sim \quad \mathcal{N}(0,\sigma^2),$$

and for the term

$$r_1(t) = \varphi_\varepsilon^{k+1} \int_0^T G\left(\frac{\tau - t}{\varphi_\varepsilon}\right) [S_\tau(X) - S_\tau(x)]\ du$$

as in (4.12) we can prove $\mathbf{E} r_1(t)^2 \to 0$. For the last term by the same Taylor's formula we have

$$r_2(t) = \varphi_\varepsilon^{k+1} \int_0^T G\left(\frac{\tau - t}{\varphi_\varepsilon}\right) [S_\tau(x) - S_t(x)]\ du = m + o(1).$$

The random variables $\eta_\varepsilon(t)$ and $\eta_\varepsilon(s)$ for $s \neq t$ are asymptotically independent, because

$$\lim_{\varepsilon\to 0} \mathbf{E}\eta_\varepsilon(t)\eta_\varepsilon(s) = 0$$

REMARK 4.3. If the initial value $X_0 = \eta$ is a random $\mathcal{F}_0$-measurable variable, then the inequality (4.4) does not change and estimator (4.7) remains consistent.

The next question, which is also interesting in this estimation problem is to understand what are the optimal estimators. This problem is similar to the problem of density function estimation, considered by R. Farrel [35]. I. Ibragimov and R. Khas'minskii [53] proposed another proof of Farrel's theorem and then applied it to the problem of signal estimation. Below, we give a variant of Farrel's bound applicable in our problem with the proof practically coinciding with the proof of theorem

7.5.1 in [53].

Theorem 4.3. *Let $l(\cdot) \in \mathbf{W}$ and $S(\cdot) \in \Theta_\beta(L)$, then for any $t_0 \in (0,T)$*

$$\liminf_{\varepsilon\to 0} \inf_{S^*_{t_0}} \sup_{S\in\Theta_\beta(L)} \mathbf{E}_S l\left(\varepsilon^{-\frac{2\beta}{2\beta+1}}(S^*_{t_0} - S_{t_0}(x))\right) > 0. \tag{4.16}$$

The inf here is taken over all possible estimators $S^*_{t_0}$ of the function $S_t(x)$ in the point t_0.

Proof. Let $S^{(0)}(X) \in \Theta_\beta\left(\frac{1}{2}L\right))$ be an arbitrary nonanticipative functional satisfying (4.3) with the constant $L/2$. Fix $t_0 \in (0,T)$ and introduce the functional $g_t(v,X) \in \Theta_\beta(\frac{1}{2}L)$, $v \in R$, such that

$$\begin{aligned} g_{t_0}(0,x) &= 1, \\ g_{t_0}(v,x) &= 0, \quad \text{if } |v| > \frac{1}{2}, \end{aligned}$$

where $x = \{x_s,\ 0 \le s \le t\}$ is a solution of Equation (4.2) as $S_t(x) = S^{(0)}_t(x)$. Denote

$$S^{(\varepsilon)}_t(\theta,X) = S^{(0)}_t(X) + \theta\varepsilon^{\frac{2\beta}{2\beta+1}} g_t\left(\kappa(t-t_0)\varepsilon^{-\frac{2}{2\beta+1}},X\right), \quad \theta \in \Theta = \{\theta\ :|\theta-\theta_0| < \kappa\} \tag{4.17}$$

the family of trend coefficients with

$$\kappa = \int_{-1/2}^{1/2} g_{t_0}(v,x)^2\, dt\ >\ 0$$

and the family of diffusion-type processes

$$dX_t = S^{(\varepsilon)}_t(\theta,X)dt + \varepsilon dW_t, \quad X_0 = x_0 \quad 0 \le t \le T, \quad \theta \in \Theta. \tag{4.18}$$

The corresponding family of measures $\{\mathbf{P}^{(\varepsilon)}_\theta,\ \theta \in \Theta\}$ induced in the measurable space $(\mathcal{C}_T,\ \mathcal{B}_T)$ has a Radon-Nikodym derivative (with $\mathbf{P}^{(\varepsilon)}_{\theta_0}$ probability 1)

$$\begin{aligned} \frac{d\mathbf{P}^{(\varepsilon)}_{\theta_0+u}}{d\mathbf{P}^{(\varepsilon)}_{\theta_0}}(X) &= \exp\left\{u\varepsilon^{-1+\frac{2\beta}{2\beta+1}} \int_0^T g_t(\kappa(t-t_0)\varepsilon^{-\frac{2}{2\beta+1}},X)dW_t - \right. \\ &\left. - \frac{u^2}{2\varepsilon^2}\varepsilon^{\frac{4\beta}{2\beta+1}} \int_0^T g_t(\kappa(t-t_0)\varepsilon^{-\frac{2}{2\beta+1}},X)^2 dt\right\}. \end{aligned}$$

Below we change the variable $v = \kappa(t-t_0)\varepsilon^{-\frac{2}{2\beta+1}}$ and introduce a *two-sided Wiener process*

$$W(v) = \sqrt{\kappa}\varepsilon^{-\frac{1}{2\beta+1}}\left[W_{t_0+v\kappa^{-1}\varepsilon^{\frac{2}{2\beta+1}}} - W_{t_0}\right],\ |v| \le \frac{1}{2}.$$

Then the likelihood ratio could be written in the form

$$\begin{aligned} \frac{d\mathbf{P}^{(\varepsilon)}_{\theta_0+u}}{d\mathbf{P}^{(\varepsilon)}_{\theta_0}}(X) &= \exp\left\{\frac{u}{\sqrt{\kappa}} \int_{-\frac{1}{2}}^{\frac{1}{2}} g_{t_0}(v,x)dW(v) - \right. \\ &\left. -\frac{u^2}{2\kappa}\int_{-\frac{1}{2}}^{\frac{1}{2}} g_{t_0}(v,x)^2 dv + \psi_\varepsilon(u,\theta_0)\right\} = \\ &= \exp\{u\,\Delta - \frac{u^2}{2} + \psi_\varepsilon(u,\theta_0)\} = Z(u)\ e^{\psi_\varepsilon(u,\theta_0)}, \end{aligned}$$

where $u \in (-\kappa, \kappa)$ and

$$\mathbf{P}_{\theta_0} - \lim_{\varepsilon \to 0} \psi_\varepsilon(u, \theta_0) = 0.$$

As was mentioned in Remark 2.3, we can choose such functions $g_{t_0}(0, x)$ and κ_0 that

$$\liminf_{\varepsilon \to 0} \sup_{|\theta - \theta_0| < \kappa_0} \mathbf{E}_\theta l(\theta_\varepsilon - \theta) \geq \frac{1}{2\sqrt{2\pi}} \int l(y) e^{-\frac{y^2}{2}} dy. \tag{4.19}$$

Since $S_t^{(\varepsilon)}(\theta, X) \in \Theta_\beta(L)$, we can write

$$\sup_{S(\cdot) \in \Theta_\beta(L)} \mathbf{E}_S l\left(\varepsilon^{-\frac{2\beta}{2\beta+1}}(S_{t_0}^* - S_{t_0}(x))\right) \geq$$

$$\geq \sup_{|\theta - \theta_0| < \kappa_0} \mathbf{E}_\theta l\left(\varepsilon^{-\frac{2\beta}{2\beta+1}}(S_{t_0}^* - S_{t_0}^{(\varepsilon)}(\theta, x))\right).$$

An estimator of parameter θ constructed by the observations (4.18) with the trend (4.17) could be obtained in the following way. Let S_t^* be some nonparametric estimator of $S_t^{(\varepsilon)}(\theta, X)$. Set

$$\theta_\varepsilon^* - [S_{t_0}^* - S_{t_0}^{(0)}(X)] g_{t_0}(0, X)^{-1} \varepsilon^{-\frac{2\beta}{2\beta+1}}.$$

Then we have

$$\sup_{|\theta - \theta_0| < \kappa_0} \mathbf{E}_\theta l\left(\varepsilon^{-\frac{2\beta}{2\beta+1}}(S_{t_0}^* - S_{t_0}^{(\varepsilon)}(\theta, x))\right) =$$

$$= \sup_{|\theta - \theta_0| < \kappa_0} \mathbf{E}_\theta l(\varepsilon^{-\frac{2\beta}{2\beta+1}}(S_{t_0}^* - S_{t_0}^{(0)}(X)) +$$

$$+ \varepsilon^{-\frac{2\beta}{2\beta+1}}(S_{t_0}^{(0)}(X) - S_{t_0}^{(0)}(x)) + \varepsilon^{-\frac{2\beta}{2\beta+1}}(S_{t_0}^{(0)}(x) - S_{t_0}^{(\varepsilon)}(\theta, x))) =$$

$$= \sup_{|\theta - \theta_0| < \kappa_0} \mathbf{E}_\theta l\left(\theta_\varepsilon^* g_{t_0}(0, X) - \theta g_{t_0}(0, x) + o(1)\right) =$$

$$= \sup_{|\theta - \theta_0| < \kappa_0} \mathbf{E}_\theta l\left(\theta_\varepsilon^* - \theta + o(1)\right).$$

Therefore, the nondegeneracy of limit (4.16) follows from (4.19). Setting $l(u) = u^2$ and using the result of Theorem 4.2, we immediately obtain the following theorem.

Theorem 4.4. *Let $S(X) \in \Theta_\beta(L)$ and $\varphi_\varepsilon = \varepsilon^{\frac{2}{2\beta+1}}$, then the estimator $\hat{S}_t$ has an asymptotically optimal rate of convergence.*

Let us consider another problem of trend estimation closely connected with the previous one. We are given an homogeneous diffusion process

$$dX_t = S(X_t)dt + \varepsilon dW_t, \quad X_0 = x_0, \quad 0 \leq t \leq T \tag{4.20}$$

the trend coefficient $S(\cdot)$ of which is unknown and should be estimated. Below we suppose that $S(\cdot)$ belongs to the class

$$\Theta(L/2) = \{S(\cdot) : \ |S(x) - S(y)| \leq L|x - y|, |S(x)| \leq L(1 + |x|)\}.$$

The limit dynamic system is described by the ordinary differential equation

$$\frac{dx_t}{dt} = S(x_t), \quad x_0, \quad 0 \le t \le T. \tag{4.21}$$

We are interesting in an estimation of function $S(x)$, $x \in [a, b]$ where the parameters a and b can as take any values between

$$a^* = \min_{0 \le t \le T} x_t, \quad \textit{and} \quad b^* = \max_{0 \le t \le T} x_t.$$

The pair $(x_t, S(x_t), 0 \le t \le T)$ also defines this function and we could consider the pair $(X_t, \hat{S}_t, 0 \le t \le T)$ as an estimator of $S(x)$ in the points $x = X_t$, $x \in [\min X_t, \max X_t]$.

The main difference between this and the previous problem is that the domain $[a, b]$ is not known to the observer, because $a = a(S)$ and $b = b(S)$ and the function $S(\cdot)$ is unknown. Particularly, if $S_1(x_0) < 0$ and $S_2(x_0) > 0$, then the solutions of Equation (4.21) as $S = S_1$ and $S = S_2$ has the only common point x_0. Below, we consider a subclass of $\Theta(L/2)$ with the positive at x_0 members, so $S(x_0) > 0$. Denote $y_0 = \inf\{x > x_0 : \ S(x) = 0\}$ as the first zero of the function $S(\cdot)$. Of course, the solution x_t of (4.21) for $t > t_0$, where $x_{t_0} = y_0$ is constant and $x_t = y_0$, $t \in [t_0, T]$. So $a = x_0$ and $b = y_0$. Note also that $y_0 = y_0(S)$. Let us introduce the other restriction of "uniform positiveness".

(P) *There exists a constant* $\Delta > 0$, *such that*

$$\inf_{x_0 \le x \le y_0 - \Delta} S(x) > \kappa. \tag{4.22}$$

We also suppose that if $S(x_t)$ *is positive for all* $t \in [0, T]$, *then* $y_0 = x_T$ *and* $t_0 = T$.

The subclass of functions from $\Theta(L/2)$ satisfying condition **(P)** is denoted by $\Theta_+(L) = \Theta_+(L, \Delta, \kappa)$.

Utilizing the known "theorem of comparison" of differential equations, we obtain that the maximal solution of (4.21) in the class $\Theta(L)$ satisfies the equation

$$\frac{dx_t}{dt} = L(1 + x_t), \quad x_0, \quad 0 \le t \le T,$$

so $x_t = (1 + x_0)e^{Lt} - 1$ and for $t = T$, we have the maximal value $d = (1 + x_0)e^{Lt} - 1$, i.e.

$$\sup_{S \in \Theta(L)} \sup_{0 \le t \le T} x_t = d.$$

Moreover, under these assumptions

$$S(x_0) = S(x_0) - S(y_0) = (x_0 - y_0)S'(\tilde{x})$$

and, for $S(\cdot) \in \Theta_+(L)$,

$$y_0 - x_0 = -\frac{S(x_0)}{S'(\tilde{x})} \ge \frac{\kappa}{L},$$

so the constant Δ in (4.22) should satisfy the inequality $\Delta < \kappa L^{-1}$.

Denote a stopping time

$$\tau_\varepsilon(x) = \inf\{t:\ X_t \geq x,\ 0 \leq t \leq T\}$$

and the values of x here should be taken from the interval $[x_0, d_\varepsilon]$, where

$$d_\varepsilon = \min\left(\max_{0\leq t\leq T} X_t, d\right) - \Delta.$$

So $T_\varepsilon(x)$ is well defined.

The estimator of function $S(\cdot)$ at point x we introduce by the equality

$$\hat{S}_\varepsilon(x) = \frac{1}{\varphi_\varepsilon}\int_0^T G\left(\frac{t-\tau_\varepsilon(x)}{\varphi_\varepsilon}\right) dX_t,$$

where the kernel $G(\cdot)$ satisfies (4.6).

Theorem 4.5. *Let $S(\cdot) \in \Theta_+(L)$, $\varphi_\varepsilon \to 0$, $\varepsilon^2\varphi_\varepsilon^{-1} \to 0$ and $c = x_0+\Delta_1$, $d = d_\varepsilon - \Delta_1$ where Δ_1 is a positive constant, then*

$$\lim_{\varepsilon\to 0} \sup_{S\in\Theta_+(L)} \mathbf{E}_S \int_c^d [\hat{S}_\varepsilon(x) - S(x)]^2 dx \chi_{\{d>c\}} = 0.$$

Proof. First, we mention that $d_\varepsilon \to d_0 = \max_{0\leq t\leq T} x_t - \Delta = y_0 - \Delta$. Then we introduce the set

$$\mathcal{A}_1 = \{\omega:\ d_\varepsilon < d_0 + \varepsilon^\delta\},$$

where $\delta \in (0, \frac{1}{2})$ and note that for the complement $\mathcal{A}_1^C$, in view of the inequality

$$|d_\varepsilon - d_0| \leq \sup_{0\leq t\leq T} |X_t - x_t| \leq C\varepsilon \sup_{0\leq t\leq T} |W_t|,$$

we have an estimate

$$\begin{aligned}\mathbf{P}(\mathcal{A}_1^C) &= \mathbf{P}(d_\varepsilon - d_0 > \varepsilon^\delta) \leq \\ &\leq \mathbf{P}\{C\varepsilon \sup_{0\leq t\leq T} |W_t| > \varepsilon^\delta\} \leq 2\exp\{-\gamma\varepsilon^{-2(1-\delta)}\},\end{aligned}$$

where $\gamma = \gamma(C,T) > 0$.

Thus, we can write

$$\begin{aligned}&\mathbf{E}_S \int_c^d [\hat{S}_\varepsilon(x) - S(x)]^2 dx \chi_{\{d>c\}} = \mathbf{E}_S \chi_{\{\mathcal{A}_1, d>c\}} \int_c^d [\hat{S}_\varepsilon(x) - S(x)]^2 dx + \\ &+ \mathbf{E}_S \chi_{\{\mathcal{A}_1^C, d>c\}} \int_c^d [\hat{S}_\varepsilon(x) - S(x)]^2 dx \leq \\ &\leq \int_c^{d_0+\varepsilon^\delta} \mathbf{E}_S[\hat{S}_\varepsilon(x) - S(x)]^2 dx + \int_c^d \mathbf{E}_S[\hat{S}_\varepsilon(x) - S(x)]^2 \chi_{\{\mathcal{A}_1^C, d>c\}} dx.\end{aligned}$$

Further

$$\mathbf{E}_S[\hat{S}_\varepsilon(x) - S(x)]^2 = \mathbf{E}_S\left[\frac{1}{\varphi_\varepsilon}\int_0^T G\left(\frac{t-\tau_\varepsilon(x)}{\varphi_\varepsilon}\right) S(X_t)dt + \right.$$

$$\left. + \frac{\varepsilon}{\varphi_\varepsilon}\int_0^T G\left(\frac{t-\tau_\varepsilon(x)}{\varphi_\varepsilon}\right) dW_t - S(x)\right]^2 \le$$

$$\le 2\mathbf{E}_S\left(\int_{-\frac{\tau_\varepsilon(x)}{\varphi_\varepsilon}}^{\frac{T-\tau_\varepsilon(x)}{\varphi_\varepsilon}} G(u)S(X_{\tau_\varepsilon(x)+\varphi_\varepsilon u})du - S(x)\right)^2 +$$

$$+2\frac{\varepsilon^2}{\varphi_\varepsilon^2}\int_0^T \mathbf{E}_S G\left(\frac{t-\tau_\varepsilon(x)}{\varphi_\varepsilon}\right)^2 dt.$$

Note also that

$$0 = X_{\tau_\varepsilon(x)} - x_{\tau(x)} = X_{\tau_\varepsilon(x)} - x_{\tau_\varepsilon(x)} + x_{\tau_\varepsilon(x)} - x_{\tau(x)}.$$

This implies that

$$|X_{\tau_\varepsilon(x)} - x_{\tau(x)}| = |\int_{\tau(x)}^{\tau_\varepsilon(x)} S(x_t)dt| \ge \kappa|\tau_\varepsilon(x) - \tau(x)|.$$

Hence, with probability 1,

$$\sup_{c\le x\le d} |\tau_\varepsilon(x) - \tau(x)| \le \kappa^{-1}C\varepsilon \sup_{0\le t\le T} |W_t|. \tag{4.23}$$

Now introduce the set

$$\mathcal{A}_2 = \{\omega : \ |\tau_\varepsilon(x) - \tau(x)| < \varepsilon^\delta\}$$

and immediately note that

$$\mathbf{P}(\mathcal{A}_2^C) \le 2\exp\left\{-\gamma_2\varepsilon^{-2(1-\delta)}\right\}$$

with $\gamma_2 = \gamma_2(\kappa, C, T) > 0$.

Set

$$\varepsilon_1 = \sup\left\{\varepsilon : -\frac{\tau(x_0+\Delta) - \varepsilon^\delta}{\varphi_\varepsilon} \le A\right\},$$

$$\varepsilon_2 = \sup\left\{\varepsilon : \frac{T - \tau(y_0-\Delta) - \varepsilon^\delta}{\varphi_\varepsilon} \ge B\right\}.$$

Then, for $\varepsilon < \varepsilon_0 = \min(\varepsilon_1, \varepsilon_2)$ and $\omega \in \mathcal{A}_2$, we have

$$A_\varepsilon = -\frac{\tau_\varepsilon(x)}{\varphi_\varepsilon} \le -\frac{\tau(x) - \varepsilon^\delta}{\varphi_\varepsilon} \le -\frac{\tau(x_0+\Delta) - \varepsilon^\delta}{\varphi_\varepsilon} \le A.$$

On the other hand,

$$B_\varepsilon = \frac{T - \tau_\varepsilon(x)}{\varphi_\varepsilon} \ge \frac{T - \tau(x) - \varepsilon^\delta}{\varphi_\varepsilon} \ge \frac{T - \tau(y_0-\Delta) - \varepsilon^\delta}{\varphi_\varepsilon} \ge B.$$

Hence, on $\mathcal{A}_2$ as $\varepsilon < \varepsilon_0$, we can write

$$\int_{A_\varepsilon}^{B_\varepsilon} G(u)S(X_{\tau_\varepsilon(x)+u\varphi_\varepsilon})du = \int_A^B G(u)S(X_{\tau_\varepsilon(x)+u\varphi_\varepsilon})du$$

and therefore

$$\mathbf{E}_S\chi_{\{\mathcal{A}_2\}}\left(\int_{A_\varepsilon}^{B_\varepsilon} G(u)S(X_{\tau_\varepsilon(x)+u\varphi_\varepsilon})du - S(x)\right)^2 \leq$$

$$\leq \mathbf{E}_S\left(\int_A^B G(u)[S(X_{\tau_\varepsilon(x)+u\varphi_\varepsilon}) - S(x_{\tau_\varepsilon(x)+u\varphi_\varepsilon})]du + \right.$$

$$\left. + \int_A^B G(u)[S(x_{\tau_\varepsilon(x)+u\varphi_\varepsilon}) - S(x_{\tau(x)})]du\right)^2 \leq$$

$$\leq 2(B-A)L^2\int_A^B G(u)^2\mathbf{E}_S|X_{\tau_\varepsilon(x)+u\varphi_\varepsilon} - x_{\tau_\varepsilon(x)+u\varphi_\varepsilon}|^2du +$$

$$+2(B-A)L^2\int_A^B G(u)^2\mathbf{E}_S|x_{\tau_\varepsilon(x)+u\varphi_\varepsilon} - x_{\tau(x)}|^2du \leq$$

$$\leq C_1\varepsilon^2 + C_2\varphi_\varepsilon^2,$$

because of

$$\mathbf{E}_S|x_{\tau_\varepsilon(x)+u\varphi_\varepsilon} - x_{\tau(x)}|^2 = \mathbf{E}_S|\int_{\tau(x)}^{\tau_\varepsilon(x)+u\varphi_\varepsilon} S(x_t)dt|^2 \leq$$

$$\leq C_1\mathbf{E}_S(\tau_\varepsilon(x) - \tau(x))^2 + C_2u^2\varphi_\varepsilon^2$$

and (4.23).

Joining all the obtained inequalities and taking into account that all the mathematical expectations on complements $\mathcal{A}_1^C$ and $\mathcal{A}_2^C$ are exponentially small, we obtain

$$\mathbf{E}_S\int_c^d[\hat{S}_\varepsilon(x) - S(x)]^2dx \leq \tilde{C}_1\varepsilon^2 + \tilde{C}_2\varphi_\varepsilon^2 + \tilde{C}_3\exp(-\gamma\varepsilon^{-2(1-\delta)}), \tag{4.24}$$

where the constants $\tilde{C}_i$, γ and δ do not depend of $S(\cdot)$ for $S(\cdot) \in \Theta_+(L)$.

For $S(\cdot) \in \Theta_\beta(L)\bigcap\Theta_+(L)$, it is possible to formulate an analog of Theorem 4.2, and the proof will follow immediately from inequality (4.24).

4.2 Linear multiplier estimation

Let us consider the problem of estimating the function $\theta(t)$, $0 \leq t \leq T$ *(linear multiplier)* by the observations of the linear diffusion process

$$dX_t = \theta(t)X_tdt + \varepsilon dW_t, \quad X_0 = x_0, \quad 0 \leq t \leq T \tag{4.25}$$

and describe the properties of the estimators as $\varepsilon \to 0$.

Below, $\Theta_\beta(L_\alpha)$ is a class of functions $\theta(t)$, $0 \leq t \leq T$ uniformly bounded and k times continuously differentiable, and the k-th derivative satisfies the Hölder condition of the order $\alpha \in (0,1]$:

$$|\theta^{(k)}(t) - \theta^{(k)}(s)| \leq L_\alpha|t-s|^\alpha, \quad \beta = k + \alpha$$

First, we introduce a minimax low bound:

Theorem 4.6. *Let $\theta(\cdot) \in \Theta_\beta(L_\alpha)$, then for any $t \in (0,T)$,*

$$\liminf_{\varepsilon\to 0} \inf_{\theta_t^*} \sup_{\theta(\cdot)\in\Theta_\beta(L_\alpha)} \mathbf{E}l\left((\theta_t^* - \theta(t))\varepsilon^{-\frac{2\beta}{2\beta+1}}\right) > 0. \tag{4.26}$$

Proof repeat the proof of Theorem 4.3 above. The vicinity of the true function $\theta_0(\cdot)$ is introduced as a family

$$\theta(t) = \theta_0(t) + \theta\varepsilon^{\frac{2\beta}{2\beta+1}} g_t\left((t-t_0)\varepsilon^{-\frac{2\beta}{2\beta+1}}\kappa\right),$$

where $\theta \in (-\kappa,\kappa)$, and so on.

The solution of the limit system

$$\frac{dx_t}{dt} = \theta(t)x_t, \quad x_0, \quad 0 \le t \le T$$

is the function

$$x_t = x_0 \exp\left\{\int_0^t \theta(t)ds\right\}.$$

This representation prompts us to search for an estimator of $\theta(t)$ as

$$\theta_t^* = \frac{d}{dt}\ln X_t, \quad 0 \le t \le T.$$

But the process X_t also takes negative values with positive probability and has the same smoothness as the Wiener process, so it is not differentiable and this estimator does not exist. Below, we try to construct an estimator on a "good set".

Denote L_0 as the constant from the condition $\theta(\cdot) \in \Theta_\beta(L_\alpha)$ such that

$$\sup_{\theta(\cdot)\in\Theta_\beta(L_\alpha)} \sup_{0\le t\le T} |\theta(t)| \le L_0. \tag{4.27}$$

Then, according to the Lemma 1.13, with probability 1, we have

$$|X_t - x_t| \le \varepsilon e^{L_0 t} \sup_{0\le t\le T} |W_s|.$$

We now introduce the set

$$\mathcal{A}_t = \left\{\omega : \inf_{0\le s\le t} X_s \ge \frac{1}{2}x_0 e^{-L_0 t}\right\}, \quad \mathcal{A} = \mathcal{A}_T$$

and the process Y_t with the differential

$$dY_t = \left[\theta(t) - \frac{\varepsilon^2}{2}X_t^{-2}\right]\chi_{\{\mathcal{A}_t\}}dt + \varepsilon X_t^{-1}\chi_{\{\mathcal{A}_t\}}dW_t.$$

Then we introduce the kernel-type estimator

$$\hat{\theta}_t = \chi_{\{\mathcal{A}\}}\frac{1}{\varphi_\varepsilon}\int_0^T G\left(\frac{t-s}{\varphi_\varepsilon}\right)dZ_s,$$

where the kernel $G(\cdot)$ satisfies the conditions imposed on $G(\cdot)$ in Theorem 4.2.

Theorem 4.7. *Let $\theta(\cdot) \in \Theta_\beta(L_\alpha)$ and $\varphi_\varepsilon = \varepsilon^{\frac{2}{2\beta+1}}$ then for any $[c,d] \subset (0,T)$,*

$$\lim_{\varepsilon\to 0} \sup_{\theta(\cdot)\in\Theta_\beta(L_\alpha)} \sup_{c\le t\le d} \mathbf{E}|\hat\theta_t - \theta(t)|^2 \varepsilon^{-\frac{4\beta}{2\beta+1}} < \infty \tag{4.28}$$

Proof. By analogy with the proofs of Theorems 4.1 and 4.2, we can write

$$\begin{aligned}
\mathbf{E}|\hat\theta_t - \theta(t)|^2 &= \mathbf{E}\Bigg|\chi_{\{\mathcal{A}\}}\frac{1}{\varphi_\varepsilon}\int_0^T G\left(\frac{t-s}{\varphi_\varepsilon}\right)[\theta(s)-\theta(t)]ds + \\
&+ \chi_{\{F^C\}}\theta(t) - \chi_{\{\mathcal{A}\}}\frac{\varepsilon^2}{2\varphi_\varepsilon}\int_0^T G\left(\frac{t-s}{\varphi_\varepsilon}\right)X_s^{-2}ds + \\
&+ \chi_{\{\mathcal{A}\}}\frac{\varepsilon}{\varphi_\varepsilon}\int_0^T G\left(\frac{t-s}{\varphi_\varepsilon}\right)X_s^{-1}dW_s\Bigg|^2 \le \\
&\le 4\mathbf{E}\chi_{\{\mathcal{A}\}}\left(\int G(u)[\theta(t+u\varphi_\varepsilon)-\theta(t)]du\right)^2 + 4\theta(t)^2\mathbf{P}\{F^C\} + \\
&+ 2\frac{\varepsilon^4}{2\varphi_\varepsilon^2}\mathbf{E}\left(\chi_{\{\mathcal{A}\}}\int_0^T G\left(\frac{t-s}{\varphi_\varepsilon}\right)X_s^{-2}ds\right)^2 + \\
&+ 4\frac{\varepsilon^2}{2\varphi_\varepsilon^2}\mathbf{E}\left(\chi_{\{\mathcal{A}\}}\int_0^T G\left(\frac{t-s}{\varphi_\varepsilon}\right)X_s^{-1}dW_s\right)^2.
\end{aligned} \tag{4.29}$$

The Taylor formula and conditions of the Theorem provide an upper estimate for the first term in the last expression as

$$\frac{4L_\alpha^2}{(k+1)!}\varphi_\varepsilon^{2\beta}\int |G(u)u^\beta|du.$$

For the second term, we have (4.27) and

$$\begin{aligned}
\mathbf{P}\{\mathcal{A}^C\} &= \mathbf{P}\left\{\inf_{0\le t\le T} X_t < \frac{1}{2}x_0e^{-L_0T}\right\} \le \\
&\le \mathbf{P}\left\{\inf_{0\le t\le T}[X_t - x_t] + \inf_{0\le t\le T} x_t < \frac{1}{2}x_0e^{-L_0T}\right\} \le \\
&\le \mathbf{P}\left\{\inf_{0\le t\le T}[X_t - x_t] < -\frac{1}{2}x_0e^{-L_0T}\right\} \le \\
&\le \mathbf{P}\left\{\sup_{0\le t\le T}|X_t - x_t| > \frac{1}{2}x_0e^{-L_0T}\right\} \le \\
&\le \mathbf{P}\left\{\varepsilon e^{L_0T}\sup_{0\le t\le T}|W_t| > \frac{1}{2}x_0e^{-L_0T}\right\} \le \\
&\le 2\exp\left\{-\frac{x_0^2\exp(-4L_0T)}{8\varepsilon^2T}\right\},
\end{aligned}$$

where we have use the evident inequality

$$x_t = x_0\exp\left\{\int_0^t \theta(s)ds\right\} \ge x_0e^{-L_0t}.$$

The third term is less than

$$8\varepsilon^4 x_0^{-2}\exp(2L_0T)\left(\int |G(u)|du\right)^2.$$

This follows from the definition of set $\mathcal{A}$. For the last term (using $\mathcal{A}_t \supset \mathcal{A}$) we have

$$\begin{aligned}
&\frac{\varepsilon^2}{\varphi_\varepsilon^2}\mathbf{E}\left(\chi_{\{\mathcal{A}\}}\int_0^T G\left(\frac{t-s}{\varphi_\varepsilon}\right)X_s^{-1}dW_s\right)^2 = \\
&= \frac{\varepsilon^2}{\varphi_\varepsilon^2}\mathbf{E}\left(\chi_{\{\mathcal{A}\}}\int_0^T G\left(\frac{t-s}{\varphi_\varepsilon}\right)X_s^{-1}\chi_{\{\mathcal{A}_s\}}dW_s\right)^2 \le \\
&\le \frac{\varepsilon^2}{\varphi_\varepsilon^2}\mathbf{E}\left(\int_0^T G\left(\frac{t-s}{\varphi_\varepsilon}\right)X_s^{-1}\chi_{\{\mathcal{A}_s\}}dW_s\right)^2 = \\
&= \frac{\varepsilon^2}{\varphi_\varepsilon^2}\int_0^T G\left(\frac{t-s}{\varphi_\varepsilon}\right)^2\mathbf{E}\left(X_s^{-2}\chi_{\{\mathcal{A}_s\}}\right)ds \le \\
&\le \frac{4\varepsilon^2}{\varphi_\varepsilon x_0^2}\exp(2L_0T)\int G(u)^2du.
\end{aligned}$$

So we obtain the final inequality

$$\mathbf{E}(\hat{\theta}_t - \theta(t))^2 \le C_1\varphi_\varepsilon^{2\beta} + C_2\exp(-c\varepsilon^{-2}) + C_3\varepsilon^4 + C_4\frac{\varepsilon^2}{\varphi_\varepsilon}.$$

Hence, if we choose $\varphi_\varepsilon = \varepsilon^{\frac{2}{2\beta+1}}$, the inequality (4.28) immediately follows on.

Comparison of (4.26) for $l(u) = u^2$ and (4.28) shows that this estimator has an asymptotically optimal rate of convergence.

REMARK 4.4. T. Phann and H. Nguen [103] have considered the problem of estimating the function $\theta(\cdot)$ by n independent observations $\{X_t^{(j)},\ 0 \le t \le T\}$, $j = 1,\dots,n$ of the process

$$dX_t^{(j)} = \theta(t)X_t^{(j)}dt + dW_t^{(j)},\quad X_0^{(j)} = x_0,\quad 0 \le t \le T \tag{4.30}$$

and by using sieves method they proposed an estimator which is consistent and asymptotically normal. Note that this model can be also reduced to the scheme (4.25) if we introduce the notations

$$X_t = \frac{1}{n}\sum_{j=1}^n X_t^{(j)},\quad W_t = \frac{1}{\sqrt{n}}\sum_{j=1}^n W_t^{(j)},\quad \varepsilon = \frac{1}{\sqrt{n}}.$$

4.3 State estimation

Let the states $x = \{x_t,\ 0 \le t \le T\}$ of the dynamic system

$$\frac{dx_t}{dt} = S(x_t),\quad x_0,\quad 0 \le t \le T \tag{4.31}$$

be unknown and we have to estimator x by the observations $X = \{X_t,\ 0 \leq t \leq T\}$ of the corresponding homogeneous diffusion process

$$dX_t = S(X_t)dt + \varepsilon dW_t, \quad X_0 = x_0, \quad 0 \leq t \leq T \tag{4.32}$$

as $\varepsilon \to 0$. The function $S(\cdot)$ is unknown to the observer, so we deal with the non-parametric problem. The states x_t, $0 \leq t \leq T$, of course, could be estimated directly by X_t, $0 \leq t \leq T$, i.e. $X_t, 0 \leq t \leq T$ is a *nonparametric estimator* of the function $x_t, 0 \leq t \leq T$ and the following question naturally appears: is it possible to improve this estimator using other $(X_s,\ s \neq t)$ observations in the estimation of x_t? Below we show that the answer is, in the appropriate sense, negative. In order to prove this, we first introduce a low minimax bound on the risk of any estimator and then we show that the estimator $\{X_t,\ 0 \leq t \leq T\}$ achieves this bound.

If $S(\cdot) \in \Theta(L)$, Equations (4.31) and (4.32) have unique solutions and all the measures $\mathbf{P}_S^{(\varepsilon)}$, $S \in \Theta(L)$ are equivalent. Here $\mathbf{P}_S^{(\varepsilon)}$ is a measure induced by the process (4.32) in the space $(\mathcal{C}_T,\ \mathcal{B}_T)$.

Let us introduce a nonparametric vicinity of a fixed trajectory x_t, $0 \leq t \leq T$ as a family of functions $x^h = (x_t^h,\ 0 \leq t \leq T)$, $h(\cdot) \in \Theta(L/2)$ defined by the family of equations

$$\frac{dx_t^h}{dt} = S(x_t^h) + \varepsilon \frac{h(x_t^h)}{S(x_t^h)}, \quad x_0^h = x_0,\ 0 \leq t \leq T. \tag{4.33}$$

We assume here that $S(\cdot) \in \Theta(L/2)$, so the functions $S(\cdot) + \varepsilon h(\cdot)S(\cdot)^{-1} \in \Theta(L)$ for all mentioned $h(\cdot)$ and sufficiently small ε. Denote X_t, $0 \leq t \leq T$ as the observations corresponding to the system (4.33):

$$dX_t^h = \left[S(X_t^h) + \varepsilon \frac{h(X_t^h)}{S(X_t^h)}\right] dt + \varepsilon dW_t, \ X_0^h = x_0,\ 0 \leq t \leq T \tag{4.34}$$

and denote $\mathbf{P}_h^{(\varepsilon)}$ as the measure of this process in $(\mathcal{C}_T,\ \mathcal{B}_T)$. So we have to estimate the trajectory x_t^h, $0 \leq t \leq T$ of (4.33) by the observations of (4.34). The loss function we choose in the form $l(y) = g(|y|)$, where $y \in \mathcal{C}_T$, $|y| = \sup_{0 \leq t \leq T} |y_t|$, and $g(\cdot)$ is nonnegative, increasing on the $[0, \infty)$ function.

In view of Lemma 1.13, this estimator is uniformly consistent on $t \in [0, T]$, with probability 1

$$\lim_{\varepsilon \to 0} \sup_{S(\cdot) \in \Theta(L)} \sup_{0 \leq t \leq T} |X_t - x_t| = 0$$

and, for continuously differentiable $S(\cdot)$, uniformly asymptotically normal:

$$\lim_{\varepsilon \to 0} \sup_{0 \leq t \leq T} |\varepsilon^{-1}(X_t - x_t) - x_t^{(1)}| = 0,$$

where $x_t^{(1)}$, $0 \leq t \leq T$ is a Gaussian linear process

$$dx_t^{(1)} = S'(x_t)x_t^{(1)}dt + dW_t, \ x_0^{(1)} = 0,\ 0 \leq t \leq T \tag{4.35}$$

(the derivative of X_t with respect to ε at point $\varepsilon = 0$).

The low bound on the risk of estimators is based on the treatment by Millar [101] of the "limits of experiments" due to L. Le Cam [92] (see also [93]).

From here on, we suppose that $x_0 > 0$. Denote

$$\Theta(L) = \{S(\cdot):\ |S(x) - S(y)| \le L|x-y|,\ |S(x) \le L(1+|x|)\}$$

and $\Theta_+(L)$ as a subclass of functions $S(\cdot)$ from $\Theta(L/2)$ such that $S(x_t) > 0,\ t \in [0,T]$.

In the theorem below, we take the following notations: *infinum* is taken over all possible estimators $\hat{x}$ of the state x^h, ξ_t, $0 \le t \le T$ is a Gaussian process defined by the equalities:

$$\xi_t = S(x_t)W(F(t)),\ \ F(t) = \int_0^t \frac{ds}{S(x_s)^2},\ \ 0 \le t \le T,$$

$W(\lambda)$, $0 \le \lambda \le F(T)$ is a Wiener process and $S(\cdot) \in \Theta_+$ corresponds to $S(\cdot) \in \Theta_+(L/2)$ with some $L > 0$.

The low bound on the risk function $\mathbf{E}_h l(\varepsilon^{-1}(\hat{x} - x^h))$ of any estimator $\hat{x}$ is given by the following theorem.

Theorem 4.8. *Let the function* $S(\cdot)$ *be continuously differentiable and* $S(\cdot) \in \Theta_+$, *then*

$$\liminf_{\varepsilon\to 0} \inf_{\hat{x}} \sup_{h\in\Theta(L/2)} \int l\left(\varepsilon^{-1}(\hat{x} - x^h)\right) d\mathbf{P}_h^{(\varepsilon)} \ge \mathbf{E}l(\xi). \tag{4.36}$$

Proof. We first establish the convergence of appropriate experiments and then use a result of P. Millar.

The likelihood ratio for the observations (4.34) with $\mathbf{P}_S^{(\varepsilon)}$ probability 1 has the representation $\left(\mathbf{P}_S^{(\varepsilon)} = \mathbf{P}_0^{(\varepsilon)},\ (h=0)\right)$

$$Z_\varepsilon(h) \equiv \frac{d\mathbf{P}_h^{(\varepsilon)}}{d\mathbf{P}_0^{(\varepsilon)}}(X) = \exp\left\{\int_0^T \frac{h(X_t)}{S(X_t)}dW_t - \frac{1}{2}\int_0^T \left(\frac{h(X_t)}{S(X_t)}\right)^2 dt\right\}.$$

In view of condition $h(\cdot)$, $S(\cdot)^{-1} \in \Theta(L/2)$ we have

$$\mathbf{P}_S - \lim_{\varepsilon\to 0} \int_0^T \frac{h(X_t)}{S(X_t)}dW_t = \int_0^T \frac{h(x_t)}{S(x_t)}dW_t$$

and

$$\mathbf{P}_S - \lim_{\varepsilon\to 0} \int_0^T \left(\frac{h(X_t)}{S(X_t)}\right)^2 dt = \int_0^T \left(\frac{h(x_t)}{S(x_t)}\right)^2 dt.$$

So for any set $h_1(\cdot),\ldots,h_j(\cdot) \in \Theta(L/2)$, the random vector $(Z_\varepsilon(h_1),\ldots,Z_\varepsilon(h_k))$ converges in distribution (and in probability) to the vector $(Z(h_1),\ldots,Z(h_k))$, where

$$Z(h) = \exp\left\{\int_0^T \frac{h(x_t)}{S(x_t)}dW_t - \frac{1}{2}\int_0^T \left(\frac{h(x_t)}{S(x_t)}\right)^2 dt\right\}.$$

Hence we have the convergence of experiments $\left\{\mathcal{C}_T,\ \mathcal{B}_T, \mathbf{P}_h^{(\varepsilon)},\ h \in \Theta(L/2)\right\}$ to the experiment $\left\{\mathcal{C}_T,\ \mathcal{B}_T,\ \mathbf{P}_h,\ h \in \Theta(\frac{1}{2}L)\right\}$ in the sense of L. Le Cam. Here, $\mathbf{P}_h$ corresponds to the Gaussian process

$$dX_t = h(x_t)S(x_t)^{-1}dt + dW_t, \quad X_0 = x_0, \quad 0 \le t \le T.$$

Put $a = x_0,\ b = x_T$ and note that for $S(\cdot) \in \Theta_+\ a \le x_t \le b$. Introduce a function

$$\begin{aligned} \lambda(x) &= \int_a^x \frac{dy}{S(y)^3}, \\ \lambda(x_t) = \int_a^{x_t} \frac{dy}{S(y)^3} &= \int_0^t \frac{dx_s}{S(x_s)^3} = \int_0^t \frac{ds}{S(x_s)^2} = F(t) \end{aligned}$$

and a Hilbert space

$$\mathcal{H} = \left\{ h(x),\ a \le x \le b : \int_a^b |h(x)|^2 d\lambda(x) = \|h\|_{\mathcal{H}}^2 < \infty \right\}.$$

The Wiener process $W(\lambda),\ 0 \le \lambda \le F(T)$ can be written as $W(\lambda(y)),\ a \le y \le b$ and the likelihood ratio of the limiting experiment has the representation

$$Z(h) = \exp\left\{ \int_a^b h(y)dW(\lambda(y)) - \frac{1}{2}\|h\|_{\mathcal{H}}^2 \right\} \tag{4.37}$$

which also corresponds to the observations

$$dX(y) = h(y)d\lambda(y) + dW(\lambda(y)),\ X(a) = 0,\ a \le x \le b. \tag{4.38}$$

Denote τ as the mapping from $\mathcal{H}$ to the space $\mathcal{C}_0$ of continuous on $[a, b]$ functions, defined by

$$\tau h(x) = \int_0^x h(y)d\lambda(y) \quad (\tau h(0) = 0).$$

Let Q be a canonical Gaussian cylindrical measure on $\mathcal{H}$ and Q_τ is its image under the mapping τ. Then the characteristic function of this measure is

$$\begin{aligned} \Phi_\tau(m) &= \int \exp\{i < z, m >_{\mathcal{C}_0}\} dQ_\tau(z) = \\ &= \int \exp\{i < z, \tau^* m >_{\mathcal{H}}\} dQ(z) = \exp\left\{ -\frac{1}{2}\|\tau^* m\|_{\mathcal{H}}^2 \right\}, \end{aligned}$$

where $m \in C^*$ and C^* is a conjugate to $\mathcal{C}_0$ space, the space of all linear, continuous functionals formed by sign measures on $[a, b]$ and the duality ratio is defined by

$$< z, m >_{\mathcal{C}_0} = \int_a^b z(x)dm(x).$$

The operator τ^* is conjugate to τ and it maps $C^* \longrightarrow \mathcal{H}$ by $\tau^*m(x) = m\{[x,b]\}$. Thus

$$\begin{aligned}
\|\tau^*m\|^2_{\mathcal{H}} &= \int_a^b \left(\int_a^b \chi_{\{[a,y]\}}(x)m(dy)\right)^2 d\lambda(x) = \\
&= \int_a^b\int_a^b\int_a^b \chi_{\{[a,y]\}}(x)\chi_{\{[a,z]\}}(x)m(dy)m(dz)\lambda(dx) = \\
&= \int_a^b\int_a^b\int_a^b \chi_{\{[a,y\wedge z]\}}(x)d\lambda(x)m(dy)m(dz) = \\
&= \int_a^b\int_a^b [\lambda(y\wedge z)-\lambda(a)]m(dy)m(dz) = \\
&= \int_a^b [\lambda(y)\wedge\lambda(z)]m(dy)m(dz),
\end{aligned}$$

where the indicator

$$\chi_{\{[a,y]\}}(x) = \begin{cases} 1, & a \le x \le y \\ 0, & x < a,\ x > y \end{cases}$$

and we have used the strict monotonicity of the function $\lambda(\cdot)$: $\lambda(y\wedge z) = \lambda(y)\wedge\lambda(z)$, $y\wedge z = \min(y,z)$. So the characteristic function of the measure Q_τ on C_0 coincides with the characteristic function of the Wiener process $W(\lambda(x))$, $a \le x \le b$. The standard Gaussian shift experiment $\mathbf{P}_h$, $h \in \mathcal{H}$ induced by the process (4.38) has the likelihood ratio (4.37) and $\mathbf{P}_0 = Q_\tau$.

The argument $\varepsilon^{-1}(\hat{x}_t - x_t^h)$ of the loss function could be written in the form $\varepsilon^{-1}(\hat{x}_t - x_t) - \varepsilon^{-1}(x_t^h - x_t) = \tilde{x}_t - Y_t^{(\varepsilon)}$, where for $Y_t^{(\varepsilon)}$, from (4.31), (4.33) we have the equation

$$\frac{dY_t^{(\varepsilon)}}{dt} = S'(\tilde{x}_t^h)Y_t^{(\varepsilon)} + \frac{h(x_t^h)}{S(x_t^h)},\quad Z_0^{(\varepsilon)} = 0$$

(here $\tilde{x}_t^h = x_t^h + \gamma(x_t - x_t^h)$, $\gamma \in (0,1)$) and the limit ($\varepsilon = 0$) equation is

$$\frac{dY_t}{dt} = S'(x_t)Y_t + \frac{h(x_t)}{S(x_t)},\quad Z_0 = 0.$$

The solution of this linear equation is

$$Y_t = S(x_t)\int_0^t \frac{h(x_s)}{S(x_s)^2}ds = S(x_t)\int_a^{x_t} h(y)d\lambda(y). \tag{4.39}$$

After a change of variables, $x_t = x$, $a \le x \le b$, this gives the function $Y_t = Y(x) = S(x)\tau h(x) = \tau_1 h(x)$. Introduced in this manner, the operator τ_1 also maps $\mathcal{H}$ into C_0 and the measure Q_τ corresponding to the Wiener process $W(\lambda(x))$, $a \le x \le b$, transforms to the measure Q_τ^0 of the process $S(x)W(\lambda(x))$, $a \le x \le b$. Let us denote $\{\mathbf{P}_h^0,\ h \in \mathcal{H}\}$ as a standard Gaussian shift family of the measure Q_τ^0. The experiments $\{\mathcal{C}_0,\ \mathcal{B}_0,\ \mathbf{P}_h,\ h \in \Theta(L/2)\}$ and $\{\mathcal{C}_0,\ \mathcal{B}_0,\ \mathbf{P}_h^0,\ h \in \Theta(L/2)\}$ are equivalent. Hence, the family of experiments $\{\mathcal{C}_T,\ \mathcal{B}_T,\ \mathbf{P}_h^{(\varepsilon)},\ h \in \Theta(L/2)\}$ converges to experiment

$\{\mathcal{C}_0, \mathcal{B}_0, \mathbf{P}_h^0, h \in \Theta(L/2)\}$. Now the proposition of the theorem follows from Theorem 4.9 of P. Millar [102] (see also p. 389 there, where the role of τ_1 is clarified).

$$\liminf_{\varepsilon\to 0} \inf_{\hat{x}} \sup_{h\in\Theta(L/2)} \int l(\varepsilon^{-1}(\hat{x} - x^h))d\mathbf{P}_h^{(\varepsilon)} =$$
$$= \liminf_{\varepsilon\to 0} \inf_{\tilde{x}} \sup_{h\in\Theta(L/2)} \int l(\tilde{x} - \tau_1 h)d\mathbf{P}_h^{(\varepsilon)} \geq$$
$$\geq \int l(x)d\mathbf{P}_0^0 = \mathbf{E}l(SW(\lambda)),$$

where we have supposed that $l(\cdot)$ is bounded and continuous. If $l(\cdot)$ is not bounded, then we can introduce the truncation $l_c(\cdot) = C \wedge l(\cdot)$ and, having inequality (4.36) for $l_c(\cdot)$, take a limit $C \longrightarrow \infty$.

In the view of (4.36), it is natural, following P. Millar[102] to call an estimator x_t^*, $0 \leq t \leq T$ of the function x_t, $0 \leq t \leq T$ *local asymptotic minimax* (LAM) for a loss function $l(\cdot)$ if

$$\lim_{\varepsilon\to 0} \sup_{h\in\Theta(L/2)} \int l(\varepsilon^{-1}(x^* - x^h))d\mathbf{P}_h^{(\varepsilon)} = \mathbf{E}l(SW(\lambda)). \tag{4.40}$$

The following theorem shows that the *natural* estimator can not be improved.

Theorem 4.9. *The estimator X_t, $0 \leq t \leq T$ of x_t, $0 \leq t \leq T$ is LAM for the loss function $l(\cdot) \in \mathbf{W}_{e,2}$.*

Proof. For the difference $y_t^{(\varepsilon)} = \varepsilon^{-1}(X_t^h - x_t^h)$, we have the equation

$$dy_t^{(\varepsilon)} = [S'(\tilde{X}_t^h)y_t^{(\varepsilon)} + h(X_t^h)S(X_t^h)^{-1} - h(x_t^h)S(x_t^h)^{-1}]dt + dW_t, \; y_0^{(\varepsilon)} = 0.$$

and as $\varepsilon \to 0$ the solution of this equation converges uniformly on $t \in [0, T]$ and $h \in \Theta(L/2)$, with probability 1, to the solution of the equation

$$dy_t = S'(x_t)y_t dt + dW_t, \;\; y_0 = 0, \;\; 0 \leq t \leq T$$

which could be presented in the form

$$\begin{aligned} y_t = S(x_t)\int_0^t \frac{dW_s}{S(x_s)} &= S(x_t)W\left(\int_0^t \frac{ds}{S(x_s)^2}\right) \\ &= S(x_t)W(\lambda(x_t)), \;\; 0 \leq t \leq T. \end{aligned}$$

So the normed difference $\varepsilon^{-1}(X_t^h - x_t^h), 0 \leq t \leq T$ converges, as $\varepsilon \to \infty$, to the process $S(x_t)W(\lambda(x_t)), 0 \leq t \leq T$ and it now remains to establish the uniform integrability of the loss function

$$l\left(\varepsilon^{-1}(X_t^h - x_t^h)\right).$$

For the difference $y_t^{(\varepsilon)}$, using standard arguments (see Lemma 1.13), we obtain the inequality

$$\sup_{h\in\Theta(L/2)} \sup_{0\leq t\leq T} \varepsilon^{-1}|X_t^h - x_t^h| \leq C \sup_{0\leq t\leq T} |W_t|.$$

This implies

$$\sup_{h\in\Theta(L/2)} \mathbf{E}_h l\left(\varepsilon^{-1}(X^h - x^h)\right) \leq \mathbf{E}\exp\left\{\gamma C \sup_{0\leq t\leq T} |W_t|^2\right\} < \infty,$$

where we have used the finiteness of the last mathematical expectation for $\gamma < (2TC)^{-1}$ (Lemma 1.14).

So the random variables $l\left(\varepsilon^{-1}(X^h - x^h)\right)$ are uniformly integrable and the passage to the limit in

$$\lim_{\varepsilon\to 0} \sup_{h\in\Theta(L/2)} \mathbf{E}_h l\left(\varepsilon^{-1}(X^h - x^h)\right) = \mathbf{E}l(SW(\lambda))$$

is valid.

Chapter 5

The Disorder Problem

The *disorder problem* is traditionally a problem of hypotheses testing using the hypothesis: did it spasmodically change some of the characteristics of the observed process during the time of observation or not? Here we describe disorder problem as the problem of estimating the moment of switching of the characteristics of an observed diffusion-type process in a situation when it is known that the switching was realized during the observation. First we consider the problem of simultaneous estimation of the moment of disorder and some other parameter when the trend of the process is a smooth function of this parameter. Then we consider the problem of disorder estimation of multidimensional observations when each observed component has its own moment of switching and a generalization of this problem, when all the moments of switching are some smooth functions of another unknown parameter. The behavior of the MLE of the switching moment when the mathematical model is wrong is considered in the final part. It is interesting to note that even in this misspecified problem, consistent estimation is possible. We will also return to disorder problems in partially observed systems (Section 6.3) and in the minimum distance estimation of switching equations (Section 7.5).

5.1 Simultaneous estimation of the smooth parameter and the moment of switching

Let the observed diffusion-type process be

$$dX_t = S_t(\theta,\tau,X)dt + \varepsilon dW_t, \quad X_0 = x_0, \quad 0 \le t \le T,$$

where the trend coefficient has the representation

$$S_t(\theta,\tau,X) = h_t(\theta,X)\chi_{\{t\le\tau\}} + g_t(\theta,X)\chi_{\{t>\tau\}}.$$

Here $\theta \in (\alpha,\beta) = \Theta, \tau \in (t_1,t_2) = \mathbf{T} \subset [0,T]$ are the unknown parameters that we have to estimate and the functions $h_t(\cdot,X)$ and $g_t(\cdot,X)$ are known and smooth with respect to θ. The function $S_t(\theta,\tau,X) \in \mathcal{P}$. So the observations are up to time τ :

$$dX_t = h_t(\theta,X)dt + \varepsilon dW_t, \quad X_0 = x_0, \quad 0 \le t \le \tau$$

and on $[\tau, T]$

$$dX_t = g_t(\theta, X)dt + \varepsilon dW_t, \quad X_\tau, \quad \tau \leq t \leq T.$$

The likelihood ratio process

$$\frac{d\mathbf{P}^{(\varepsilon)}_{\theta,\tau}}{d\mathbf{P}^{(\varepsilon)}_{\theta_0,\tau_0}}(X), \quad \theta \in \Theta, \quad \tau \in \mathbf{T}$$

is not differentiable with respect to τ and Theorem 2.2 is not applicable here.

We are interested in the asymptotic behavior of the two-dimensional maximum-likelihood $(\hat{\theta}_\varepsilon, \hat{\tau}_\varepsilon)$ and Bayes $(\tilde{\theta}_\varepsilon, \tilde{\tau}_\varepsilon)$ estimators as $\varepsilon \to 0$.

Let us denote $x = (x_t, 0 \leq t \leq T)$ as the solution of the differential equation

$$\frac{dx_t}{dt} = \begin{cases} h_t(\theta, x), & x_0, \quad 0 \leq t \leq \tau, \\ g_t(\theta, x), & \tau < t \leq T, \end{cases}$$

the difference

$$V_t(\theta, x) = h_t(\theta, x) - g_t(\theta, x),$$

and

$$\sigma(\theta,\tau)^2 = \int_0^\tau \dot{h}_t(\theta, x)^2 dt + \int_\tau^T \dot{g}_t(\theta, x)^2 dt = \int_0^T \dot{S}_t(\theta, \tau, x_0)^2 dt,$$

where $\dot{h}(\cdot)$ and $\dot{g}(\cdot)$ are the derivatives of $h(\cdot)$ and $g(\cdot)$ with respect to θ.

We need several conditions

I. *The functionals $h(\cdot)$ and $g(\cdot)$ satisfy the inequalities*

$$\begin{aligned} |h_t(\theta, X) - h_t(\theta, Y)| &+ |g_t(\theta, X) - g_t(\theta, Y)| \leq \\ &\leq L_1|X_t - Y_t| + L_2 \int_0^t |X_s - Y_s| ds \end{aligned}$$

and

$$|h_t(\theta, X)| + |g_t(\theta, X)| \leq L_1(1 + |X_t|) + L_2 \int_0^t (1 + |X_s|) ds,$$

where the constants L_1 and L_2 do not depend on θ.

II. *The functions $h(\theta, \cdot)$ and $g(\theta, \cdot)$ have two continuous bounded derivatives with respect to θ and the first derivatives $\dot{h}_t(\theta, X)$ and $\dot{g}_t(\theta, X)$ are uniformly on θ continuous on X in $L_2[0, T]$ at point $x = (x_t, 0 \leq t \leq T)$.*

III. *For any compact $K_T = [\tau_1, \tau_2] \subset (t_1, t_2)$*

$$\inf_{\theta \in \Theta} \left[\int_0^{\tau_1} h_t(\theta, x)^2 dt + \int_{\tau_2}^T g_t(\theta, x)^2 dt \right] > 0,$$

$$\inf_{\theta \in \Theta, \tau \in K_T} V_\tau(\theta, x)^2 > 0,$$

and

$$\inf_{\theta \in \Theta, \tau \in K_t} \sigma(\theta, \tau)^2 > 0.$$

Let us also introduce a random field

$$Z(u,v) = \exp\left\{u\xi - \frac{1}{2}u^2\sigma(\theta,\tau)^2 + \right.$$
$$\left. + V_\tau(\theta,x)W(v) - \frac{1}{2}|v|V_\tau(\theta,x)^2\right\}, \qquad u,v \in R^2,$$

where $W(v)$ is a two-sided Wiener process

$$W(v) = \begin{cases} W_+(v), & v \geq 0, \\ W_-(-v), & v < 0, \end{cases}$$

$W^{(1)}, W^{(2)}$ are independent Wiener processes and ξ is a Gaussian random variable

$$\mathcal{L}\{\xi\} = \mathcal{N}\left(0, \sigma(\theta,\tau)^2\right).$$

Denote also $\xi^* = \sigma(\theta,\tau)^{-2}\xi$ and

$$\zeta^* = \arg\sup_{v\in R}\left\{V_\tau(\theta,x)W(v) - \frac{1}{2}|v|V_\tau(\theta,x)^2\right\}.$$

So, the random vector (ξ^*,ζ^*) is defined by

$$(\xi^*,\zeta^*) = \arg\sup_{(u,v)\in R^2} Z(u,v).$$

The role of Fisher information will be played by the matrix

$$I_\varepsilon(\theta,\tau) = \begin{pmatrix} \varepsilon^{-2}\sigma(\theta,\tau)^2 & 0 \\ 0 & \varepsilon^{-4}V_\tau(\theta,x)^2. \end{pmatrix}$$

Theorem 5.1. *Under conditions I–III, the MLE $(\hat\theta_\varepsilon;\hat\tau_\varepsilon)$ is uniformly on $(\theta;\tau) \in K_\Theta \times K_\tau$,*

(a) consistent

$$\mathbf{P}_{\theta,\tau} - \lim_{\varepsilon\to 0}(\hat\theta_\varepsilon;\hat\tau_\varepsilon) = (\theta;\tau);$$

(b) converges in distribution

$$\mathcal{L}_{\theta,\tau}\left\{I_\varepsilon(\theta,\tau)^{\frac{1}{2}}(\hat\theta_\varepsilon - \theta;\hat\tau_\varepsilon - \tau)\right\} \Longrightarrow \mathcal{L}\{(\xi^*,\zeta^*)\};$$

(c) the moments also converge

$$\lim_{\varepsilon\to 0}\mathbf{E}_{\theta,\tau}\left|I_\varepsilon(\theta,\tau)^{\frac{1}{2}}(\hat\theta_\varepsilon - \theta;\hat\tau_\varepsilon - \tau)\right|^p = \mathbf{E}|(\xi^*,\zeta^*)|^p.$$

Proof. The likelihood ratio $Z_\varepsilon(u,v)$ has the representation

$$\begin{aligned}\ln Z_\varepsilon(u,v) &= \frac{1}{\varepsilon}\int_0^T [S_t(\theta+u\varepsilon,\tau+v\varepsilon^2,X) - S_t(\theta,\tau,X)]dW_t - \\ &- \frac{1}{2\varepsilon^2}\int_0^T [S_t(\theta+u\varepsilon,\tau+v\varepsilon^2,X) - S_t(\theta,\tau,X)]^2 dt.\end{aligned}$$

Denote by $\Delta_t = S_t(\theta+u\varepsilon,\tau+v\varepsilon^2,X) - S_t(\theta,\tau,X)$, then we can write for $v>0$

$$\int_0^T \Delta_t^2 dt = \int_0^\tau \Delta_t^2 dt + \int_\tau^{\tau+v\varepsilon^2} \Delta_t^2 dt + \int_{\tau+v\varepsilon^2}^T \Delta_t^2 dt.$$

The first and third integrals can be transformed, as in the proof of Lemma 2.1. For example

$$\begin{aligned}\int_0^\tau \Delta_t^2 dt &= \int_0^\tau [h_t(\theta+u\varepsilon,X) - h_t(\theta,X)]^2 = \\ &= \int_0^\tau [h_t(\theta+u\varepsilon,X) - h_t(\theta,X) - u\varepsilon\dot h_t(\theta,X)]^2 dt + \\ &+ 2u\,\varepsilon\int_0^\tau (h_t(\theta+u\varepsilon,X) - h_t(\theta,X))\dot h_t(\theta,X)dt - \\ &-u^2\varepsilon^2\int_0^\tau \dot h_t(\theta,X)^2 dt.\end{aligned}$$

Hence

$$\begin{aligned}&\varepsilon^{-2}\int_0^\tau |\Delta_t - u\varepsilon\dot h_t(\theta,x)|^2 dt \le \\ &\le c\varepsilon^2 + 2|u|^2\int_0^\tau |\dot h_t(\theta,X) - \dot h_t(\theta,x)|^2 dt = o(1).\end{aligned}$$

In a similar way, we establish

$$\begin{aligned}&\varepsilon^{-2}\int_{\tau+v\varepsilon^2}^T |\Delta_t - u\varepsilon\dot g_t(\theta,x)|^2 dt \le \\ &\le c\varepsilon^2 + 2|u|^2\int_0^T |\dot g_t(\theta,X) - \dot g_t(\theta,x)|^2 dt = o(1).\end{aligned}$$

Further

$$\begin{aligned}\varepsilon^{-2}\int_\tau^{\tau+v\varepsilon^2} \Delta_t^2 dt &= \varepsilon^{-2}\int_\tau^{\tau+v\varepsilon^2} [h_t(\theta+u\varepsilon,X) - g_t(\theta,X)]^2 dt = \\ &= \varepsilon^{-2}\int_\tau^{\tau+v\varepsilon^2} [h_t(\theta+u\varepsilon,x) - g_t(\theta,x)]^2 dt + o(1) = \\ &= \varepsilon^{-2}\int_\tau^{\tau+v\varepsilon^2} V_t(\theta,x)^2 dt + o(1) \longrightarrow v\,V_\tau(\theta,x)^2.\end{aligned}$$

Hence, with $\mathbf{P}_{\theta,\tau}$ probability 1,

$$\varepsilon^{-2}\int_0^T \Delta_t^2 dt \longrightarrow u^2 I(\theta,\tau) + vV_\tau(\theta,x)^2.$$

Similar arguments allow us to prove

$$\begin{aligned}\varepsilon^{-1}\int_0^T \Delta_t dW_t &= \varepsilon^{-1}\int_0^\tau \Delta_t dW_t + \varepsilon^{-1}\int_\tau^{\tau+v\varepsilon^2} \Delta_t dW_t + \\ &\quad +\varepsilon^{-1}\int_{\tau+v\varepsilon^2}^T \Delta_t dW_t = u\int_0^\tau \dot{h}_t(\theta,x)dW_t + u\int_{\tau+v\varepsilon^2}^T \dot{g}_t(\theta,x)dW_t + \\ &\quad +\varepsilon^{-1}\int_\tau^{\tau+v\varepsilon^2} V_t(\theta,x)dW_t + o(1) = \\ &= u\xi + V_\tau(\theta,x)\left[\frac{W_{\tau+v\varepsilon^2}-W_\tau}{\varepsilon}\right] + o(1) = u\xi + V_\tau(\theta,x)\tilde{W}_v + o(1),\end{aligned}$$

where $\tilde{W}_v$ is a two-sided Wiener process with the same distribution as $W(v)$, but its realizations depend on ε. The same limit we will have if we consider any sequences $u_\varepsilon \to u$ and $v_\varepsilon \to v$.

The analogy convergences for $v < 0$ can be derived in the same way.

So the convergence uniform, with respect to $\theta, \tau \in K_\Theta \times K_\tau$ of one-dimensional distributions of $Z_\varepsilon(u,v)$ is established. The convergence of finite-dimensional distributions could be proved in a similar way.

Lemma 5.1. *Let conditions I–II be fulfilled, then for any compacts K_Θ and K_τ, we have*

$$\begin{aligned}\sup_{(\theta,\tau)\in K_\Theta\times K_\tau, |u_i|<R, |v_i|<R} &[(u_2-u_1)^2+(v_2-v_1)^2]^{-1} \times \\ \times \mathbf{E}_{\theta,\tau}[Z\varepsilon^{\frac{1}{4}}(u_1,v_1) - Z_\varepsilon^{\frac{1}{4}}(u_2,v_2)]^4 &\leq B(1+R^6).\end{aligned}$$

Proof. Let us denote by $\Delta_t = S_t(\theta+u_1\varepsilon, \tau+v_1\varepsilon^2, x) - S_t(\theta+u_2\varepsilon, \tau+v_2\varepsilon^2, x)$, then repeating the derivation of (2.13), we obtain

$$\begin{aligned}&\mathbf{E}_{\theta,\tau}|Z_\varepsilon^{\frac{1}{4}}(u_1,v_1) - Z_\varepsilon^{\frac{1}{4}}(u_2,v_2)|^4 \leq \\ &\leq \frac{8\cdot 3^4}{32^4\varepsilon^8}\mathbf{E}_{\theta_2,\tau_2}\left|\int_0^T \Delta_t^2 Y_t dt\right|^4 + \frac{8\cdot 3^4}{4^4\varepsilon^4}\mathbf{E}_{\theta_2,\tau_2}\left|\int_0^T \Delta_t Y_t dW_t\right|^4.\end{aligned}$$

Here, with $\mathbf{P}^{(\varepsilon)}_{\theta_2,\tau_2}$ probability 1,

$$Y_t = \exp\left\{\frac{1}{4\varepsilon}\int_0^t \Delta_s dW_s - \frac{1}{8\varepsilon^2}\int_0^t \Delta_s^2 ds\right\} = \left(\frac{d\mathbf{P}^{(\varepsilon)}_{\theta_1,\tau_1}}{d\mathbf{P}^{(\varepsilon)}_{\theta_2,\tau_2}}\right)_t^{\frac{1}{4}},$$

where $(\cdot)_t$ means that the measures $\mathbf{P}^{(\varepsilon)}_{\theta_i,\tau_i}, i = 1,2$ are defined on $(\mathcal{C}_t, \mathcal{B}_t)$ and $\theta_i = \theta + u_i\varepsilon$, $\tau_i = \tau + v_i\varepsilon^2$, $i = 1,2$. Note also that

$$\begin{aligned}\mathbf{E}_{\theta_2,\tau_2}\int_0^T (\Delta_t Y_t)^4 dt &= \int_0^T \mathbf{E}_{\theta_2,\tau_2}\left(\frac{d\mathbf{P}^{(\varepsilon)}_{\theta_1,\tau_1}}{d\mathbf{P}^{(\varepsilon)}_{\theta_2,\tau_2}}\right)_t \Delta^4 dt = \\ &= \mathbf{E}_{\theta_1,\tau_1}\int_0^T \Delta_t^4 dt < \infty.\end{aligned}$$

By (1.8) we have

$$\mathbf{E}_{\theta,\tau}\left|\int_0^T \Delta_t Y_t dW_t\right|^4 < 6T^3\int_0^T E_{\theta_1,\tau_1}\Delta_t^4 dt < \infty.$$

So, for $v > 0$

$$\mathbf{E}_{\theta_2,\tau_2}\left|\int_0^T \Delta_t^2 Y_t dt\right|^4 =$$

$$= \mathbf{E}_{\theta_2,\tau_2}\left[\left(\int_0^\tau + \int_{\tau+v\varepsilon^2}^T\right)\Delta_t^2 Y_t dt + \int_\tau^{\tau+v\varepsilon^2}\Delta_t^2 Y_t dt\right]^4 \leq$$

$$\leq 8\mathbf{E}_{\theta_2,\tau_2}\left|\left(\int_0^\tau + \int_{\tau+v\varepsilon^2}^T\right)\Delta_t^2 Y_t dt\right|^4 + 8\mathbf{E}_{\theta_2,\tau_2}\left|\int_\tau^{\tau+v\varepsilon^2}\Delta_t^2 Y_t dt\right|^4,$$

where $v = v_1 - v_2$. Schwarz inequality allows us to write

$$\mathbf{E}_{\theta_2,\tau_2}\left(\int_\tau^{\tau+v\varepsilon^2}\Delta_t^2 Y_t dt\right)^4 \leq v^3\varepsilon^6\int_\tau^{\tau+v\varepsilon^2}\mathbf{E}_{\theta_1,\tau_1}\Delta_t^8 dt \leq Cv^4\varepsilon^8,$$

because, we have

$$\sup_{\theta,\tau,t}\mathbf{E}_{\theta,\tau}|\Delta_t|^8 < \infty.$$

Moreover

$$\mathbf{E}_{\theta_2,\tau_2}\left(\int_0^\tau \Delta_t^2 Y_t dt\right)^4 \leq C\tau^3\int_0^\tau \mathbf{E}_{\theta_1,\tau_1}\Delta_t^8 dt \leq C\varepsilon^8(u_2-u_1)^8.$$

Hence

$$\varepsilon^{-8}\mathbf{E}_{\theta_2,\tau_2}\left|\int_0^T \Delta_t^2 Y_t dt\right|^4 \leq C_1|u_2-u_1|^8 + C_2|v_2-v_1|^4.$$

For the stochastic integral, by analogy we have

$$\mathbf{E}_{\theta_2,\tau_2}\left|\int_0^T \Delta_t Y_t dW_t\right|^4 \leq$$

$$\leq 8\mathbf{E}\left|\left(\int_0^\tau + \int_{\tau+v\varepsilon^2}^T\right)\Delta_t Y_t dW_t\right|^4 + 8\mathbf{E}\left(\int_\tau^{\tau+v\varepsilon^2}\Delta_t Y_t dW_t\right)^4 \leq$$

$$\leq C_3\varepsilon^4|u_2-u_1|^4 + C_4\varepsilon^4(v_2-v_1)^2.$$

Finally, for $|u_i|, |v_i| < R$ we obtain

$$\sup_{\theta,\tau\in K_\Theta\otimes K_\tau}\left(|u_2-u_1|^2+|v_2-v_1|^2\right)^{-1}\times$$

$$\times\mathbf{E}_{\theta,\tau}\left|Z_\varepsilon^{\frac{1}{4}}(u_1,v_1) - Z_\varepsilon^{\frac{1}{4}}(u_2,v_2)\right|^4 \leq B(1+R^6).$$

Lemma 5.2. *Let conditions I–III be fulfilled, then for any compacts K_Θ and K_τ, we have*

$$\sup_{\theta,\tau \in K_\Theta \otimes K_\tau} \mathbf{E}_{\theta,\tau} Z_\varepsilon^{\frac{1}{2}}(u,v) \leq C \exp\{-g(u,v)\},$$

where $g(u,v) = \kappa_1 u^2 + \kappa_2 |v|$.

Proof. As this follows from the proof of Lemma 2.3, it is sufficient to consider the behavior of the (deterministic) integral

$$\int_0^T [S_t(\theta + u\theta, \tau + v\varepsilon^2, x) - S_t(\theta, \tau, x)]^2 dt =$$
$$= \int_0^T [h_t(\theta + u\varepsilon, x) - h_t(\theta, x)]^2 dt + \int_{\tau + v\varepsilon^2}^T [g_t(\theta + u\varepsilon, x) -$$
$$- g_t(\theta, x)]^2 dt + \int_\tau^{\tau + v\varepsilon^2} V_t(\theta, x)^2 dt.$$

Conditions II and III allow us to write (by analogy with (2.3))

$$\int_0^T [h_t(\theta + u, x) - h_t(\theta, x)]^2 dt + \int_{\tau + v\varepsilon^0}^T [g_t(\theta + u, x) -$$
$$- g_t(\theta, x)]^2 dt \geq \kappa_1 u^2,$$

and

$$\int_\tau^{\tau + v} V_t(\theta, x)^2 dt \geq \kappa_2 v \qquad (v > 0).$$

Hence,

$$\varepsilon^{-2} \int_0^T [S_t(\theta + u\varepsilon, \tau + v\varepsilon^2, x) - S_t(\theta, \tau, x)]^2 dt \geq$$
$$\geq \kappa_1 u^2 + \kappa_2 |v|.$$

Repeating the corresponding derivation of Lemma 2.3, we complete the proof of this lemma.

The properties of the likelihood ratio $Z_\varepsilon(u,v)$ established in Lemmas 5.1 and 5.2 allow us to refer to Theorem 1.4. Note only that Theorem 1.4 remains true with this function $g(\cdot)$ (see, [53], Theorem 1.10.1).

To describe the properties of the Bayes estimators of (θ, τ), we now suppose that (θ, τ) is a random variable with the prior density $\pi(r)$, $r \in \Theta \times \mathbf{T}$ and suppose also that the loss function is quadratic.

Theorem 5.2. *Let the function $\pi(r)$ be positive, continuous and bounded on Θ and conditions I–III be fulfilled, then the Bayes estimator $(\tilde{\theta}_\varepsilon, \tilde{\tau}_\varepsilon)$, uniformly on $(\theta, \tau) \in K_\Theta \times K_\tau$,*

(a) is consistent

$$\mathbf{P}_{\theta,\tau} - \lim_{\varepsilon \to 0} (\tilde{\theta}_\varepsilon; \tilde{\tau}_\varepsilon) = (\theta; \tau);$$

(b) converges in distribution

$$\mathcal{L}_{\theta,\tau}\left\{I_\varepsilon(\theta,\tau)^{\frac{1}{2}}(\tilde\theta_\varepsilon-\theta;\tilde\tau_\varepsilon-\tau)\right\}\Longrightarrow\mathcal{L}\{(\xi^*,\tilde\zeta)\};$$

where

$$\tilde\zeta=\left(\int_{-\infty}^{\infty}Z(v)dv\right)^{-1}\int_{-\infty}^{\infty}vZ(v)dv$$

(c) the convergence of moments also takes place.

The properties of the likelihood ratio process established in Lemmas 5.1 and 5.2 allow us to use Theorem 1.5 and immediately obtain the mentioned properties of the estimators.

REMARK 5.1. As follows from the theorem 1.9.1 of [53], for any estimator $(\theta_\varepsilon,\tau_\varepsilon)$ the inequality

$$\lim_{\delta\to0}\lim_{\varepsilon\to0}\sup_{|\theta-\theta_0|+|\tau-\tau_0|<\delta}\mathbf{E}_{\theta,\tau}l\left(I_\varepsilon(\theta,\tau)^{\frac{1}{2}}(\theta_\varepsilon-\theta;\tau_\varepsilon-\tau)\right)\geq\mathbf{E}l(\xi^*,\tilde\zeta),\qquad(5.1)$$

holds, so the LAM estimators can be introduced (as estimators with equality here for all $(\theta_0;\tau_0)\in\Theta\times\mathbf{T}$) and, for polynomial loss functions $l(\cdot)$ the LAM of Bayes estimators can be proved.

REMARK 5.2. The existence of the r.v. ζ^* follows from lemma 7.2.5 of [53]. A. Terent'yev [110] found the value

$$\mathbf{E}(\zeta^*)^2=26\;V_\tau(\theta,x)^{-1}$$

(see also [53], section 7.3).

5.2 Multidimensional disorder

Below, we consider two problems of parameter estimation by the observations of diffusion-type processes with switching in every component. We suppose that a k-dimensional diffusion-type process $X_t=(X_t^{(1)},\ldots,X_t^{(k)})$, $0\leq t\leq T$ is observed and the components satisfy the equations

$$\begin{aligned}dX_t^{(i)}&=h_t^{(i)}(X)dt+\varepsilon dW_t^{(i)},\quad X_0^{(i)}=x_0^{(i)},\quad 0\leq t\leq\tau^{(i)},\\ dX_t^{(i)}&=g_t^{(i)}(X)dt+\varepsilon dW_t^{(i)},\quad X_{\tau^{(i)}},\quad \tau^{(i)}\leq t\leq T,\end{aligned}\qquad(5.2)$$

where $W_t^{(i)}$, $i=1,\ldots,k$ are independent Wiener processes and the vector $\tau=$ $=(\tau^{(1)},\ldots,\tau^{(k)})$ is unknown to the observer. The functions $h_t^{(i)}(X)$, $g_t^{(i)}(X)$ and the values $x_0^{(i)}$, $i=1,\ldots,k$ are known.

Such statements can, for example, arise in the problem of detecting a signal by k different detectors arranged in the space when this signal attains an i-th detector at a proper moment $\tau^{(i)}$. So the first problem is to estimate this vector of switching τ.

The second problem could be explained as follows. Suppose now that the source of this signal is situated at point $(\theta^{(1)},\theta^{(2)})$ on the plane and it is possible to calculate the time of propagation $t_{(i)}(\theta^{(1)}$ of this signal from the source to the i-th detector, then $\tau^{(i)}=\theta^{(3)}+t_{(i)}(\theta^{(1)},\theta^{(2)})$ (here $\theta^{(3)}$ is the moment of beginning the transmission). So all the values $\tau^{(1)},\ldots,\tau^{(k)}$ are smooth functions of $\theta=(\theta^{(1)},\theta^{(2)},\theta^{(3)})$ and can be calculated if the vector θ is known. The second problem corresponds to this situation. We have the same observations (5.2) where τ is a known (smooth) function of another (unknown) vector-parameter θ and we have to estimate θ by $(X_t,\ 0\le t\le T)$ (without a preliminary estimation of τ). This problem can be motivated by the detection of seismic signals and estimation of the epicenter's position. We call these two problems Problem 1 and Problem 2, correspondingly.

We denote $x_t = x_t(\tau) = (x_t^{(1)},\ldots,x_t^{(k)})$, $0\le t\le T$ as the vector with the components $(i=1,\ldots,k)$

$$\frac{dx_t^{(i)}}{dt}=\begin{cases} h_t^{(i)}, & 0\le t\le \tau^{(i)},\quad x_0^{(i)},\\ g_t^{(i)}, & \tau^{(i)}\le t\le T,\end{cases}$$

The analog of conditions $\mathcal{L}$ here are the conditions

I. *The functionals* $h_t(X)=(h_t^{(1)}(X),\ldots,h_t^{(k)}(X))$ *and* $g_t(X)=(g_t^{(1)}(X),\ldots,$ $g_t^{(k)}(X))$ *satisfy the inequalities*

$$|h_t(X)-h_t(Y)|+|g_t(X)-g_t(Y)|\le L_1\int_0^t |X_s-Y_s|ds+L_2|X_t-Y_t|$$

and

$$|h_t(X)|+|g_t(X)|\le L_1\int_0^t (1+|X_s|)ds+L_2(1+|X_t|),$$

where $|\cdot|$ *is* R^k*-norm, so, for example,*

$$|h_t(X)-h_t(Y)|^2=\sum_{i=1}^{k}|h_t^{(i)}(X)-h_t^{(i)}(Y)|^2$$

and X_t, Y_t, $0\le t\le T$ *are* k*-dimensional continuous vector-functions.*

Using the same arguments as in the proof of Lemma 1.13, we can obtain the inequality

$$\sup_{0\le t\le T}|X_t-x_t(\tau)|\le C\varepsilon\sup_{0\le t\le T}|W_t|, \tag{5.3}$$

where $C=C(L_1,L_2,T)>0$ and $|W_t|^2=\sum_{i=1}^{k}(W_t^{(i)})^2$, which is true with $\mathbf{P}_\tau^{(\varepsilon)}$ probability 1. $\mathbf{P}_\tau^{(\varepsilon)}$ is the measure induced by the process (5.2).

If τ is a random variable with a prior density $\pi(y)$, $y\in\mathbf{T}$, then we can introduce the Bayes estimator $\tilde\tau_\varepsilon$ of τ as

$$\tilde\tau=\arg\inf_{s\in\mathbf{T}}\int_{\mathbf{T}} l(s-y)L(y,X)\pi(y)dy,$$

where $l(\cdot)$ is a loss function and $L(y,X)$ is the likelihood ratio.

Problem 1 **(τ-estimation).** The maximum likelihood estimator $\hat{\tau}_\varepsilon$ is defined by

$$\hat{\tau}_\varepsilon = \arg\sup_{\tau\in\mathbf{T}} L(\tau, X),$$

where $\mathbf{T} = (\tau_1, \tau_2),\ , \tau_1 = (\tau_1^{(1)}, \dots, \tau_1^{(k)}), \tau_2 = (\tau_2^{(1)}, \dots, \tau_2^{(k)})$ is the set of possible values of τ, $0 < \tau_1^{(i)} < \tau_2^{(i)} < T$.

$$L(\tau, X) \equiv \frac{d\mathbf{P}_\tau^{(\varepsilon)}}{d\mathbf{P}^{(\varepsilon)}}(X) = \exp\left\{\frac{1}{\varepsilon^2}\int_0^\tau (h_t(X), dX_t) + \right.$$
$$\left. + \frac{1}{\varepsilon^2}\int_\tau^T (V_t(X), dX_t) - \frac{1}{2\varepsilon^2}\int_0^\tau |h_t(X)|^2 dt - \frac{1}{2\varepsilon^2}\int_\tau^T |V_t(X)|^2 dt\right\}$$

Here we denote $\mathbf{P}^{(\varepsilon)}$ the measure induced by the vector-process $\{x_0^{(i)} + \varepsilon W_t^{(i)},\ 0 \le t \le T,\ i = 1, \dots, k\}$ and $V_t(X) = h_t(X) - g_t(X)$. The stochastic integral

$$\int_0^\tau (h_t(X), dX_t) \equiv \sum_{i=1}^k \int_0^{\tau^{(i)}} h_t^{(i)}(X) dX_t^{(i)}$$

and so on.

This expression for the likelihood ratio of the multidimensional process can be found, for example, in [97].

To formulate the main result we need additional notations.

Let us introduce

(a) $2k$ independent Wiener processes

$$W_+^{(i)}(v), \quad W_-^{(i)}(v), \quad v \ge 0,\ i = 1, \dots, k,$$

(b) stochastic processes

$$Z^{(j)}(v) = \exp\left\{r^{(j)}\bar{W}^{(j)}(v) - \frac{1}{2}(r^{(j)})^2|v|\right\},\ v \in R,$$

where $r^{(j)} = V_{\tau^{(j)}}^{(j)}(x) = h_{\tau^{(j)}}^{(j)}(x) - g_{\tau^{(j)}}^{(j)}(x)$ and

$$\bar{W}^{(j)}(v) = \begin{cases} W_+^{(j)}(v), & v \ge 0 \\ W_-^{(j)}(-v), & v < 0, \end{cases}$$

(c) random field

$$Z(u) = \prod_{j=1}^k Z^{(j)}(u^{(j)}), \qquad u = (u^{(1)}, \dots, u^{(k)}) \in R^k,$$

(d) random vector $\xi = (\xi^{(1)}, \ldots, \xi^{(k)})$, where

$$\xi^{(j)} = \arg \sup_{v \in R^k} Z^{(j)}(v),$$

(e) random vector

$$\zeta = \arg \inf_{s \in R^k} \int_{R^k} l(s-y) Z(y) dy.$$

Note also that the vector ξ could be represented as

$$\xi = \arg \sup_{u \in R^k} Z(u).$$

We introduce now the condition of identifiability

II.

$$\min_{j=1,\ldots,k} \inf_{\tau \in \mathbf{T}} |V^{(j)}_{\tau(j)}(x(\tau))| > 0$$

Below, $\mathbf{K}$ denotes an arbitrary compact in $\mathbf{T}$.

Theorem 5.3. *Let conditions I and II be fulfilled, then, uniformly on $\tau \in \mathbf{K}$, the MLE $\hat{\tau}_\varepsilon$ is*

$$\mathbf{P}_\tau - \lim_{\varepsilon \to 0} \hat{\tau}_\varepsilon = \tau,$$

$$\mathcal{L}_\tau\{\varepsilon^{-2}(\hat{\tau}_\varepsilon - \tau)\} \Longrightarrow \mathcal{L}\{\xi\},$$

$$\lim_{\varepsilon \to 0} \mathbf{E}_\tau |\varepsilon^{-2}(\hat{\tau}_\varepsilon - \tau)|^p = \mathbf{E}|\xi|^p$$

for any $p \geq 1$.

The proof of this theorem is based on Theorem 1.4. Below we check the conditions of that theorem by three lemmas.

Lemma 5.3. *Under condition I, the finite-dimensional distributions of the normed likelihood ratio field*

$$Z_\varepsilon(u) = \frac{d\mathbf{P}^{(\varepsilon)}_{\tau+\varepsilon^2 u}}{d\mathbf{P}^{(\varepsilon)}_\tau}(X), \quad u \in U_\varepsilon = \{u : \tau + \varepsilon^2 u \in \mathbf{T}\}$$

converge to the finite-dimensional distributions of the field $Z(u), u \in R^k$ uniformly on $\tau \in \mathbf{K}$.

Proof. The substitution of observations (5.2) in the expression

$$Z_\varepsilon(u) = \frac{L(\tau + \varepsilon^2 u, X)}{L(\tau, X)}$$

for the likelihood ratio, implies the representation

$$\begin{aligned}\ln Z_\varepsilon(u) &= \frac{1}{\varepsilon}\sum_{j=1}^{k}\int_{\tau^{(j)}}^{\tau^{(j)}+\varepsilon^2u^{(j)}}[h_t^{(j)}(X)-g_t^{(j)}(X)]dW_t^{(j)} - \\ &- \frac{1}{2\varepsilon^2}\sum_{j=1}^{k}\int_{\tau^{(j)}}^{\tau^{(j)}+\varepsilon^2u^{(j)}}[h_t^{(j)}(X)-g_t^{(j)}(X)]^2dt.\end{aligned}$$

Note that the constant C in (5.2) does not depend on τ, so the convergence below is uniform on $\tau \in \mathbf{K}$:

$$\begin{aligned}&\frac{1}{2\varepsilon^2}\int_{\tau^{(j)}}^{\tau^{(j)}+\varepsilon^2u^{(j)}}[h_t^{(j)}(X)-g_t^{(j)}(X)]^2dt = \\ &= \frac{1}{2\varepsilon^2}\int_{\tau^{(j)}}^{\tau^{(j)}+\varepsilon^2u^{(j)}}V_t^{(j)}(x)^2dt + \sum_{j=1}^{k}u^{(j)}o(\varepsilon) = \\ &= \left(V_{\tau^{(j)}}^{(j)}(x)\right)^2u^{(j)}(1+o(1))\end{aligned}$$

and

$$\begin{aligned}\frac{1}{\varepsilon}\int_{\tau^{(j)}}^{\tau^{(j)}+\varepsilon^2u^{(j)}}V_t^{(j)}(X)dW_t^{(j)} &= \frac{1}{\varepsilon}\int_{\tau^{(j)}}^{\tau^{(j)}+\varepsilon^2u^{(j)}}V_t^{(j)}(x)dW_t^{(j)} + o(1) = \\ &= V_{\tau^{(j)}}^{(j)}(x)\varepsilon^{-1}[W_{\tau^{(j)}+\varepsilon^2u^{(j)}}^{(j)} - W_{\tau^{(j)}}^{(j)}] + o(1) = \\ &= V_{\tau^{(j)}}^{(j)}(x)W_+^{(j)}(u^{(j)}) + o(1)\end{aligned}$$

for $u^{(j)} \geq 0$ and we have similar equalities for $u^{(j)} < 0$.

Thus, the likelihood ratio $Z_\varepsilon(u)$ converges in $\mathbf{P}_\tau^{(\varepsilon)}$ probability:

$$\ln Z_\varepsilon(u) \longrightarrow \sum_{j=1}^{k}\{V_{\tau^{(j)}}^{(j)}\bar{W}^{(j)}(u^{(j)}) - \frac{1}{2}(V_{\tau^{(j)}}^{(j)})^2|u^{(j)}|\} = \ln Z(u)$$

and this convergence is uniform on $\tau \in \mathbf{K}$. In the same way, we can prove the convergence of the vectors $\{Z_\varepsilon(u_1),\ldots,Z_\varepsilon(u_m)\}$ to the vector $\{Z(u_1),\ldots,Z(u_m)\}$, if we consider the sum

$$\sum_{l=1}^{m}\lambda_l \ln Z_\varepsilon(u_l) \longrightarrow \sum_{l=1}^{m}\lambda_l \ln Z(u_l),$$

where $\lambda \in R^m$ is an arbitrary vector.

Lemma 5.4. *Under condition I, for any integer $m > 1$, we have*

$$\begin{aligned}&\sup_{\tau\in\mathbf{K}}\ \sup_{|u_1|+|u_2|<R_0}\left(\sum_{j=1}^{k}|u_2^{(j)}-u_1^{(j)}|^m\right)^{-1}\times \\ &\times\mathbf{E}_\tau\left|Z_\varepsilon^{\frac{1}{2m}}(u_2)-Z_\varepsilon^{\frac{1}{2m}}(u_1)\right|^{2m} \leq C(1+R_0^m). \qquad (5.4)\end{aligned}$$

Proof. Let us denote

$$Y_t = \exp\left\{\frac{1}{2m\varepsilon}\int_0^t (S_s(\tau,X),dW_s) - \frac{1}{2m\varepsilon^2}\int_0^t |S_s(\tau,X)|^2 ds\right\},$$

where

$$\begin{aligned} S_t(\tau,X) &= h_t(X)\chi_{\{t<\tau_{u_2}\}} + g_t(X)\chi_{\{t\geq\tau_{u_2}\}} - \\ &- h_t(X)\chi_{\{t<\tau_{u_1}\}} - g_t(X)\chi_{\{t\geq\tau_{u_1}\}},\ \tau_u = \tau+\varepsilon^2 u, \end{aligned}$$

and note that

$$Y_T = \left(\frac{d\mathbf{P}^{(\varepsilon)}_{\tau+\varepsilon^2 u_2}}{d\mathbf{P}^{(\varepsilon)}_{\tau+\varepsilon^2 u_1}}(X)\right)^{\frac{1}{2m}} = \left(\frac{Z_\varepsilon(u_2)}{Z_\varepsilon(u_1)}\right)^{\frac{1}{2m}}.$$

Using Itô formula, we rewrite Y_t as

$$\begin{aligned} Y_T = 1 &- \frac{3}{8m^2\varepsilon^2}\int_0^T Y_t|S_t(\tau,X)|^2 dt + \\ &+ \frac{1}{2m\varepsilon}\int_0^T Y_t(S_t(\tau,X),dW_t). \end{aligned}$$

Hence (as was done in (2.13)), we derive

$$\begin{aligned} &\mathbf{E}_\tau|Z_\varepsilon^{\frac{1}{2m}}(u_2) - Z_\varepsilon^{\frac{1}{2m}}(u_1)|^{2m} = \mathbf{E}_{\tau+\varepsilon^2 u_1}|1-Y_t|^{2m} = \\ &= \mathbf{E}_{\tau_1}\left|\frac{3}{8m^2\varepsilon^2}\int_0^T Y_t|S_t(\tau,X)|^2 dt - \frac{1}{2m\varepsilon}\int_0^T Y_t(S_t(\tau,X),dW_t)\right|^{2m} \leq \\ &\leq C_{m,1}\mathbf{E}_{\tau_1}\varepsilon^{-4m}\left|\int_0^T Y_t|S_t(\tau,X)|^2 dt\right|^{2m} + \\ &+C_{m,2}\mathbf{E}_{\tau_1}\varepsilon^{-2m}\left|\int_0^T Y_t(S_t(\tau,X),dW_t)\right|^{2m}, \end{aligned} \tag{5.5}$$

where we denoted $\tau_1 = \tau + \varepsilon^2 u_1$ and used the inequality $(a+b)^n \leq 2^{n-1}(a^n+b^n)$.

Further, by definition,

$$\begin{aligned} &\int_0^T Y_t|S_t(\tau,X)|^2 dt = \\ &\qquad = \int_0^T Y_t\sum_{j=1}^k [h_t^{(j)}(X) - g_t^{(j)}(X)]\chi_{\{\tau_1^{(j)}\leq t\leq\tau_2^{(j)}\}}]^2 dt = \\ &\qquad = \sum_{j=1}^k \int_{\tau^{(j)}+\varepsilon^2 u_1^{(j)}}^{\tau^{(j)}+\varepsilon^2 u_2^{(j)}} Y_t(V_t^{(j)}(X))^2 dt, \end{aligned} \tag{5.6}$$

$$\int_0^T (Y_t(S_t(\tau,X),dW_t) = \sum_{j=1}^k \int_{\tau^{(j)}+\varepsilon^2 u_1^{(j)}}^{\tau^{(j)}+\varepsilon^2 u_2^{(j)}} Y_t V_t^{(j)}(X) dW_t^{(j)}.$$

This yields

$$\varepsilon^{-4m}\mathbf{E}_{\tau_1}\left|\int_0^T Y_t|S_t(\tau,X)|^2dt\right|^{2m} =$$

$$= \mathbf{E}_{\tau_1}\left|\sum_{j=1}^k \varepsilon^{-2}\int_{\tau^{(j)}+\varepsilon^2u_1^{(j)}}^{\tau^{(j)}+\varepsilon^2u_2^{(j)}} Y_t[V_t^{(j)}(X)]^2dt\right|^{2m} \leq$$

$$\leq C_{m,3}\sum_{j=1}^k \mathbf{E}_{\tau_1}\left(\varepsilon^{-2}\int_{\tau^{(j)}+\varepsilon^2u_1^{(j)}}^{\tau^{(j)}+\varepsilon^2u_2^{(j)}} Y_t[V_t^{(j)}(X)]^2dt\right)^{2m} \leq$$

$$\leq C_{m,3}\sum_{j=1}^k |u_2^{(j)}-u_1^{(j)}|^{2m-1}\mathbf{E}_{\tau_1}\varepsilon^{-2}\int_{\tau^{(j)}+\varepsilon^2u_1^{(j)}}^{\tau^{(j)}+\varepsilon^2u_2^{(j)}} Y_t^{2m}[V_t^{(j)}(X)]^{4m}dt =$$

$$= C_{m,3}\sum_{j=1}^k |u_2^{(j)}-u_1^{(j)}|^{2m-1}\int_{\tau^{(j)}+\varepsilon^2u_1^{(j)}}^{\tau^{(j)}+\varepsilon^2u_2^{(j)}} \mathbf{E}_{\tau_2}[V_t^{(j)}(X)]^{4m}dt \leq$$

$$\leq C\sum_{j=1}^k |u_2^{(j)}-u_1^{(j)}|^{2m}. \tag{5.7}$$

In view of condition I

$$\mathbf{E}_\tau|V_t(X)|^{4m} \leq \mathbf{E}_\tau\left(L_1\int_0^t(1+|X_s|)ds + L_2(1+|X_t|)\right)^{4m} \leq$$

$$\leq 2^{4m-1}L_1^{4m}\mathbf{E}_\tau\left(\int_0^t(1+|X_s|)ds\right)^{4m} + 2^{4m-1}L_2^{4m}\mathbf{E}_\tau(1+|X_t|)^{4m},$$

and the moments $\mathbf{E}_\tau|X_t|^{4m}$ are uniformly bounded on $t\in[0,T]$ (Theorem 1.1).

A similar argument shows that

$$\mathbf{E}_{\tau_1}\varepsilon^{-2m}\left(\int_0^T Y_t(S_t(\tau,X),dW_t)\right)^{2m} \leq C\sum_{j=1}^k |u_2^{(j)}-u_1^{(j)}|^m.$$

So for $|u_2|+|u_1|<R_0$

$$\sup_{\tau\in\mathbf{K}}\mathbf{E}_\tau|Z_\varepsilon^{\frac{1}{2m}}(u_2)-Z_\varepsilon^{\frac{1}{2m}}(u_1)|^{2m} \leq C\sum_{j=1}^k |u_2^{(j)}-u_1^{(j)}|^m(1+(2R_0)^m).$$

Lemma 5.5. *If conditions I and II are fulfilled, then for some $m>2, \tilde\kappa>0$ and $b>0$ for all $|u|>b$,*

$$\sup_{\tau\in\mathbf{K}}\mathbf{E}_\tau Z_\varepsilon^{\frac{1}{m}}(u) \leq 2\exp\{-\tilde\kappa\sum_{j=1}^k|u^{(j)}|\}. \tag{5.8}$$

Proof. First, we note that, by condition II, for $u^{(j)}>0$

$$\sum_{j=1}^k \varepsilon^{-2}\int_{\tau^{(j)}}^{\tau^{(j)}+\varepsilon^2u^{(j)}} V_t^{(j)}(x)^2dt \geq \kappa\sum_{j=1}^k u^{(j)},$$

and for $u^{(j)} < 0$

$$\sum_{j=1}^{k} \varepsilon^{-2} \int_{\tau^{(j)}+\varepsilon^2 u^{(j)}}^{\tau^{(j)}} V_t^{(j)}(x)^2 dt \geq -\kappa \sum_{j=1}^{k} u^{(j)}.$$

Further (for $u^{(j)} > 0$)

$$\varepsilon^{-2} \int_{\tau}^{\tau+\varepsilon^2 u} |V_t(X)|^2 dt = \varepsilon^{-2} \sum_{j=1}^{k} \int_{\tau^{(j)}}^{\tau^{(j)}+\varepsilon^2 u^{(j)}} V_t^{(j)}(X)^2 dt =$$

$$= \sum_{j=1}^{k} \varepsilon^{-2} \int_{\tau^{(j)}}^{\tau^{(j)}+\varepsilon^2 u^{(j)}} V_t^{(j)}(x)^2 dt + \sum_{j=1}^{k} \varepsilon^{-2} \times$$

$$\times \int_{\tau^{(j)}}^{\tau^{(j)}+\varepsilon^2 u^{(j)}} [V_t^{(j)}(X)^2 - V_t^{(j)}(x)^2] dt \geq \kappa \sum_{j=1}^{k} u^{(j)} - C\varepsilon \sum_{j=1}^{k} u^{(j)}.$$

So, as in Lemma 2.3, we have

$$\left(\mathbf{E}_\tau Z_\varepsilon^{\frac{1}{m}}(u)\right)^2 \leq \mathbf{E}_\tau \exp\left\{\frac{m-2}{m} \sum_{j=1}^{k} \int_{\tau^{(j)}}^{\tau^{(j)}+\varepsilon^2 u^{(j)}} |V_t^{(j)}(X)|^2 dt\right\} \leq$$

$$\leq \exp\left\{\frac{-\kappa(m-2)}{m^2} \sum_{j=1}^{k} |u^{(j)}|\right\} \times$$

$$\times \mathbf{E} \exp\left\{\frac{m-2}{m^2} \varepsilon C \sum_{j=1}^{k} |u^{(j)}| \sup_{0 \leq t \leq T} |W_t|\right\} \leq$$

$$\leq \exp\left\{-\kappa \frac{m-2}{m^2} \sum_{j=1}^{k} |u^{(j)}|\right\} (1 + \sqrt{8\pi T} A \exp\left\{\frac{1}{2} A^2 T\right\},$$

where $A = \frac{m-2}{m^2} \varepsilon C \sum_{j=1}^{k} |u^{(j)}|$ and we have used (1.50).

Consequently,

$$\left(\mathbf{E}_\tau Z_\varepsilon^{\frac{1}{m}}(u)\right)^2 \leq 2 \exp\left\{-\kappa \frac{m-2}{m^2} \sum_{j=1}^{k} |u^{(j)}| (1 - \right.$$

$$\left. -\frac{1}{2} \frac{(m-2)}{\kappa m^2} T \varepsilon^2 C^2 \sum_{j=1}^{k} |u^{(j)}|) + \frac{m^2}{\kappa(m-2)} \ln(A\sqrt{8\pi T})\right\}. \tag{5.9}$$

But the value $\varepsilon^2 \sum_{j=1}^{k} |u^{(j)}|$ is bounded (together with $\mathbf{T}$) by the quantity $k\mathbf{T}$, hence choosing sufficiently large m and b, we can always obtain for $|u| > b$ the inequality

$$\mathbf{E}_\tau Z_\varepsilon^{\frac{1}{m}}(u) \leq 2 \exp\left\{-\frac{\kappa(m-2)}{4m^2} \sum_{j=1}^{k} |u^{(j)}|\right\}.$$

The properties of the likelihood ratio field $Z_\varepsilon(u)$ established in Lemmas 5.3 - 5.5, allow us to formulate the asymptotic behavior of the Bayes estimator.

Theorem 5.4. *Let conditions I and II be fulfilled and the prior density $\pi(y), y \in \mathbf{T}$ be positive, continuous and bounded on $\mathbf{T}$, then uniformly on $\tau \in \mathbf{K}$*

$$\mathbf{P}_\tau - \lim_{\varepsilon \to 0} \tilde{\tau}_\varepsilon = \tau,$$

$$\mathcal{L}_\tau\{\varepsilon^{-2}(\tilde{\tau}_\varepsilon - \tau)\} \Longrightarrow \mathcal{L}\{\xi\},$$

$$\lim_{\varepsilon \to 0} \mathbf{E}_\tau |\varepsilon^{-2}(\tilde{\tau}_\varepsilon - \tau)|^p = \mathbf{E}|\xi|^p$$

for any $p \geq 1$.

Proof follows from Lemmas 5.3-5.5 and Theorem 1.5.

Problem 2 (θ-estimation). We are given the same observations (5.2), where the disorder parameter τ is a function of another (unknown) parameter θ and we have to estimate θ by the observations $\{X_t, 0 \leq t \leq T\}$. Let us denote Θ as the set of all possible values of θ, Θ is an open, convex bounded subset of $\mathbf{R}^d$, and the set $\{\tau(\theta), \theta \in \Theta\} \subset (0,T)^k$; $\mathbf{P}_\theta^{(\varepsilon)}$ is a measure induced by the process (5.2) with $\tau = \tau(\theta)$. The MLE $\hat{\theta}_\varepsilon$ is introduced by the formula

$$\hat{\theta}_\varepsilon = \arg \sup_{\theta \in \Theta} \frac{d\mathbf{P}_\theta^{(\varepsilon)}}{d\mathbf{P}^{(\varepsilon)}}(X).$$

We introduce

(a) the matrix

$$\dot{\tau} = \{\dot{\tau}^{(j,i)}\} = \left\{ \frac{\partial \tau^{(j)}}{\partial \theta^{(i)}}, i = 1, \ldots, d;\ j = 1, \ldots, k \right\}$$

and the vector

$$\dot{\tau}^{(j)} = (\dot{\tau}^{(j,1)}, \ldots, \dot{\tau}^{(j,k)})$$

(b) random functions

$$Y(u) = \sum_{j=1}^k r_j W_j((\dot{\tau}^{(j)}(\theta), u)), \qquad u \in R^d,$$

$$Z^\circ(u) = \exp\left\{ Y(u) - \frac{1}{2} \sum_{j=1}^k r_j^2 |(\dot{\tau}^{(j)}(\theta), u)| \right\},$$

where

$$r_j \equiv V_{\tau(j)(\theta)}^{(j)}(x(\theta)) = h_{\tau(j)(\theta)}(x(\theta)) - g_{\tau(j)(\theta)}^{(j)}(x(\theta)),$$

$$(\dot{\tau}^{(j)}(\theta), u) = \sum_{i=1}^d \dot{\tau}^{(j,i)}(\theta) u^{(i)}$$

and

$$W_j(v) = \begin{cases} W_+^{(j)}(v), & v \geq 0 \\ W_-^{(j)}(-v), & v < 0 \end{cases}$$

with independent Wiener processes $W_+^{(j)}(v), W_-^{(j)}(v), u > 0$, $j = 1, \ldots, k$.

So the random functions $Y(u), u \in R^d$ are Gaussian with the properties

$$\mathbf{E}Y(u) = 0, \qquad \mathbf{E}Y(u)^2 = \sum_{j=1}^{k} r_j^2 |(\dot{\tau}^{(j)}(\theta), u)|,$$

$$\mathbf{E}Y(u)Y(v) = \sum_{j=1}^{k} r_j^2 (\dot{\tau}^{(j)}(\theta)u) \wedge (\dot{\tau}^{(j)}(\theta), v)),$$

where $a \wedge b = \min(a, b)$.

The limit random vectors ξ° and ζ° are defined by the equations

$$\xi^\circ = \arg \sup_{u \in R^d} Z^\circ(u),$$

$$\zeta^\circ = \arg \inf_{s \in R^d} \int_{R^d} l(s - y) Z^\circ(y) dy.$$

Here $l(\cdot)$ is a loss function of the polynomial type $l(y) = |y|^a, a > 1$.

The condition of identifiability here is

III. *The components $\tau^{(j)}(\theta)$ of the vector $\tau(\theta)$ have continuous bounded derivatives with respect to θ and*

$$\text{(a)} \quad \min_{1 \le j \le k} \inf_{\theta \in \Theta} \left| V^{(j)}_{\tau^{(j)}(\theta)}(x(\theta)) \right| > 0,$$

$$\text{(b)} \quad \min_{1 \le j \le k} \inf_{\theta \in \Theta} |(\dot{\tau}^{(j)}(\theta), \lambda)| \ge \kappa |\lambda|,$$

where $\kappa > 0$.

Theorem 5.5. *Let conditions I and III be fulfilled then, uniformly on $\theta \in \mathbf{K}$ ($\mathbf{K}$ is a compact in Θ)*

$$\mathbf{P}_\theta - \lim_{\varepsilon \to 0} \hat{\theta}_\varepsilon = \theta,$$

$$\mathcal{L}_\theta\{\varepsilon^{-2}(\hat{\theta}_\varepsilon - \theta)\} \Longrightarrow \mathcal{L}\{\xi^\circ\},$$

$$\lim_{\varepsilon \to 0} \mathbf{E}_\theta |\varepsilon^{-2}(\hat{\theta}_\varepsilon - \theta)|^p = \mathbf{E}|\xi^\circ|^p,$$

for any $p \ge 1$.

Proof. First, we establish the uniform convergence of finite-dimensional distributions of the process (random field)

$$Z_\varepsilon(u) = \frac{d\mathbf{P}^{(\varepsilon)}_{\theta+\varepsilon^2 u}}{d\mathbf{P}^{(\varepsilon)}_\theta}(X), \qquad u \in U_{\theta,\varepsilon} = \{u : \theta + \varepsilon^2 u \in \Theta\}$$

to the finite-dimensional distributions of the process $Z^\circ(u), u \in R^d$ and then we derive inequalities like (5.4) and (5.8).

The above-mentioned convergence follows from the representation

$$\begin{aligned}\ln Z_\varepsilon(u_\varepsilon) &= \frac{1}{\varepsilon}\sum_{j=1}^{k} r_j\left[W^{(j)}_{\tau^{(j)}(\theta+\varepsilon^2 u)} - W^{(j)}_{\tau^{(j)}(\theta)}\right] - \\ &\quad -\frac{1}{2\varepsilon^2}\sum_{j=1}^{k} r_j^2|\tau^{(j)}(\theta+\varepsilon^2 u) - \tau^{(j)}(\theta)| + \psi_1(\theta,u_\varepsilon,\varepsilon) = \\ &= Y(u) - \frac{1}{2}\sum_{j=1}^{k} r_j^2|(\dot\tau^{(j)}(\theta),u)| + \psi_2(\theta,u_\varepsilon,\varepsilon),\end{aligned} \tag{5.10}$$

where $u_\varepsilon \to u$ and the function $\psi_2(\theta,u_\varepsilon,\varepsilon)$ converges to zero uniformly on $\theta\in\mathbf{K}$ and $|u_\varepsilon|<C$ for any compact $\mathbf{K}\subset\Theta$ and $C>0$ i.e., for any $\delta>0$ and $u_\varepsilon\to u$,

$$\lim_{\varepsilon\to 0}\sup_{\theta\in\mathbf{K}} \mathbf{P}^{(\varepsilon)}_\theta\{|\psi_2(\theta,u_\varepsilon,\varepsilon)|>\delta\} = 0.$$

This representation could be obtained using the same arguments as in Lemma 5.3.

From condition III which gives the inequality

$$|\tau(\theta_2)-\tau(\theta_1)| \le C|\theta_2-\theta_1|$$

with some $C>0$, and repeating (5.5), (5.6) and (5.7), we obtain

$$\begin{aligned}\mathbf{E}_\theta\left|Z_\varepsilon^{\frac{1}{2m}}(v_2) - Z_\varepsilon^{\frac{1}{2m}}(v_1)\right|^{2m} &\le C_{R_0}\sum_{j=1}^{k}|(\dot\tau^{(j)}(\tilde\theta),(v_2-v_1))|^m \le \\ &\le \tilde C_{R_0}\|v_2-v_1\|^m, \qquad (|v_1|+|v_2|<R_0).\end{aligned}$$

The last inequality, similar to (5.8), is derived by repeating the calculus of Lemma 5.5, where we use condition III :

$$\varepsilon^{-2}\int_{\tau^{(j)}(\theta)}^{\tau^{(j)}(\theta+\varepsilon^2 v)} V_t^{(j)}(x)^2 dt \ge \frac{\kappa}{\varepsilon^2}|\tau^{(j)}(\theta+\varepsilon^2 v)-\tau^{(j)}(\theta)| \ge \kappa_1|v|.$$

Here the constant $\kappa_1>0$ does not depend of $\theta\in\mathbf{K}\subset\Theta$. It now only remains to make a reference to Theorem 1.4.

Suppose now that θ is a random vector with positive, bounded and continuous prior density $\pi(y), y\in\bar\Theta$, then the Bayes estimators $\tilde\theta_\varepsilon$ have uniformly on $\theta\in\mathbf{K}$, the following properties :

$$\mathbf{P}_\theta - \lim_{\varepsilon\to 0}\tilde\theta_\varepsilon = \theta,$$

$$\mathcal{L}_\theta\{\varepsilon^{-2}(\tilde\theta_\varepsilon-\theta)\} \Longrightarrow \mathcal{L}\{\zeta^\circ\},$$

$$\lim_{\varepsilon\to 0}\mathbf{E}_\theta|\varepsilon^{-2}(\tilde\theta_\varepsilon-\theta)|^p = \mathbf{E}|\zeta^\circ|^p, \qquad p\ge 1.$$

This follows immediately from the above-mentioned properties of $Z_\varepsilon(\cdot)$ and Theorem 1.5.

The minimax low bound on risks of estimators like (5.1) could be also introduced and the LAM of the Bayes estimators could be proved.

Example 5.1. The dependence $\tau = \tau(\theta)$ could be defined by the deterministic equation. Let, for example, $d = 1, k = 1$, and the moment of switching τ be the first exit time of the function $x_s, s \geq 0$ from the region $[x_0, \theta]$. So

$$\frac{dx_t}{dt} = \begin{cases} h_t(x), & x_0 \leq x_t \leq \theta, \\ g_t(x), & \theta < x_t \leq x_T, \end{cases}$$

where $h_t(x) > 0, g_t(x) > 0, 0 \leq t \leq T$, $\theta \in (\alpha, \beta)$ and $x_0 < \alpha$.

Then

$$\theta = x_0 + \int_0^\tau h_s(x)ds$$

and

$$\frac{d\theta}{d\tau} = h_\tau(x), \quad \frac{d\tau}{d\theta} = h_\tau(x)^{-1}, \quad \alpha < \theta < \beta.$$

Hence for $u > 0$

$$\frac{d\mathbf{P}^{(\varepsilon)}_{\theta+\varepsilon^2 u}}{d\mathbf{P}^{(\varepsilon)}_\theta}(X) =$$

$$= \exp\left\{\frac{1}{\varepsilon}\int_{\tau(\theta)}^{\tau(\theta+\varepsilon^2 u)} [h_t(X) - g_t(X)]dW_t - \frac{1}{2\varepsilon^2}\int_{\tau(\theta)}^{\tau(\theta+\varepsilon^2 u)} [h_t(X) - g_t(X)]^2 dt\right\} =$$

$$= \exp\left\{V_{\tau(\theta)}(x)\frac{W(\tau(\theta+\varepsilon^2 u)) - W(\tau(\theta))}{\varepsilon} - \frac{1}{2}V_{\tau(\theta)}(x)^2\frac{\tau(\theta+\varepsilon^2) - \tau(\theta)}{\varepsilon^2} + \psi_\varepsilon\right\} =$$

$$= \exp\left\{V_{\tau(\theta)}(x)\tilde{W}(h^{-1}_{\tau(\theta)}(x)u) - \frac{V_{\tau(\theta)}(x)^2 u}{2h_{\tau(\theta)}(x)} + \tilde{\psi}_\varepsilon\right\} = \exp\left\{r_\theta \hat{W}(u) - \frac{1}{2}r_\theta^2 u + \psi_\varepsilon\right\},$$

where $r_\theta = V_{\tau(\theta)}(x)h_{\tau(\theta)}(x)^{-1/2}$. So we have the same problem as earlier on, but with $\dot{\tau} = h_\tau(x)^{-1}$.

5.3 Misspecified disorder

Let the observed diffusion-type process $X_t, 0 \leq t \leq T$ be

$$\begin{aligned} dX_t &= \left[h_t(X)\chi_{\{t<\tau_0\}} + g_t(X)\chi_{\{t\geq\tau_0\}}\right] dt + \varepsilon dW_t, \\ &\qquad X_0 = x_0, \quad 0 \leq t \leq T, \end{aligned} \tag{5.11}$$

where $\tau_0 \in (\tau_1, \tau_2), 0 < \tau_1 < \tau_2 < T$ is a time of switching and we have to estimate τ, but the functions $h(\cdot)$ and $g(\cdot)$ are not known to the observer. The observer knows that $h_t(x) < g_t(x)$ for all $t \in [\tau_1, \tau_2]$. He has to suppose that the process $X_t, 0 \leq t \leq T$ satisfies the equation

$$\begin{aligned} dX_t &= \left[a_t(X)\chi_{\{t<\tau_0\}} + b_t(X)\chi_{\{t\geq\tau_0\}}\right] dt + \varepsilon dW_t, \\ &\qquad X_0 = x_0, \quad 0 \leq t \leq T, \end{aligned} \tag{5.12}$$

with some known functions $a(\cdot)$ and $b(\cdot)$ such that $a_t(\cdot) < b_t(\cdot)$ for all $t \in [\tau_1, \tau_2]$ and then construct the maximum likelihood estimator

$$\hat{\tau}_\varepsilon = \arg \sup_{\tau_1<\tau<\tau_2} \frac{d\mathbf{P}_\tau^{(\varepsilon)}}{d\mathbf{P}^{(\varepsilon)}}(X)$$

and corresponding Bayes estimators $\tilde{\tau}_\varepsilon$ on the basis of the parametric model (5.12), but substitute there observations (5.11). So, as in section 2.6, we have a misspecified model. What is really surprising here is that under not so restrictive conditions, the MLE and BE are consistent.

We introduce several conditions. The first one is

I. *The functions $h(\cdot), g(\cdot), a(\cdot)$ and $b(\cdot)$ satisfy the condition $\mathcal{L}$.*

Let us denote $x_t = x_t(\tau_0), 0 \leq t \leq T$ the solution of the deterministic switching system

$$\frac{dx_t}{dt} = \begin{cases} h_t(x), & 0 \leq t \leq \tau_0, \quad x_0, \\ g_t(x), & \tau_0 < t \leq T, \end{cases} \tag{5.13}$$

then the second condition will be

II. *The functions $h_t(x), g_t(x), a_t(x), b_t(x)$ are continuous on $t \in [\tau_1, \tau_2]$ and*

$$h_t(x) < \frac{a_t(x) + b_t(x)}{2} < g_t(x),$$

where x_t is the solution of equation (5.13).

Theorem 5.6. *Let conditions I and II be fulfilled, then for any $\nu > 0$, there exist constants $c > 0$ and $\gamma > 0$ such that*

$$\mathbf{P}_{\tau_0}^{(\varepsilon)}\{|\hat{\tau}_\varepsilon - \tau_0| > \nu\} \leq C \exp\left\{-\gamma \varepsilon^{-\frac{2}{3}}\right\}.$$

Proof. By definition of MLE we have

$$\hat{\tau}_\varepsilon = \arg \sup_{\tau_1<\tau<\tau_2} \frac{d\mathbf{P}_\tau^{(\varepsilon)}}{d\mathbf{P}^{(\varepsilon)}}(X) = \arg \sup_{\tau_1<\tau<\tau_2} \frac{d\mathbf{P}_\tau^{(\varepsilon)}}{d\mathbf{P}_{\tau_0}^{(\varepsilon)}}(X).$$

Hence

$$\mathbf{P}_{\tau_0}^{(\varepsilon)}\{|\hat{\tau}_\varepsilon - \tau_0| > \nu\} = \mathbf{P}_{\tau_0}^{(\varepsilon)}\left\{ \sup_{|\tau-\tau_0|\leq\nu} \frac{d\mathbf{P}_\tau^{(\varepsilon)}}{d\mathbf{P}_{\tau_0}^{(\varepsilon)}}(X) < \sup_{|\tau-\tau_0|>\nu} \frac{d\mathbf{P}_\tau^{(\varepsilon)}}{d\mathbf{P}_{\tau_0}^{(\varepsilon)}}(X) \right\} \leq$$

$$\leq \mathbf{P}_{\tau_0}^{(\varepsilon)}\left\{ \sup_{|\tau-\tau_0|>\nu} \frac{d\mathbf{P}_\tau^{(\varepsilon)}}{d\mathbf{P}_{\tau_0}^{(\varepsilon)}}(X) > 1 \right\} =$$

$$= \mathbf{P}_{\tau_0}^{(\varepsilon)}\left\{ \sup_{|\tau-\tau_0|>\nu} \left(\frac{1}{\varepsilon^2} \int_{\tau\wedge\tau_0}^{\tau\vee\tau_0} [a_t(X) - b_t(X)] dX_t - \right.\right.$$

$$
- \frac{1}{2\varepsilon^2} \int_{\tau \wedge \tau_0}^{\tau \vee \tau_0} [a_t(X)^2 - b_t(X)^2] dt \Big) > 0 \Big\} \leq
$$

$$
\leq \mathbf{P}_{\tau_0}^{(\varepsilon)} \Big\{ \sup_{\tau > \tau_0 + \nu} \Big(\int_{\tau_0}^{\tau} [a_t(X) - b_t(X)] dW_t -
$$

$$
- \frac{1}{2\varepsilon} \int_{\tau_0}^{\tau} ([a_t(X) - g_t(X)]^2 - [b_t(X) - g_t(X)]^2) dt \Big) > 0 \Big\} +
$$

$$
+ \mathbf{P}_{\tau_0}^{(\varepsilon)} \Big\{ \sup_{\tau < \tau_0 - \nu} \Big(\int_{\tau_0}^{\tau} [b_t(X) - a_t(X)] dW_t -
$$

$$
- \frac{1}{2\varepsilon} \int_{\tau_0}^{\tau} ([b_t(X) - h_t(X)]^2 - [a_t(X) - h_t(X)]^2) dt \Big) > 0 \Big\}.
$$

Denote

$$
V_t(X) = [a_t(X) - g_t(X)]^2 - [b_t(X) - g_t(X)]^2
$$

and consider the last probability

$$
\mathbf{P}_{\tau_0}^{(\varepsilon)} \Big\{ \sup_{\tau > \tau_0 + \nu} \Big(\int_{\tau_0}^{\tau} [a_t(X) - b_t(X)] dW_t -
$$

$$
- \frac{1}{2\varepsilon} \int_{\tau_0}^{\tau} V_t(X) dt - \frac{1}{4\varepsilon} \int_{\tau_0}^{\tau} V_t(x) dt + \frac{1}{4\varepsilon} \int_{\tau_0}^{\tau} V_t(x) dt \Big) > 0 \Big\} \leq
$$

$$
\leq \mathbf{P}_{\tau_0}^{(\varepsilon)} \Big\{ \sup_{\tau_0 + \nu < \tau < \tau_2} \Big(\int_{\tau_0}^{\tau} [a_t(X) - b_t(X)] dW_t -
$$

$$
- \frac{1}{4\varepsilon} \int_{\tau_0}^{\tau} V_t(x) dt \Big) > 0 \Big\} +
$$

$$
+ \mathbf{P}_{\tau_0}^{(\varepsilon)} \Big\{ \sup_{\tau_0 + \nu < \tau < \tau_2} \Big(\frac{1}{2\varepsilon} \int_{\tau_0}^{\tau} [V_t(x) - V_t(X)] dt -
$$

$$
- \frac{1}{4\varepsilon} \int_{\tau_0}^{\tau} V_t(x) dt \Big) > 0 \Big\} \leq
$$

$$
\leq \mathbf{P}_{\tau_0}^{(\varepsilon)} \Big\{ \sup_{\tau_0 + \nu < \tau < \tau_2} \int_{\tau_0}^{\tau} [a_t(X) - b_t(X)] dW_t >
$$

$$
> \frac{1}{4\varepsilon} \inf_{\tau_0 + \nu < \tau < \tau_2} \int_{\tau_0}^{\tau} V_t(x) dt \Big\} +
$$

$$
+ \mathbf{P}_{\tau_0}^{(\varepsilon)} \Big\{ \frac{1}{\varepsilon} \int_{\tau_0}^{\tau_2} |V_t(X) - V_t(x)| dt > \frac{1}{2\varepsilon} \inf_{\tau_0 + \nu < \tau < \tau_2} \int_{\tau_0}^{\tau} V_t(x) dt \Big\}.
$$

Denote

$$
\kappa_1 = \inf_{\tau_1 < t < \tau_2} V_t(x)
$$

and remember that by condition II, $\kappa_1 > 0$ because

$$
\begin{aligned}
V_t(x) &= [a_t(x) - b_t(x)]\,[a_t(x) + b_t(x) - 2g_t(x)] = \\
&= 2\,[b_t(x) - a_t(x)] \left(g_t(x) - \frac{a_t(x) + b_t(x)}{2} \right).
\end{aligned}
$$

Hence, continuing this chain of inequalities, we have

$$
\leq \mathbf{P}_{\tau_0}^{(\varepsilon)} \Big\{ \sup_{\tau_0 + \nu < \tau < \tau_2} \int_{\tau_0}^{\tau} [a_t(X) - b_t(X)] dW_t > \frac{\kappa_1 \nu}{4\varepsilon} \Big\} +
$$

$$+\mathbf{P}\left\{CT\sup_{0\le t\le T}|W_t|>\frac{\kappa_1\nu}{2\varepsilon}\right\}\le$$

$$\le C_1\exp\left\{-\frac{\kappa_1^2\nu^{\frac{2}{3}}}{\varepsilon^{\frac{2}{3}}}\right\}+2\exp\left\{-\frac{\kappa_1^2\nu^2}{8C^2T^3\varepsilon^2}\right\},$$

where we have used inequality (1.19). Note also that

$$\mathbf{E}\exp\left\{\left(\int_0^T[a_t(X)-b_t(X)]^2dt\right)^{\frac{1}{2}}\right\}\le$$

$$\le\mathbf{E}\exp\left\{\left(\int_0^T(A+B\sup_{0\le s\le t}|W_s|^2)dt\right)^{\frac{1}{2}}\right\}<\infty$$

and the condition of Lemma 1.7 is fulfilled with $p=2$.

REMARK 5.3. Introduce the function

$$\begin{aligned}G(\tau) &= \int_\tau^{\tau_0}\left[(b_t(x)-h_t(x))^2-(a_t(x)-h_t(x))^2\right]\,dt\,\chi_{\{\tau<\tau_0\}}+\\ &+\int_{\tau_0}^{\tau}\left[(a_t(x)-g_t(x))^2-(b_t(x)-g_t(x))^2\right]\,dt\,\chi_{\{\tau\ge\tau_0\}}.\end{aligned}$$

The consistency of the MLE is a consequence of the equality

$$\tau_0=\arg\inf_{\tau\in(\tau_1,\tau_2)}G(\tau).$$

In general case if Condition II is not satisfied the MLE will converge to the value

$$\tau^*=\arg\inf_{\tau\in(\tau_1,\tau_2)}G(\tau).$$

To study the asymptotic behavior of the normed difference $\varepsilon^{-2}(\hat\tau_\varepsilon-\tau_0)$ we need an additional condition

III. *The functions $h_t(x), g_t(x), a_t(x)$ and $b_t(x)$ are continuous on $t\in[\tau_1,\tau_2]$ and*

$$h_t(x)\le a_t(x)<b_t(x)\le g_t(x)$$

for all $t\in[0,T]$.

Introduce the notations:

(a) process

$$Z(u)=\begin{cases}\exp\left\{rW_+(u)-\frac{1}{2}r_1^2u\right\}, & u\ge 0\\ \exp\left\{-rW_-(-u)+\frac{1}{2}r_2^2u\right\}, & u<0,\end{cases}$$

where $W_+(\cdot)$ and $W_-(\cdot)$ are independent Wiener processes;

(b) constants

$$\kappa_1 = \inf_{\tau_1\leq t\leq\tau_2}[g_t(x) - b_t(x)],$$
$$\kappa_2 = \inf_{\tau_1\leq t\leq\tau_2}[b_t(x) - a_t(x)],$$
$$\kappa_3 = \inf_{\tau_1\leq t\leq\tau_2}[a_t(x) - h_t(x)],$$
$$r = a_{\tau_0}(x) - b_{\tau_0}(x),$$
$$r_1^2 = [a_{\tau_0}(x) - g_{\tau_0}(x)]^2 - [b_{\tau_0}(x) - g_{\tau_0}(x)]^2,$$
$$r_2^2 = [b_{\tau_0}(x) - h_{\tau_0}(x)]^2 - [a_{\tau_0}(x) - h_{\tau_0}(x)]^2;$$

(c) random variable

$$\xi = \arg\sup_u Z(u).$$

Theorem 5.7. *Let conditions I and III be fulfilled, then*

$$\mathbf{P}_{\tau_0} - \lim_{\varepsilon\to 0}\hat{\tau}_\varepsilon = \tau_0,$$

$$\mathcal{L}_{\tau_0}\left\{\varepsilon^{-2}(\hat{\tau}_\varepsilon - \tau)\right\} \Longrightarrow \mathcal{L}\{\xi\}.$$

Proof. The likelihood ratio process $Z_\varepsilon(u)$ on observations (5.11) could be written in the form $(u \geq 0)$

$$\begin{aligned} Z_\varepsilon(u) &= \frac{d\mathbf{P}^{(\varepsilon)}_{\tau_0+\varepsilon^2 u}}{d\mathbf{P}^{(\varepsilon)}_{\tau_0}}(X) = \exp\left\{\frac{1}{\varepsilon}\int_{\tau_0}^{\tau_0+\varepsilon^2 u}[a_t(X) - b_t(X)]dW_t - \right. \\ &\quad \left. -\frac{1}{2\varepsilon^2}\int_{\tau_0}^{\tau_0+\varepsilon^2 u}\left([a_t(X) - g_t(X)]^2 - [b_t(X) - g_t(X)]^2\right)dt\right\} = \\ &= \exp\left\{\frac{1}{\varepsilon}\int_{\tau_0}^{\tau_0+\varepsilon^2 u}[a_t(X) - b_t(X)]dW_t - \right. \\ &\quad \left. -\frac{1}{2\varepsilon^2}\int_{\tau_0}^{\tau_0+\varepsilon^2 u}\left([a_t(x) - g_t(x)]^2 - [b_t(x) - g_t(x)]^2\right)dt + \tilde{r}_\varepsilon(u)\right\} = \\ &= \exp\left\{\frac{r}{\varepsilon}(W_{\tau_0+\varepsilon^2 u} - W_{\tau_0}) - \frac{1}{2}r_1^2 u + r_\varepsilon(u)\right\}, \end{aligned} \tag{5.14}$$

where $r_\varepsilon(u_\varepsilon)$ converges to zero in probability for any bounded sequence u_ε. So the convergence of finite-dimensional distributions of $Z_\varepsilon(u), u \geq 0$ to the finite-dimensional distributions of $Z(u), u \geq 0$ is established. A similar argument shows the same convergence for $u < 0$ also.

Lemma 5.6. *Under established assumptions, there exists a constant $p \in (0, \frac{1}{2})$ such that*

$$\sup_{\varepsilon\in(0,1]}\sup_{\tau\in(\tau_1,\tau_2)}\mathbf{E}_{\tau_0}\left(\frac{d\mathbf{P}^{(\varepsilon)}_\tau}{d\mathbf{P}^{(\varepsilon)}_{\tau_0}}\right)^{2p} < \infty.$$

Proof. Denote $f_t = 2p\varepsilon^{-1}[a_t(X) - b_t(X)]$ and prove this inequality for $\tau > \tau_0$. We have

$$
\begin{aligned}
\mathbf{E}_{\tau_0}\left(\frac{d\mathbf{P}^{(\varepsilon)}_\tau}{d\mathbf{P}^{(\varepsilon)}_{\tau_0}}\right)^p &= \mathbf{E}_{\tau_0}\exp\left\{\frac{p}{\varepsilon}\int_{\tau_0}^{\tau}[a_t(X) - b_t(X)]dW_t - \right.\\
&\left. -\frac{p}{2\varepsilon^2}\int_{\tau_0}^{\tau}\left([a_t(X) - g_t(X)]^2 - [b_t(X) - g_t(X)]^2\right)dt\right\} =\\
&= \mathbf{E}_{\tau_0}\exp\left\{\frac{1}{2}\int_{\tau_0}^{\tau} f_t dW_t - \frac{1}{4}\int_{\tau_0}^{\tau} f_t^2 dt - \right.\\
&-\frac{p}{2\varepsilon^2}\int_{\tau_0}^{\tau}\left([a_t(X) - g_t(X)]^2 - [b_t(X) - g_t(X)]^2\right)dt +\\
&\left. +\frac{p^2}{\varepsilon^2}\int_{\tau_0}^{\tau}[a_t(X) - b_t(X)]^2 dt\right\} \le\\
&\le \left(\mathbf{E}_{\tau_0}\exp\left\{\int_{\tau_0}^{\tau} f_t dW_t - \frac{1}{2}\int_{\tau_0}^{\tau} f_t^2 dt\right\}\right)^{\frac{1}{2}} \times\\
&\times\left(\mathbf{E}_{\tau_0}\exp\left\{-\frac{p}{\varepsilon^2}\int_{\tau_0}^{\tau}\left([a_t(X) - g_t(X)]^2 - \right.\right.\right.\\
&\left.\left.\left. - [b_t(X) - g_t(X)]^2 - 2p[a_t(X) - b_t(X)]^2\right)dt\right\}\right)^{\frac{1}{2}}.
\end{aligned}
$$

The first mathematical expectation in the last expression is equal to 1, so we have to consider the last one where we change $2p$ by 1. Hence,

$$
\begin{aligned}
&\mathbf{E}_{\tau_0}\exp\left\{-\frac{p}{\varepsilon^2}\int_{\tau_0}^{\tau}[a_t(X)^2 + g_t(X)^2 - 2a_t(X)g_t(X) - b_t(X)^2 + \right.\\
&\qquad\left. + 2b_t(X)g_t(X) - g_t(X)^2 - a_t(X)^2 + 2a_t(X)b_t(X) - b_t(X)^2]dt\right\} =\\
&= \mathbf{E}_{\tau_0}\exp\left\{-\frac{2p}{\varepsilon^2}\int_{\tau_0}^{\tau}[b_t(X) - a_t(X)][g_t(X) - b_t(X)]dt\right\} \le\\
&\le \mathbf{E}_{\tau_0}\exp\left\{-\frac{2p}{\varepsilon^2}\int_{\tau_0}^{\tau}[b_t(x) - a_t(x)][g_t(x) - b_t(x)]dt + \right.\\
&\qquad\left. + p\frac{C_1(\tau - \tau_0)}{\varepsilon}\sup_{0\le t\le T}|W_t| + C_2(\tau - \tau_0)p\sup_{0\le t\le T}|W_t|^2\right\} \le\\
&\le \exp\left\{-\frac{2p}{\varepsilon^2}\int_{\tau_0}^{\tau}[b_t(x) - a_t(x)][g_t(x) - b_t(x)]dt\right\} \times\\
&\qquad\times\left(\mathbf{E}\exp\left[\frac{2C_1(\tau - \tau_0)p}{\varepsilon}\sup_{0\le t\le T}|W_t|\right]\right)^{\frac{1}{2}} \times\\
&\qquad\times\left(\mathbf{E}\exp[2C_2(\tau - \tau_0)p\sup_{0\le t\le T}|W_t|^2]\right)^{\frac{1}{2}} \le\\
&\le \exp\left\{-\frac{2p}{\varepsilon^2}\int_{\tau_0}^{\tau}[b_t(x) - a_t(x)][g_t(x) - b_t(x)]dt + \frac{2C_1^2(\tau_2 - \tau_1)^2p^2}{T\varepsilon^2} + K_1\right\} \le\\
&\le \exp\left\{-\frac{2p}{\varepsilon^2}(\tau - \tau_0)\kappa_1\kappa_2 + \frac{2C_1^2(\tau_2 - \tau_1)^2p^2}{T\varepsilon^2} + K_1\right\},
\end{aligned}
$$

where we have used the Schwarz inequality and the properties (1.50), (1.51) of the Wiener process. The constant $K_1 = K_1(p) < \infty$ for sufficiently small p (see (1.51). So, if we set $p \in (0, p_0)$, where

$$p_0 = (2\sqrt{2}C_2T\sqrt{T})^{-1},$$

then the assertion of Lemma 5.6 for $\tau \geq \tau_0$ follows if we take $p_1 < 2p_0$.

Analogously, for $\tau < \tau_0$ we have

$$\mathbf{E}_{\tau_0}\left(\frac{d\mathbf{P}_\tau^{(\varepsilon)}}{d\mathbf{P}_{\tau_0}^{(\varepsilon)}}(X)\right)^p = \mathbf{E}_{\tau_0}\exp\left\{\frac{p}{\varepsilon}\int_\tau^{\tau_0}[b_t(X)-a_t(X)]dW_t- \right.$$
$$\left. -\frac{p}{2\varepsilon^2}\int_{\tau_0}^{\tau}\left([b_t(X)-h_t(X)]^2-[a_t(X)-h_t(X)]^2\right)dt\right\} \leq$$
$$\leq \exp\left\{-\frac{2p}{\varepsilon^2}\int_\tau^{\tau_0}[b_t(x)-a_t(x)][a_t(x)-h_t(x)]dt+\right.$$
$$\left.+\frac{2C_1^2(\tau_2-\tau_1)^2}{T\varepsilon^2}p^2+K_2\right\}$$

and the constant p_2 is chosen in a similar way. Therefore, the assertion of Lemma 5.3 is valid for $p = \min(p_1, p_2)$.

Let us set $Z_\varepsilon^*(u) = Z_\varepsilon(u)^p$, where p is taken from Lemma 5.3.

Lemma 5.7. *Under previously made assumptions,*

$$\sup_{|u_1|+|u_2|<R_0} |u_2-u_1|^2\mathbf{E}_{\tau_0}|Z_\varepsilon^*(u_2)^{\frac{1}{4}} - Z_\varepsilon^*(u_1)^{\frac{1}{4}}|^4 \leq C(1+R_0^2)$$

for $u_2 > u_1 > 0$ and $0 > u_2 > u_1$.

Proof. As in the proof of Lemma 5.4, we introduce a process

$$Y_{\tau_2} = \left(Z_\varepsilon^*(u_2)Z_\varepsilon^*(u_1)^{-1}\right)^{\frac{1}{4}}$$

and for $Y_t, 0 \leq t \leq T$ by Itô formula, we have the representation

$$dY_t = \left(-\frac{p}{8\varepsilon^2}V_t(X)Y_t + \frac{p^2}{32\varepsilon^2}[a_t(X)-b_t(X)]Y_t\right)dt +$$
$$+Y_t\frac{p}{4\varepsilon}[a_t(X)-b_t(X)]dW_t, \quad Y_0 = 1.$$

This implies

$$Y_{\tau_2} = 1 + \int_{\tau_1}^{\tau_2} Y_s\left[-\frac{p}{8\varepsilon^2}V_t(X) + \frac{p^2}{32\varepsilon^2}[a_t(X)-b_t(X)]\right]dt +$$
$$+\frac{p}{4\varepsilon}\int_{\tau_1}^{\tau_2} Y_t[a_t(X)-b_t(X)]dW_t.$$

Hence

$$\mathbf{E}_{\tau_0}|Z_\varepsilon^*(u_2)^{\frac14} - Z_\varepsilon^*(u_1)^{\frac14}|^4 = \mathbf{E}_{\tau_0} Z_\varepsilon^*(u_2)|1-Y_{\tau_2}|^4 =$$
$$= \frac{p}{8}\mathbf{E}_{\tau_0} Z_\varepsilon^*(u_2)\left|\frac{1}{\varepsilon^2}\int_{\tau_1}^{\tau_2} Y_t[V_t(X) - \frac{p}{4}[a_t(X)-b_t(X)]]dt + \right.$$
$$\left. + \frac{2}{\varepsilon}\int_{\tau_1}^{\tau_2} Y_t[a_t(X)-b_t(X)]dW_t\right|^4 \le$$
$$\le C_1\mathbf{E}_{\tau_0} Z_\varepsilon^*(u_2)\left|\frac{1}{\varepsilon^2}\int_{\tau_1}^{\tau_2} Y_t R_t(X)dt\right|^4 +$$
$$+ C_2\mathbf{E}_{\tau_0} Z_\varepsilon^*(u_2)\left|\frac{1}{\varepsilon}\int_{\tau_1}^{\tau_2} Y_t H_t(X)dW_t\right|^4,$$

where we took the obvious notations. Below, several times we will use the Schwarz inequality

$$\left(\mathbf{E}_{\tau_0} Z_\varepsilon^*(u_2)\left|\frac{1}{\varepsilon^2}\int_{\tau_1}^{\tau_2} Y_t R_t(X)dt\right|^4\right)^2 \le$$
$$\le \mathbf{E}_{\tau_0} Z_\varepsilon^*(u_2)^2 \mathbf{E}_{\tau_0}\left(\frac{1}{\varepsilon^2}\int_{\tau_1}^{\tau_2} Y_t R_t(X)dt\right)^8 \le$$
$$\le \frac{C(\tau_2-\tau_1)^7}{\varepsilon^{14}}\frac{1}{\varepsilon^2}\int_{\tau_1}^{\tau_2}\mathbf{E}_{\tau_0} Y_t^8 R_t(X)^8 dt \le C|u_2-u_1|^8,$$
$$\left(\mathbf{E}_{\tau_0} Z_\varepsilon^*(u_2)\left(\frac{1}{\varepsilon}\int_{\tau_1}^{\tau_2} Y_t H_t(X)dW_t\right)^4\right)^2 \le$$
$$\le \mathbf{E}_{\tau_0} Z_\varepsilon^*(u_2)^2 \mathbf{E}_{\tau_0}\left(\frac{1}{\varepsilon}\int_{\tau_1}^{\tau_2} Y_t H_t(X)dW_t\right)^8 \le$$
$$\le \frac{C(\tau_2-\tau_1)^3}{\varepsilon^{6}}\frac{1}{\varepsilon^2}\int_{\tau_1}^{\tau_2}\mathbf{E}_{\tau_0} Y_t^8 H_t(X)^8 dt \le C|u_2-u_1|^4.$$

And, finally for $u_2 \ge u_1 \ge 0$, we have

$$\mathbf{E}_{\tau_0}|Z_\varepsilon^*(u_2)^{\frac14} - Z_\varepsilon^*(u_1)^{\frac14}|^4 \le C_1|u_2-u_1|^4 + C_2|u_2-u_1|^2.$$

This yields the assertion of Lemma 5.7 for $u_2 > u_1 > 0$. Analogously, we obtain the same inequality for $u_1 < u_2 < 0$.

Consider now the process $Z_\varepsilon^*(u)$ for $u \ge 0$ and $u \le 0$ separately. According to the representation (5.14) and Lemmas 5.6 and 5.7, the conditions of Theorem 1.4 are fulfilled for every component ($\{Z_\varepsilon^*(u), u \ge 0\}$ and $\{Z_\varepsilon^*(u), u \le 0\}$). As follows from the proof of Theorem 1.4, the stochastic processes $\{Z_\varepsilon^*(u), u \ge 0\}$ and $\{Z_\varepsilon^*, u \le 0\}$ weakly converge in $\mathcal{C}[0,\infty)$ and $\mathcal{C}(-\infty,0]$ correspondingly to the random processes $\{Z^p(u), u \ge 0\}$ and $\{Z^p(u), u \le 0\}$. Hence, $Z_\varepsilon^*(\cdot)$ converges weakly to $Z^p(\cdot)$ in $\mathcal{C}(-\infty,+\infty)$. This implies

$$\mathbf{P}_{\tau_0}^{(\varepsilon)}\{\varepsilon^{-2}(\hat\tau_\varepsilon - \tau_0) < x\} = \mathbf{P}_{\tau_0}^{(\varepsilon)}\{\sup_{u<x} Z_\varepsilon^*(u) > \sup_{u>x} Z_\varepsilon^*(u)\} \longrightarrow$$
$$\longrightarrow \mathbf{P}\{\sup_{u<x} Z(u)^p > \sup_{u>x} Z(u)^p\} = \mathbf{P}\{\xi < x\}.$$

and the theorem is proved.

REMARK 5.4. Suppose that τ is a random variable with positive continious on $[\tau_1, \tau_2]$ density function $\pi(\cdot)$. The Bayes estimator $\tilde{\tau}_\varepsilon$ for quadratic loss function admits the representation

$$\tilde{\tau}_\varepsilon = \tau_0 + \varepsilon^2 \frac{\int_{U_\varepsilon} u\, \pi(\tau_0 + \varepsilon^2 u) Z_\varepsilon(u)\, du}{\int_{U_\varepsilon} \pi(\theta_0 + \varepsilon^2 u) Z_\varepsilon(u)\, du}$$

and its consistence could be established under the same conditions as in Theorem 5.6.

Chapter 6

Partially Observed Systems

We consider several problems of two-dimensional system identification when only one component of the system is observed. For linear diffusion-type processes, it is possible (using Kalman filtration) to construct the MLE and BE of unknown parameters and to describe their properties in regular (Section 6.1) and disorder (Section 6.3) cases. For nonlinear systems, calculation of the likelihood ratio is difficult, so we use some approximations of the original system and then apply the approach of a "misspecified model". This allows us to construct pseudo-MLE's and pseudo-BE's and to prove their consistence and asymptotic normality (Section 6.2). We shall return to the identification of partially observed systems in Section 7.5, where we will use the minimum distance approach.

6.1 Kalman filter identification

We are given a complete probability space $\{\Omega, \mathcal{F}, \mathbf{P}\}$, family $\hat{\mathcal{F}}$, an open, convex, bounded set $\Theta \in \mathbf{R}^d$, two independent Wiener processes $\{W_t, \mathcal{F}_t, 0 \leq t \leq T\}$ and $\{V_t, \mathcal{F}_t, 0 \leq t \leq T\}$ and two-dimensional diffusion-type process $\{X_t, Y_t, 0 \leq t \leq T\}$ with the observed component

$$dX_t = f_t(\theta)Y_t dt + \varepsilon dW_t, \quad X_0 = 0, \quad 0 \leq t \leq T. \tag{6.1}$$

Here the process $Y_t, 0 \leq t \leq T$ is unobserved and also has a linear representation

$$dY_t = b_t(\theta)Y_t dt + \varepsilon\sigma_t(\theta)dV_t, \quad Y_0 = y_0 \neq 0, 0 \leq t \leq T. \tag{6.2}$$

The measurable functions $f_t(\cdot), b_t(\cdot)$ and $\sigma_t(\cdot)$ are known to the observer.

According to the "innovation theorem" (1.32), the process $X_t, 0 \leq t \leq T$ also admits the representation as a diffusion-type process

$$dX_t = S_t(\theta, X)dt + \varepsilon d\bar{W}_t, \quad X_0 = 0, \quad 0 \leq t \leq T, \tag{6.3}$$

where $S_t(\theta, X) = f_t(\theta)m_t(\theta)$ and $m_t(\theta)$ is the conditional mathematical expectation

$$m_t(\theta) = \mathbf{E}_\theta\{Y_t | X_s, \ 0 \leq s \leq t\}.$$

The Wiener process $\bar{W}_t, 0 \leq t \leq T$ is defined by (6.3). So the representation (6.3) is similar to that considered in this work for diffusion-type processes, but the realizations of the Wiener process $\bar{W}_t, 0 \leq t \leq T$ depend on $\varepsilon \in (0,1]$.

We are interested in the asymptotic properties of MLE $\hat{\theta}_\varepsilon$ and BE $\tilde{\theta}_\varepsilon$ as $\varepsilon \to 0$.

We begin with the local asymptotic normality of the corresponding family of measures $\{\mathbf{P}_\theta^{(\varepsilon)}, \theta \in \Theta\}$ induced by the realizations of (6.1) and then prove the consistence, asymptotic normality, and LAM of the mentioned estimators.

Let us introduce the following notations:

function $y_t = y_t(\theta)$:

$$y_t = y_0 \exp\left\{\int_0^t b_s(\theta) ds\right\}, \quad 0 \leq t \leq T;$$

vector-function:

$$\begin{aligned}\dot{y}_t(\theta) &= \int_0^t [\dot{b}_s(\theta) - f_s(\theta)\dot{f}_s(\theta)z_s(\theta)]y_s(\theta) \times \\ &\quad \times \exp\left\{\int_s^t [b_v(\theta) - f_v(\theta)^2 z_v(\theta)]dv\right\} ds, \qquad (6.4)\end{aligned}$$

where $\dot{b}(\cdot)$ and $\dot{f}(\cdot)$ are the derivatives of $b(\cdot)$ and $f(\cdot)$ with respect to θ and the function $z_t(\theta)$ satisfies the differential *Riccati equation*

$$\frac{\partial}{\partial t} z_t(\theta) = 2b_t(\theta)z_t(\theta) - f_t(\theta)^2 z_t(\theta)^2 + \sigma_t(\theta)^2, \quad z_0(\theta) = 0; \qquad (6.5)$$

matrix:

$$I(\theta) = \int_0^T (\dot{f}_t(\theta)y_t(\theta) + f_t(\theta)\dot{y}_t(\theta))(\dot{f}_t(\theta)y_t(\theta) + f_t(\theta)\dot{y}_t(\theta))^T dt.$$

Note, that here $\dot{y}_t(\theta) \neq \frac{\partial}{\partial\theta} y_t(\theta)$.

Theorem 6.1. *Let the functions $f_t(\theta), b_t(\theta), \sigma_t(\theta), 0 \leq t \leq T, \theta \in \Theta$ be continuous and bounded with their first two derivatives with respect to θ and the matrix $I(\theta)$ be positive definite, then the family of measures $\{\mathbf{P}_\theta^{(\varepsilon)}, \theta \in \Theta\}$ is LAN in Θ with the normalizing matrix $\varphi_\varepsilon(\theta) = \varepsilon I(\theta)^{-\frac{1}{2}}$ and the vector*

$$\Delta(\theta, x) = I(\theta)^{-\frac{1}{2}} \int_0^T [\dot{f}_t(\theta)y_t(\theta) + f_t(\theta)\dot{y}_t(\theta)]d\bar{W}_t. \qquad (6.6)$$

Proof. The processes $X_t, 0 \leq t \leq T$ and $Y_t, 0 \leq t \leq T$ are Gaussian and continuous, so all the measures $\{\mathbf{P}_\theta^{(\varepsilon)}, \theta \in \Theta\}$ are equivalent. The condition $\mathcal{L}$ follows from the linearity of these processes. We now check the other conditions of Theorem 2.1.

According to the theory of optimal filtration, the process $m_t(\theta), 0 \leq t \leq T$ satisfies the equation

$$\begin{aligned}dm_t(\theta) &= b_t(\theta)m_t(\theta)dt + \varepsilon^{-2}\gamma_t(\theta)f_t(\theta) \times \\ &\quad \times [dX_t - f_t(\theta)m_t(\theta)dt], \quad m_0(\theta) = y_0, \quad 0 \leq t \leq T, \qquad (6.7)\end{aligned}$$

where the function $\gamma_t(\theta) = \mathbf{E}_\theta(m_t(\theta) - Y_t(\theta))^2$ is a solution of the Riccati equation

$$\begin{aligned}\frac{\partial \gamma_t(\theta)}{\partial t} &= 2b_t(\theta)\gamma_t(\theta) - \varepsilon^{-2}\gamma_t(\theta)^2 f_t(\theta)^2 + \varepsilon^2\sigma_t(\theta)^2,\\ &\gamma_0 = 0, \qquad 0 \le t \le T.\end{aligned} \tag{6.8}$$

This couple of equations (6.7) and (6.8) is called Kalman-Bucy or *Kalman filter*. So if we divide this equation by ε^2 and denote $z_t(\theta) = \gamma_t(\theta)\varepsilon^{-2}$, then we obtain, for $z_t(\theta)$, Equation (6.5) and for $m_t(\theta), 0 \le t \le T$ the representations

$$dm_t(\theta) = b_t(\theta)m_t(\theta)dt + z_t(\theta)f_t(\theta)[dX_t - f_t(\theta)m_t(\theta)dt], \tag{6.9}$$

$$dm_t(\theta) = b_t(\theta)m_t(\theta)dt + \varepsilon z_t(\theta)f_t(\theta)d\bar{W}_t, \quad m_0(\theta) = y_0. \tag{6.10}$$

Denote

$$\begin{aligned}Q_t(\theta) &= b_t(\theta) - z_t(\theta)f_t(\theta)^2\\ R_t(\theta) &= z_t(\theta)f_t(\theta)\end{aligned}$$

and

$$\begin{aligned}\Delta Q_t(h) &= Q_t(\theta + h) - Q_t(\theta) - (h, \dot{Q}_t(\theta)),\\ \Delta R_t(h) &= R_t(\theta + h) - R_t(\theta) - (h, \dot{R}_t(\theta)),\end{aligned}$$

where $\dot{Q}_t(\theta)$ and $\dot{R}_t(\theta)$ are the vectors of derivatives of $Q_t(\theta)$ and $R_t(\theta)$ with respect to θ.

Then we can write

$$dm_t(\theta) = Q_t(\theta)m_t(\theta)dt + R_t(\theta)dX_t, \quad m_0(\theta) = y_0$$

and

$$dm_t(\theta + h) = Q_t(\theta + h)m_t(\theta + h)dt + R_t(\theta + h)dX_t, \quad m_0(\theta + h) = y_0$$

and for the difference

$$\Delta m_t(h) = m_t(\theta + h) - m_t(\theta) - (h, \dot{m}_t(\theta)),$$

where the vector-process $\dot{m}_t(\theta), 0 \le t \le T$ is a solution of the system

$$d\dot{m}_t(\theta) = [\dot{Q}_t(\theta)m_t(\theta) + Q_t(\theta)\dot{m}_t(\theta)]dt + \dot{R}_t(\theta)dX_t, \dot{m}_t(\theta) = 0. \tag{6.11}$$

We obtain the representation

$$\begin{aligned}d\Delta m_t(h) = {} & [Q_t(\theta + h)m_t(\theta + h) - Q_t(\theta)m_t(\theta) -\\ & -(h, \dot{Q}_t(\theta))m_t(\theta) - Q_t(\theta)(\dot{m}_t(\theta), h)]dt + [R_t(\theta + h) -\\ & -R_t(\theta) - (h, \dot{R}_t(\theta))]dX_t.\end{aligned}$$

Hence

$$d\Delta m_t(h) = [Q_t(\theta+h)\Delta m_t(h) + \Delta Q_t(h)m_t(\theta) + \\ +(Q_t(\theta+h) - Q_t(\theta))(h, \dot{m}_t(\theta))]dt + \Delta R_t(h)dX_t. \qquad (6.12)$$

The boundedness of the derivatives allows us to derive estimates

$$\sup_{0\le t\le T} |\Delta Q_t(h)| \le C_1|h|^2, \quad \sup_{0\le t\le T} |\Delta R_t(h)| \le C_2|h|^2$$

with some constants $C_i > 0$, and so, using Lemma 1.12 (or a direct solution of Equation (6.12)), we obtain

$$\sup_{0\le t\le T} \mathbf{E}_\theta(\Delta m_t(h))^2 \le C|h|^4.$$

We have similar inequality for the process $S_t(\theta, X) = f_t(\theta)m_t(\theta), 0 \le t \le T$. So

$$\sup_{0\le t\le T} \mathbf{E}_\theta[S_t(\theta+h, X) - S_t(\theta, X) - (h, \dot{S}_t(\theta, X))]^2 \le C|h|^4$$

with some other constant $C > 0$.

Thus, the process $S_t(\theta, X), 0 \le t \le T$ is differentiable in $L_2[0,T]$ with respect to θ in the following sense ($\|\cdot\|$ is a norm in $L_2[0,T]$)

$$\mathbf{E}_\theta\|S(\theta+h, X) - S(\theta, X) - (h, \dot{S}(\theta, X))\|^2 = O(|h|^4).$$

Put

$$q_t(\theta, X) = \dot{f}_t(\theta)m_t(\theta) + f_t(\theta)\dot{m}_t(\theta), \quad 0 \le t \le T.$$

Then

$$\mathbf{E}_\theta\|(S(\theta+\varepsilon v, X) - S(\theta, X))\varepsilon^{-1} - (v, q(\theta, X))\|^2 \le C\varepsilon^2|v|^4.$$

From the linearity of Equation (6.11), it follows that the process $\dot{m}_t(\theta), 0 \le t \le T$ converges to the function $\dot{y}_t(\theta), 0 \le t \le T$ as $\varepsilon \to 0$.

The limit equation for (6.11) is

$$\left.\frac{d\dot{m}_t(\theta)}{dt}\right|_{\varepsilon=0} = \dot{Q}_t(\theta)m_t(\theta)\Big|_{\varepsilon=0} + Q_t(\theta)\dot{m}_t(\theta)\big|_{\varepsilon=0} + \\ +\dot{R}_t(\theta)f_t(\theta)y_t(\theta), \quad \dot{m}_t(\theta)\big|_{\varepsilon=0} = 0$$

and $\dot{y}_t(\theta)$ is a solution of this equation.

So the process $q_t(\theta, X), 0 \le t \le T$ converges to the deterministic function $q_t(\theta, x)$, $0 \le t \le T$ and the matrix

$$I(\theta, X) = \langle q(\theta, X), q(\theta, X)^T\rangle$$

is continuous with respect to X at the point $x = \{x_t, 0 \le t \le T\}$ with

$$\frac{dx_t}{dt} = f_t(\theta)y_t(\theta).$$

Here, $\langle f, g\rangle$ means the inner product in $L_2[0,T]$. It only remains to change the scale $u = I(\theta)^{-\frac{1}{2}}v$. The LAN of the family $\{\mathbf{P}_\theta^{(\varepsilon)}, \theta \in \Theta\}$ yields the inequality of Hajek-Le Cam (Theorem 2.1)

$$\lim_{\delta\to 0}\liminf_{\varepsilon\to 0}\sup_{|\theta-\theta_0|<\delta} \mathbf{E}_\theta l(\varphi_\varepsilon(\theta_0)^{-1}(\theta_\varepsilon^* - \theta)) \geq \mathbf{E}l(\xi)$$

for any estimator θ_ε^* and $l(\cdot) \in \mathbf{W}_{e,2}$. Here $\xi \sim \mathcal{N}(0, I)$ and I is a unit $d \times d$ matrix. This inequality allows us to define the locally asymptotically minimax estimators in this problem.

We need additional conditions to describe the properties of the MLE $\hat\theta_\varepsilon$ and the BE $\tilde\theta_\varepsilon$ of parameter θ.

Denote $y_t(\theta, \theta_0)$ as the solution of the equation

$$\frac{\partial y_t(\theta,\theta_0)}{\partial t} = Q_t(\theta)y_t(\theta,\theta_0) + R_t(\theta)f_t(\theta_0)$$

and introduce the conditions

I. *The functions $f_t(\theta)$, $b_t(\theta)$ and $\sigma_t(\theta)$ are continuous and bounded with their first two derivatives w.r.t. θ uniformly on $t \in [0, T]$.*

II. *The Fisher's information matrix $I(\theta)$ is positive definite in the following sense:*

$$\inf_{\theta\in\Theta}\inf_{|\lambda|=1}(I(\theta)\lambda, \lambda) > 0.$$

III. *For any $h > 0$ and any compact $\mathbf{K} \subset \Theta$,*

$$\inf_{\theta\in\mathbf{K}}\inf_{|\theta-\theta_0|>h}\|f(\theta)y(\theta,\theta_0) - f(\theta_0)y(\theta_0)\| > 0.$$

Theorem 6.2. *If conditions I-III hold, then the MLE $\hat\theta_\varepsilon$ is, uniformly on $\theta \in \mathbf{K}$,*

(a) consistent

$$\mathbf{P}_\theta - \lim_{\varepsilon\to 0}\hat\theta_\varepsilon = \theta;$$

(b) asymptotically normal

$$\mathcal{L}_\theta\left\{\varepsilon^{-1}(\hat\theta_\varepsilon - \theta)\right\} \Longrightarrow \mathcal{N}(0, I(\theta)^{-1});$$

(c) LAM for the loss functions $l(\cdot) \in \mathbf{W}_p$.

Proof of this theorem consists of checking the conditions of Theorem 2.2. We have already checked conditions I-III of this theorem in a nonuniform on $\theta \in \mathbf{K}$ way. The uniform on $\theta \in \mathbf{K}$ version of LAN with $u_\varepsilon \to u$ and u_ε bounded has no additional difficulties because all estimates could be chosen uniformly on θ and $|u_\varepsilon| < C$. The IV-th condition immediately follows from the boundedness of the derivatives of the functions $f_t(\theta)$, $b_t(\theta)$ and $\sigma_t(\theta)$ and the finiteness of all corresponding mathematical expectations of the equation

$$\begin{aligned} d\dot{m}_t(\theta) &= [\dot{Q}_t(\theta)m_t(\theta) + Q_t(\theta)\dot{m}_t(\theta) + \dot{R}_t(\theta)f_t(\theta_1)Y_t(\theta_1)]dt + \\ &+\varepsilon\dot{R}_t(\theta)dW_t, \quad \dot{m}_0(\theta) = 0, \quad 0 \le t \le T. \end{aligned}$$

The last condition of Theorem 2.2 coincides with our III if we put $\varepsilon = 0$ in the integral

$$\int_0^T [f_t(\theta)m_t(\theta) - f_t(\theta_0)m_t(\theta_0)]^2 dt.$$

The continuity of the derivatives (condition IV) can be checked in the following way:

$$\begin{aligned} &\lim_{\theta\to\theta_0,\varepsilon\to 0} \mathbf{E}_{\theta_0}\|\dot{S}(\theta,X) - \dot{S}(\theta_0,x)\|^2 \le \\ &\le 2\limsup_{\varepsilon\to 0}\sup_{\theta} \mathbf{E}_{\theta_0}\|\dot{S}(\theta,X) - \dot{S}(\theta,x)\|^2 + 2\lim_{\theta\to\theta_0}\|\dot{S}(\theta,x) - \dot{S}(\theta_0,x)\|^2 = 0. \end{aligned}$$

So all the conditions of Theorem 2.2 are fulfilled and the MLE $\hat{\theta}_\varepsilon$ has the mentioned properties.

REMARK 6.1. If θ is a random variable $\mathcal{F}_0$-measurable with the prior density $\pi(y), y \in \Theta$ and the loss function is polynomial, then the BE $\tilde{\theta}_\varepsilon$ of parameter θ is also uniformly consistent, asymptotically normal with the same parameters and LAM.

REMARK 6.2. The generalization of these results to the schemes with multidimensional observed and unobserved processes and to the case

$$dX_t = [h_t(\theta) + f_t(\theta)Y_t]dt + \varepsilon r_t dW_t, \quad X_0 = x_0,$$

$$dY_t = [g_t(\theta) + b_t(\theta)Y_t]dt + \varepsilon\sigma_t(\theta)dV_t, \quad Y_0 = y_0$$

could be done without additional difficulties.

REMARK 6.3. It is also possible to consider the problem of estimating the initial value of an unobserved process, say

$$dX_t = f_tY_tdt + \varepsilon dW_t, \quad X_0,$$

$$dY_t = b_tY_tdt + \varepsilon\sigma_t dV_t, \quad Y_0 = \theta, \quad 0 \le t \le T,$$

where unknown parameter $\theta \in (\alpha,\beta), \alpha > 0$. In this case, we can denote $\tilde{Y}_t = Y_t - \theta$ and reduce the original model to

$$dX_t = (f_t\tilde{Y}_t + \theta f_t)dt + \varepsilon dW_t, \quad X_0 = 0, \quad 0 \le t \le T,$$

$$d\tilde{Y}_t = (b_tY_t + \theta b_t)dt + \varepsilon\sigma_t dV_t, \quad \tilde{Y}_0 = 0$$

and then check the conditions of Theorem 2.2 as it was done earlier.

6.2 Nonlinear systems

1. Conditionally Gaussian process. We begin with a nonlinear partially observed system of the type

$$dX_t = b_t(\theta, X)Y_t dt + \varepsilon dW_t, \quad X_0 = 0, \quad 0 \le t \le T,$$

$$dY_t = a_t(\theta, X)Y_t dt + \varepsilon\sigma_t(\theta, X)dV_t, \quad Y_0 = y_0, \quad 0 \le t \le T,$$

which is linear w.r.t. the unobserved component Y_t. The unknown parameter $\theta \in \Theta, \Theta$ is an open, bounded subset of $\mathbf{R}^d$.

We suppose that the following condition holds.

I. *The functions* $b_t(\theta, X), a_t(\theta, X), \ \sigma_t(\theta, X) \in \mathcal{P}$, *satisfy the condition* $\mathcal{L}$ *and are bounded.*

Then, for the conditional mathematical expectation $m_t(\theta) = \mathbf{E}_\theta(Y_t|X_s, 0 \le s \le t)$ R. Liptser and A. Shiryayev [97] proved the following theorem.

Theorem 6.3. *If condition I holds, then the process* $m_t(\theta), 0 \le t \le T$ *is the only continuous solution of the equation*

$$\begin{aligned} dm_t(\theta) &= a_t(\theta, X)m_t(\theta)dt + \varepsilon^{-2}\gamma_t(\theta)b_t(\theta, X)[dX_t - \\ &\quad -b_t(\theta, X)m_t(\theta)dt], \quad m_0(\theta) = 0, \quad 0 \le t \le T \end{aligned} \tag{6.13}$$

and the process $\gamma_t(\theta) = \gamma_t(\theta, X)$ *solves the equation*

$$\frac{\partial \gamma_t(\theta)}{\partial t} = 2a_t(\theta, X)\gamma_t(\theta) - \varepsilon^{-2}\gamma_t(\theta)b_t(\theta, X) + \varepsilon^2\sigma_t(\theta, X),$$
$$\gamma_0(\theta) = 0, \quad 0 \le t \le T.$$

REMARK 6.4. Theorem 12.1 in [97] is a more general result. Here we give a simplified version.

This result allows us to calculate the likelihood ratio

$$\begin{aligned} L(\theta, X) &= \exp\left\{\frac{1}{\varepsilon^2}\int_0^T b_t(\theta, X)m_t(\theta)dX_t - \right. \\ &\quad \left. -\frac{1}{2\varepsilon^2}\int_0^T b_t(\theta, X)^2 m_t(\theta)^2 dt\right\} \end{aligned}$$

and to study the properties of the MLE and BE as $\varepsilon \to 0$.

This situation is similar to the preceding one in Section 6.1 and uses the same technique.

Stochastic process $Y_t, 0 \le t \le T$ uniformly on $t \in [0, T]$ converges to the function

$$\tilde{y}_t(\theta) = y_0 \exp\left\{\int_0^t a_s(\theta, x)ds\right\},$$

where $x = \{x_t, 0 \le t \le T\}$ is a solution of the equation

$$\frac{dx_t}{dt} = b_t(\theta, x)\tilde{y}_t(\theta), \quad x_0 = 0, \quad 0 \le t \le T \tag{6.14}$$

and is a uniform on $t \in [0,T]$ limit of the processes $X_t, 0 \le t \le T$.

Let us denote $S_t(\theta, X) = b_t(\theta, X)m_t(\theta)$ and the derivative w.r.t. θ

$$\dot{S}_t(\theta, X) = \dot{b}_t(\theta, X)m_t(\theta) + b_t(\theta, X)\dot{m}_t(\theta),$$

where the vector-process $\dot{m}_t(\theta)$ satisfies the following equation which is similar to (6.11):

$$\begin{aligned} d\dot{m}_t(\theta) &= [\dot{Q}_t(\theta, X)m_t(\theta) + Q_t(\theta, X)\dot{m}_t(\theta)]dt + \\ &\quad +\dot{R}_t(\theta, X)dX_t, \qquad \dot{m}_0(\theta) = 0, \quad 0 \le t \le T. \end{aligned}$$

Here

$$\begin{aligned} Q_t(\theta, X) &= b_t(\theta, X) - z_t(\theta, X)b_t(\theta, X)^2, \\ R_t(\theta, X) &= z_t(\theta, X)b_t(\theta, X) \end{aligned}$$

and the process $z_t(\theta, X) = \varepsilon^{-2}\gamma_t(\theta)$ is taken from the corresponding Riccati equation.

Denote also $\tilde{y}_t(\theta, \theta_0)$ as the solution of the equation

$$\frac{\partial \tilde{y}_t(\theta, \theta_0)}{\partial t} = Q_t(\theta, x)\tilde{y}_t(\theta, \theta_0) + R_t(\theta, x)b_t(\theta_0, x)\tilde{y}_t(\theta_0), \quad \tilde{y}_0(\theta, \theta_0) = 0$$

and introduce the conditions.

II. *The functions $b_t(\theta, X), a_t(\theta, X)$ and $\sigma_t(\theta, X)$ have two continuous bounded derivatives w.r.t. θ and the first derivatives are Lipschitzian w.r.t. X.*

For example,

$$|\dot{b}_t(\theta, X) - \dot{b}_t(\theta, Y)| \le L\int_0^t |X_s - Y_s|ds + L|X_t - Y_t|,$$

where X and Y are some functions from C_T.

III. *For any $h > 0$ and any $\mathbf{K} \subset \Theta$,*

$$\inf_{\theta \in \mathbf{K}} \inf_{|\theta - \theta_0| > h} \|b(\theta, x)\tilde{y}(\theta, \theta_0) - b(\theta_0, x)\tilde{y}(\theta_0)\| > 0.$$

The matrix of information is now

$$I(\theta) = \int_0^T \dot{S}_t(\theta, x)\dot{S}_t(\theta, x)^T dt$$

and we suppose that

IV. *$I(\theta)$is uniformly nondegenerate*

$$\inf_{\theta \in \Theta} \inf_{|e|=1} (I(\theta)e, e) > 0.$$

Theorem 6.4. *If conditions I-IV hold, then the MLE $\hat{\theta}_\varepsilon$ is, uniformly on $\theta \in \mathbf{K}$:*

(a) *consistent*

$$\mathbf{P}_\theta - \lim_{\varepsilon \to 0} \hat{\theta}_\varepsilon = 0;$$

(b) *asymptotically normal*

$$\mathcal{L}_\theta\{\varepsilon^{-1}(\hat{\theta}_\varepsilon - \theta)\} \Longrightarrow \mathcal{N}(0, I(\theta)^{-1});$$

(c) *LAM for the loss functions $l(\cdot) \in \mathbf{W}_p$.*

Proof of this theorem is virtually a repeat of the proof of Theorem 6.1 because the likelihood ratio process has a very similar form.

2. General case. Let us now consider the two-dimensional nonlinear system

$$dX_t = f(\theta, Y_t)dt + \varepsilon dW_t, \quad X_0 = 0, \quad 0 \le t \le T, \tag{6.15}$$

$$dY_t = b(\theta, Y_t)dt + \varepsilon dV_t, \quad Y_0 = y_0, \quad 0 \le t \le T, \tag{6.16}$$

where X_t is an observed and Y_t is an unobserved components. Parameter θ is one-dimensional, $\theta \in \Theta = (\alpha, \beta)$. The Wiener processes W_t and V_t are independent.

The investigation of MLE estimator $\hat{\theta}_\varepsilon$ is now difficult because the likelihood ratio

$$L(\theta, X) = \exp\left\{\frac{1}{\varepsilon^2}\int_0^T \hat{f}_t(\theta)dX_t - \frac{1}{2\varepsilon^2}\int_0^T \hat{f}_t(\theta)^2 dt\right\}$$

depends on the conditional expectation $\hat{f}_t(\theta) = \mathbf{E}_\theta(f(\theta, Y_t)|X_s, 0 \le s \le t)$ and there is no closed finite system of equations describing $\hat{f}_t(\theta)$ (details of existing difficulties can be found in [97] or in any other work on nonlinear filtration).

Below, we propose some consistent and asymptotically normal estimators of the parameter θ based on the linearization of the original system in the vicinity of a deterministic limit system.

We suppose that

I. *The coefficients $f(\theta, x)$ and $b(\theta, x)$ are continuous and bounded with their first four derivatives w.r.t. θ and x.*

Then the functions $f(\theta, x)$ and $b(\theta, x)$ are also Lipschitzian w.r.t. x and, using standard arguments, we obtain the inequality

$$\sup_{0 \le t \le T} |Y_t - y_t| \le \varepsilon e^{LT} \sup_{0 \le t \le T} |V_t|,$$

where L is a constant from the inequality $|b(\theta, x) - b(\theta, y)| \le L|x - y|$ (because $|b'_x(\theta, x)| < L$) and y_t is a solution of differential equation

$$\frac{dy_t}{dt} = b(\theta, y_t), \quad y_0, \quad 0 \le t \le T.$$

For the process X_t the analog inequality is

$$\sup_{0\leq t\leq T}|X_t-x_t|\leq C\varepsilon\sup_{0\leq t\leq T}|V_t|+\varepsilon\sup_{0\leq t\leq T}|W_t|$$

and

$$x_t=x_t(\theta)=\int_0^t f(\theta,y_s)ds.$$

Let us introduce the linearized system

$$dX_t=[f(\theta,y_t)+f'(\theta,y_t)(Y_t-y_t)]dt+\varepsilon dW_t,\quad X_0=0, \tag{6.17}$$

$$\begin{aligned}dY_t&=[b(\theta,y_t)+b'(\theta,y_t)(Y_t-y_t)]dt+\varepsilon dV_t,\\ Y_0&=y_0,\quad 0\leq t\leq T,\end{aligned} \tag{6.18}$$

where $f'(\theta,y)=\frac{\partial}{\partial y}f(\theta,y)$ and we try to estimate θ by observations (6.15) but assuming that the observed system is (6.17). So we have a "misspecified model" as in Section 2.6.

On the basis of the linearized system (6.17), (6.18), we introduce a "Kalman filter" for the "conditional mathematical expectation" $m_t(\theta)=\mathbf{E}_\theta(Y_t|X_s,0\leq s\leq t)$

$$\begin{aligned}dm_t(\theta)&=[b(\theta,y_t)+b'(\theta,y_t)(m_t(\theta)-y_t)]dt+\\ &+z_t(\theta)f'(\theta,y_t)[dX_t-(f(\theta,y_t)+f'(\theta,y_t))(m_t(\theta)-y_t)dt],\end{aligned} \tag{6.19}$$

where $z_t(\theta)$ is a solution of

$$\frac{\partial z_t(\theta)}{\partial t}=2b'(\theta,y_t)z_t(\theta)-f'(\theta,y_t)^2z_t(\theta)^2+1,\quad z_0(\theta)=0. \tag{6.20}$$

Observations $X_t, 0\leq t\leq T$ in (6.19) are taken from (6.15), so all the processes are non-Gaussian and the solution of (6.19) is not a conditional expectation (if the system is nonlinear, of course).

Denote $S_t(\theta,X)=f(\theta,y_t)+f'(\theta,y_t)[m_t(\theta)-y_t]$ and introduce the function

$$\begin{aligned}L(\theta,\theta_0,X)&=\frac{1}{\varepsilon^2}\int_0^T[S_t(\theta,X)-S_t(\theta_0,X)]dX_t-\\ &-\frac{1}{2\varepsilon^2}\int_0^T[S_t(\theta,X)^2-S_t(\theta_0,X)^2]dt.\end{aligned}$$

The pseudo-MLE $\hat\theta_\varepsilon$ is defined by the relation

$$\sup_{\theta\in\Theta}L(\theta,\theta_0,X)=L(\hat\theta_\varepsilon,\theta_0,X) \tag{6.21}$$

and we describe its properties as $\varepsilon\to 0$. This situation, as we mentioned above, is similar to that considered in Section 2.6 and the difference between them is that for $\varepsilon=0$, Equations (6.15), (6.16) and (6.17), (6.18) coincide in our case (see Remark 2.9).

The condition of identifiability is now

II. *For any $h > 0$ and any $\mathbf{K} \subset \Theta$*

$$\inf_{\theta \in \mathbf{K}} \inf_{|\theta-\theta_0|>h} \|S(\theta, x) - S(\theta_0, x)\| > 0. \tag{6.22}$$

We recall that in our notation

$$S(\theta, x) = f(\theta, y_t(\theta)) + f'(\theta, y_t(\theta))[y_t(\theta, \theta_0) - y_t(\theta)]$$

and $y_t(\theta, \theta_0)$ solves the equation

$$\begin{aligned} \frac{\partial y_t(\theta, \theta_0)}{\partial t} &= b(\theta, y_t) + b'(\theta, y_t)[y_t(\theta, \theta_0) - y_t(\theta)] + \\ &+ z_t(\theta) f'(\theta, y_t(\theta))[f(\theta_0, y_t(\theta_0)) - f(\theta, y_t(\theta)) - \\ &- f'(\theta, y_t)(y_t(\theta, \theta_0) - y_t(\theta))], \quad y_0(\theta, \theta_0) = 0. \end{aligned}$$

So $y_t(\theta_0, \theta_0) = y_t(\theta_0)$ and condition (6.22) could be rewritten in the form

$$\begin{aligned} \inf_{\theta \in \mathbf{K}} \inf_{|\theta-\theta_0|>h} & \|f(\theta, y(\theta)) - f(\theta_0, y(\theta_0)) + f'(\theta, y(\theta)) \times \\ \times & [y(\theta, \theta_0) - y(\theta)]\| > 0. \end{aligned}$$

To introduce the "Fisher information" $I(\theta)$ in this problem, we first differentiate Equation (6.19) w.r.t. θ, then put $\theta = \theta_0$ and $\varepsilon = 0$. The result of this procedure we denote $\dot{n}_t(\theta_0)$ (deterministic function satisfying the equation introduced below). So

$$I(\theta_0) = \int_0^T [\dot{f}(\theta_0, y_t(\theta_0)) + f'(\theta_0, y_t(\theta_0))\dot{n}_t(\theta_0)]^2 dt$$

and the corresponding condition is

III.

$$\inf_{\theta \in \Theta} I(\theta) > 0.$$

Finally, define the Gaussian random variable

$$\xi = I(\theta_0)^{-1} \left\{ \int_0^T \dot{S}_t(\theta_0, x) dW_t - \int_0^T \dot{S}_t(\theta_0, x) f_t'(\theta_0, y_t) n_t^{(1)} dt \right\},$$

where the Gaussian process $n_t^{(1)}$ is defined below in (6.24).

Theorem 6.5. *If conditions I-III hold, then the pseudo-MLE $\hat{\theta}_\varepsilon$ is consistent and asymptotically normal:*

$$\mathcal{L}_\theta\{\varepsilon^{-1}(\hat{\theta}_\varepsilon - \theta)\} \Longrightarrow \{\xi\}.$$

Proof is similar to the proof of Theorem 2.16 in Section 2.6. So we only mention here the main differences in the calculation of the corresponding derivatives.

Denote θ_0 as the true value and introduce a localizing set

$$\mathcal{A}_1 = \{\omega : \sup_{|\theta-\theta_0|>\nu} L(\theta,\theta_0,X) < 0\},$$

then, as before, for $\omega \in \mathcal{A}_1$ we have

$$|\hat{\theta}_\varepsilon - \theta_0| < \nu$$

on the set $\mathcal{A}_1$ the pseudo-MLE $\hat{\theta}_\varepsilon = \theta_0 + \hat{u}_\varepsilon$, where $\hat{u}_\varepsilon$ is one of the solutions of the equation

$$\dot{L}(\theta_0 + u, \theta_0, X) = 0, \quad -\nu < u < \nu,$$

which can be written as

$$\begin{aligned}\varepsilon \int_0^T \dot{S}_t(\theta_0 + u, X)dW_t - \int_0^T \dot{S}_t(\theta_0 + u, X)[S_t(\theta_0 + u, X) - \\ -b(\theta_0, Y_t)]dt = 0, \quad u \in (-\nu, \nu).\end{aligned} \tag{6.23}$$

Let us denote the left side of this equation as $F(\varepsilon, u)$ and introduce the set

$$\mathcal{A}_2 = \left\{\omega : \sup_{|u|<\nu} \frac{\partial F}{\partial u} < -\frac{1}{2}I(\theta_0)\right\},$$

providing the uniqueness of the solution of equation

$$F(\varepsilon, u) = 0, \quad -\nu < u < \nu.$$

Then on $\mathcal{A}_1 \cap \mathcal{A}_2$ this equation defines an implicit function $u = u(\varepsilon)$ and we can calculate the expansion of u by the powers of ε.

First, note that as follows from the condition of identifiability, $u(0) = 0$. The derivative

$$\begin{aligned}\dot{S}_t(\theta, X) &= \dot{f}(\theta, y_t(\theta)) + [\dot{f}'(\theta, y_t(\theta) + f''(\theta, y_t(\theta))\dot{y}_t(\theta)] \times \\ &\quad \times [m_t(\theta) - y_t(\theta)] + f'(\theta, y_t(\theta))\dot{m}_t(\theta)\end{aligned}$$

and $\dot{S}_t(\theta_0, x) = \dot{f}(\theta_0, y_t) + f'(\theta_0, y_t)\dot{n}_t$, where $y_t = y_t(\theta_0)$, $\dot{n}_t = \dot{n}_t(\theta_0)$ and the function $\dot{n}_t$ satisfies the equation

$$\begin{aligned}\frac{d\dot{n}_t}{dt} &= -z_t(\theta_0)f'(\theta_0, y_t)\dot{f}(\theta_0, y_t) + \\ &\quad +(b'(\theta_0, y_t) - z_t(\theta_0)f'(\theta_0, y_t)^2)\dot{n}_t, \quad \dot{n}_0 = 0.\end{aligned}$$

The first derivatives F'_u and F'_ε at the point $\varepsilon = 0$ are obtained by a direct differentiating of the equation (6.23):

$$\begin{aligned}F'_u(\varepsilon, u)|_{\varepsilon=0} &= -\int_0^T (\dot{f}(\theta_0, y_t) + f'(\theta_0, y_t)\dot{n}_t)^2 dt, \\ F'_\varepsilon(\varepsilon, u)|_{\varepsilon=0} &= \int_0^T (\dot{f}(\theta_0, y_t) + f'(\theta_0, y_t)\dot{n}_t)dW_t - \\ &\quad -\int_0^T (\dot{f}(\theta_0, y_t) + f'(\theta_0, y_t)\dot{n}_t)f'(\theta_0, y_t(\theta_0))n_t^{(1)} dt,\end{aligned}$$

where the term $f'(\theta_0, y_t)n_t^{(1)}$ is obtained by differentiating $f'(\theta, y_t(\theta))m_t(\theta)-$ $-f(\theta_0, Y_t)$ on ε, so

$$\begin{aligned}\frac{\partial}{\partial\varepsilon}[S_t(\theta, X) - f(\theta_0, Y_t)]|_{\varepsilon=0} &= f'(\theta_0, y_t)m_t'(\theta_0)|_{\varepsilon=0} - \\ -f'(\theta_0, y_t)y_t^{(1)} &= f'(\theta_0, y_t)n_t^{(1)}\end{aligned}$$

and $n_t^{(1)}$ is a solution of the equation

$$\begin{aligned}dn_t^{(1)} &= [[b'(\theta_0, y_t) - z_t(\theta_0)f'(\theta_0, y_t)^2]n_t^{(1)} + \\ &\quad +z_t(\theta_0)f'(\theta_0, y_t)^2 y_t^{(1)}]dt + z_t(\theta_0)f'(\theta_0, y_t)dW_t \qquad (6.24)\end{aligned}$$

with

$$dy_t^{(1)} = b'(\theta_0, y_t)y_t^{(1)}dt + dV_t, \quad y_0^{(1)} = 0,\ 0 \le t \le T.$$

So the first derivative $u'(0)$ is

$$\begin{aligned}u'(0) &= -\frac{F_\varepsilon''}{F_u'}\bigg|_{\varepsilon=0} = \\ &= I(\theta_0)^{-1}\left\{\int_0^T \dot S_t(\theta_0, x)dW_t - \int_0^T \dot S(\theta_0, y_t)f'(\theta_0, y_t)n_t^{(1)}dt\right\}\end{aligned}$$

and $u'(0) = \xi$ in our notation.

We can now represent the pseudo-MLE

$$\hat\theta_\varepsilon = \theta_0 + (\varepsilon\xi + \varepsilon^{\frac{3}{2}}\eta_1)\chi_{\{\mathcal{A}\}} + \zeta_\varepsilon\chi_{\{\mathcal{A}^C\}},$$

where $\mathcal{A} = \cap_{i=1}^3 \mathcal{A}_i$ and $\mathcal{A}_3 = \{\omega : |\eta_1| \le 1\}$ and then estimate the probabilities $\mathbf{P}_{\theta_0}^{(\varepsilon)}(\mathcal{A}_i^C)$, $i = 1, 2, 3$ and $\mathbf{P}_{\theta_0}^{(\varepsilon)}(|\zeta_\varepsilon| > x)$ as was done in Section 2.6.

REMARK 6.5. It is also possible to construct a more simple, consistent and asymptotically normal pseudo-MLE. Let us consider the equations

$$\begin{aligned}dX_t &= f(\theta, y_t(\theta))dt + \varepsilon dW_t, \quad X_0 = 0, \\ \frac{\partial y_t(\theta)}{\partial t} &= a(\theta, y_t(\theta)), \quad y_0,\end{aligned}$$

as a "misspecified model" for the observed process (6.15) with the unobserved component (6.16).

The "loglikelihood ratio" is now

$$L(\theta, \theta_0, X) = \frac{1}{\varepsilon^2}\int_0^T [f_t(\theta) - f_t(\theta_0)]dX_t - \frac{1}{2\varepsilon^2}\int_0^T [f_t(\theta)^2 - f_t(\theta_0)^2]dt,$$

where we denote $f_t(\theta) = f(\theta, y_t(\theta))$.

The pseudo-MLE $\hat\theta_\varepsilon$ is defined by the equation

$$L(\hat\theta_\varepsilon, \theta_0, X) = \sup_{\theta\in\bar\Theta} L(\theta, \theta_0, X)$$

and the observations X are taken from Equation (6.15).

The condition of identifiability is now

$$\inf_{|\theta-\theta_0|>h} \int_0^T [f(\theta, y_t(\theta)) - f(\theta_0, y_t(\theta_0))]^2 dt > 0.$$

Under the regularity conditions (including this identifiability) it is possible to prove the consistence and asymptotic normality

$$\mathcal{L}_\theta\{\varepsilon^{-1}(\hat{\theta}_\varepsilon - \theta)\} \Longrightarrow \mathcal{N}(0, \sigma^2) = \mathcal{L}\{\xi\},$$

where

$$\xi = I(\theta)^{-1}\left\{\int_0^T \dot{f}_t(\theta)dW_t + \int_0^T \dot{f}_t(\theta) f'(\theta, y_t(\theta)) y_t^{(1)} dt\right\}.$$

The Gaussian process $y_t^{(1)}$ satisfies the same equation as before and $\dot{f}_t(\theta) = \dot{f}(\theta, y_t(\theta)) + f'(\theta, y_t(\theta))\dot{y}_t(\theta)$.

The proof is not presented here because this result is a particular case of the previous theorem if we set $f' = 0$ and $b' = 0$ in Equations (6.17), (6.18).

Extended Kalman filter. If the observed process is linear

$$dX_t = f_t(\theta) Y_t dt + \varepsilon\, dW_t,\ X_0 = 0,\ 0 \le t \le T$$

with respect to the unobserved component

$$dY_t = b(\theta, Y_t)\, dt + \varepsilon\, dV_t,\ Y_0 = y_0,\ 0 \le t \le T$$

then we can substitute in the likelihood ratio formula in the place of conditional expectation, the solution $m_t(\theta)$ of the , so-called, *extended Kalman filter*

$$dm_t(\theta) = b(\theta, m_t(\theta))dt + [dX_t - f_t(\theta) m_t(\theta) dt],\ m_0(\theta) = y_0.$$

The process $m_t(\theta)$ converges uniformly on $\theta \in \Theta$ to the function $y_t(\theta, \theta_0)$, solving the equation

$$\frac{\partial y_t(\theta, \theta_0)}{\partial t} = b(\theta, y_t(\theta, \theta_0)) + f_t(\theta_0) y_t(\theta_0) - f_t(\theta) y_t(\theta, \theta_0)$$

with $y_t(\theta_0) = y_t(\theta_0, \theta_0)$. So the condition of identifiability is

$$\inf_{|\theta-\theta_0|>\nu} \int_0^T [f_t(\theta) y_t(\theta, \theta_0) - f_t(\theta_0) y_t(\theta_0)]^2 dt > 0$$

and if this and other conditions of regularity are fulfilled, then using standard arguments (as was done before), one can prove the uniform exponential consistence of the MLE and its asymptotic normality. This problem was considered by F. Konecny [60].

6.3 Disorder problem for Kalman filter

Disorder in the equation of observations. First, we have a switching in the equation of observations

$$dX_t = f_t(\tau)Y_t dt + \varepsilon dW_t, \quad X_0 = 0, \quad 0 \le t \le T, \tag{6.25}$$

where

$$f_t(\tau) = \chi_{\{t \le \tau\}} h_t + \chi_{\{t > \tau\}} g_t$$

and $\tau \in (\tau_1, \tau_2) \subset [0,T]$ of a partially observed linear system with the state equation

$$dY_t = b_t Y_t dt + \varepsilon \sigma_t dV_t, \quad Y_0 = y_0 \ne 0, \quad 0 \le t \le T. \tag{6.26}$$

All the functions h_t, g_t, b_t and σ_t are measurable, bounded, and known. So we have to estimate the moment of switching τ by the observations $\{X_t, 0 \le t \le T\}$.

According to the equations of optimal filtration, the conditional mathematical expectation $m_t(\tau) = \mathbf{E}_\tau(Y_t | X_s, 0 \le s \le t)$ (with known τ) satisfies the equation

$$dm_t(\tau) = b_t m_t(\tau)dt + z_t(\tau)h_t[dX_t - h_t m_t(\tau)dt], \ m_0 = Y_0, \tag{6.27}$$

where

$$\frac{\partial z_t(\tau)}{\partial t} = 2b_t z_t(\tau) - z_t(\tau)^2 h_t^2 + \sigma_t^2, \ z_0(\tau) = 0 \tag{6.28}$$

for $t \in [0, \tau]$ and

$$dm(\tau) = b_t m_t(\tau)dt + z_t(\tau)g_t[dX_t - g_t m_t(\tau)dt], \ m_\tau(\tau,) \tag{6.29}$$

with

$$\frac{\partial z_t(\tau)}{\partial t} = 2b_t z_t(\tau) - z_t(\tau)^2 g_t^2 + \sigma_t^2, \ z_\tau(\tau). \tag{6.30}$$

for $t \in [\tau, T]$, where $m_\tau(\tau)$ and $z_\tau(\tau)$ are taken from (6.27), (6.28).

The stochastic processes X_t and Y_t correspondingly converge as $\varepsilon \to 0$ to the functions

$$\begin{aligned} x_t(\tau) &= \int_0^t f_s(\tau) y_s ds, \quad t \in [0,T], \\ y_t &= y_0 \exp\left\{\int_0^t b_s ds\right\}, \quad t \in [0,T]. \end{aligned}$$

Let us denote

$$v_t = (h_t - g_t)x_t$$

and introduce (as in Section 5.1) the stochastic process

$$Z(u) = \exp\left\{v_\tau W(u) - \frac{1}{2}|u|v_\tau^2\right\}, \ u \in (-\infty, \infty),$$

where

$$W(u) = \begin{cases} W_+(u), & u \ge 0, \\ W_-(-u), & u < 0. \end{cases}$$

and $W_+(\cdot)$,$W_-(\cdot)$ are independent Wiener processes.

Also denote ζ as the solution of the equation

$$Z(\zeta) = \sup_u Z(u).$$

Let $\mathbf{P}_\tau^{(\varepsilon)}$ be a measure induced by the observations (6.25) in $(\mathcal{C}_T, \mathcal{B}_T)$. The measures $\{\mathbf{P}_\tau^{(\varepsilon)}, \tau \in (\tau_1, \tau_2)\}$ are equivalent ($\varepsilon \neq 0$) and the MLE $\hat{\tau}_\varepsilon$ is defined by

$$\frac{d\mathbf{P}_\tau^{(\varepsilon)}}{d\mathbf{P}^{(\varepsilon)}}(X) = \sup_{\tau\in(\tau_1,\tau_2)} \frac{d\mathbf{P}_{\hat{\tau}_\varepsilon}^{(\varepsilon)}}{d\mathbf{P}^{(\varepsilon)}}(X),$$

where $\mathbf{P}^{(\varepsilon)}$ is a measure of the process $\{\varepsilon W_t, 0 \leq t \leq T\}$.

We suppose that

$$\inf_{\tau_1<t<\tau_2} |v_t| > 0.$$

Theorem 6.6. *Under the previously made assumptions, the MLE $\hat{\theta}_\varepsilon$, as $\varepsilon \to 0$, is uniformly on the compacts $\mathbf{K}_\tau \subset (\tau_1, \tau_2)$*

(a) consistent

$$\mathbf{P}_\tau - \lim_{\varepsilon\to 0} \hat{\tau}_\varepsilon = \tau;$$

(b) converges in distribution

$$\mathcal{L}_\tau\left\{\varepsilon^{-2}(\hat{\tau}_\varepsilon - \tau)\right\} \Longrightarrow \mathcal{L}_\tau\{\xi\};$$

(c) the moments of the MLE also converge ($p > 0$)

$$\lim_{\varepsilon\to 0} \mathbf{E}_\tau|\hat{\tau}_\varepsilon - \tau|^p \varepsilon^{-2p} = \mathbf{E}_\tau|\xi|^p.$$

Proof. We are going to use Theorem 1.4, so first we need to prove the following three lemmas.

Lemma 6.1. *The finite-dimensional distributions of the likelihood ratio process*

$$Z_\varepsilon(u_\varepsilon) = \frac{d\mathbf{P}_{\tau+\varepsilon^2 u_\varepsilon}^{(\varepsilon)}}{d\mathbf{P}_\tau^{(\varepsilon)}}(X), \quad u_\varepsilon \in \left(\frac{\tau_1-\tau}{\varepsilon^2}, \frac{\tau_2-\tau}{\varepsilon^2}\right),$$

as $\varepsilon \to 0$ converge uniformly on $\tau \in \mathbf{K}_\tau$ for any bounded sequence $u_\varepsilon \to u$ to the finite-dimensional distributions of the process $Z(u), u \in (-\infty, \infty)$.

Proof. Let $\tau_u = \tau + \varepsilon^2 u$ and $u > 0$, then $Z_\varepsilon(u)$ has the representation ($\tilde{W}_t$ below is an innovation process)

$$\begin{aligned}\ln Z_\varepsilon(u) &= \frac{1}{\varepsilon}\int_0^T [f_t(\tau_u)m_t(\tau_u) - f_t(\tau)m_t(\tau)]d\tilde{W}_t - \\ &\quad -\frac{1}{2\varepsilon^2}\int_0^T [f_t(\tau_u)m_t(\tau_u) - f_t(\tau)m_t(\tau)]^2 dt =\end{aligned}$$

$$
\begin{aligned}
&= \frac{1}{\varepsilon}\int_{\tau}^{\tau_u}[h_t m_t(\tau_u) - g_t m_t(\tau)]d\tilde{W}_t + \frac{1}{\varepsilon}\int_{\tau_u}^{T} g_t[m_t(\tau_u) - \\
&\quad - m_t(\tau)]d\tilde{W}_t - \frac{1}{2\varepsilon^2}\int_{\tau}^{\tau_u}[h_t m_t(\tau_u) - g_t m_t(\tau)]^2 dt - \\
&\quad - \frac{1}{2\varepsilon^2}\int_{\tau_u}^{T} g_t^2[m_t(\tau_u) - m_t(\tau)]^2 dt,
\end{aligned}
$$

where $m_t(\tau)$ satisfies Equations (6.27) and (6.29) and the process $m_t(\tau_u)$ coincides with the process $m_t(\tau)$ up to time $t = \tau$. At the interval $[\tau, \tau_u]$, it is a solution of the equation

$$
\begin{aligned}
dm_t(\tau_u) &= b_t m_t(\tau_u)dt + z_t(\tau_u)h_t \times \\
&\quad \times [dX_t - h_t m_t(\tau_u)dt], \quad m_\tau(\tau_u) = m_\tau(\tau)
\end{aligned} \tag{6.31}
$$

and the function $z_t(\tau_u)$ up to time $t = \tau_u$ coincides with the solution of Equation (6.31). Further, the equations for $m_t(\tau)$ and $m_t(\tau_u)$ for $t \in [\tau_u, T]$ coincide with (6.29) but the initial values $m_t(\tau)$ and $m_t(\tau_u)$ are different and, consequently, the difference $m_t(\tau) - m_t(\tau_u) \neq 0$ with $\mathbf{P}_\tau^{(\varepsilon)}$ probability 1. We obtain the value $m_{\tau_u}(\tau)$ by solving Equations (6.27), (6.28) at the interval $[0, \tau]$ and (6.29), (6.30) at$[\tau, \tau_u]$ and $m_{\tau_u}(\tau_u)$ is equal to the value of the solution of Equation (6.27) at the point where $t = \tau_u$.

Let us consider the integral

$$
\int_{\tau}^{\tau_u}[h_t m_t(\tau_u) - g_t m_t(\tau)]^2 dt.
$$

Substituting the observations

$$
dX_t = g_t m_t(\tau)dt + \varepsilon d\bar{W}_t, \quad t \in [\tau, \tau_u],
$$

in the filter of $m_t(\tau_u)$, we obtain

$$
\begin{aligned}
dm_t(\tau_u) &= b_t m_t(\tau_u)dt + z_t(\tau_u)h_t[dX_t - h_t m_t(\tau_u)dt] = \\
&= b_t m_t(\tau_u)dt + z_t(\tau_u)h_t[g_t m_t(\tau) - h_t m_t(\tau_u)]dt + \\
&\quad + \varepsilon z_t(\tau_u)h_t d\bar{W}_t, \quad m_\tau(\tau_u) = m_\tau(\tau),
\end{aligned} \tag{6.32}
$$

where $z_t(\tau_u)$ is a solution of (6.28). Hence, we have the convergence

$$
\sup_{\tau \le t \le \tau_u} \left(|m_t(\tau_u) - x_t(\tau_u)| + |m_t(\tau) - x_t(\tau)|\right) \longrightarrow 0
$$

and, for any sequence $u_\varepsilon \to u$,

$$
\frac{1}{\varepsilon^2}\int_{\tau}^{\tau+\varepsilon^2 u_\varepsilon}[h_t m_t(\tau_{u_\varepsilon}) - g_t m_t(\tau)]^2 dt \longrightarrow u v_\tau^2. \tag{6.33}
$$

This convergence allows us to prove the representation

$$
\begin{aligned}
&\frac{1}{\varepsilon}\int_{\tau}^{\tau_u}[h_t m_t(\tau_u) - g_t m_t(\tau)]d\tilde{W}_t = \\
&\quad = \tilde{\tilde{W}}\left(\frac{1}{\varepsilon^2}\int_{\tau}^{\tau_u}[h_t m_t(\tau_u) - g_t m_t(\tau)]^2 dt\right) \Longrightarrow \\
&\quad \Longrightarrow \tilde{\tilde{W}}(u v_\tau^2) = v_\tau W_+(u).
\end{aligned} \tag{6.34}
$$

Then we consider the integral

$$\int_{\tau}^{T} g_t^2[m_t(\tau_u) - m_t(\tau)]^2 dt. \tag{6.35}$$

The difference $\Delta_t = m_t(\tau_u) - m_t(\tau)$ is a solution of the equation

$$d\Delta_t = [b_t - z_t(\tau_u)g_t^2]\Delta_t dt + \varepsilon[z_t(\tau_u) - z_t(\tau)]g_t d\bar{W}_t,$$
$$t \in [\tau_u, T], \quad \Delta_{\tau_u} = m_{\tau_u}(\tau_u) - m_{\tau_u}(\tau)$$

and using Lemma 1.11, we can obtain the estimate

$$\sup_{\tau_u \le t \le T} |\Delta_t| \le C_1|\Delta_{\tau_u}| +$$
$$+ \varepsilon C_2 \sup_{\tau_u \le t \le T} \left| \int_{\tau_u}^{t} [z_s(\tau_u) - z_s(\tau)] g_s \exp\left\{ \int_s^t [b_v - z_v(\tau_u) g_v^2] dv \right\} d\bar{W}_s \right|. \tag{6.36}$$

We recall that

$$\int_{\tau_u}^{T} [z_s(\tau_u) - z_s(\tau)]^2 ds \le H, \tag{6.37}$$

hence

$$\sup_{\tau_u \le t \le T} |\Delta_t| \le C_1|\Delta_{\tau_u}| + C\varepsilon \sup_{\tau_u \le t \le T} \left| \int_0^t q(s) d\bar{W}_s \right| \tag{6.38}$$

with bounded $q(\cdot)$. To estimate the value Δ_{τ_u}, we consider the difference of the solutions Equations (6.27) and (6.31) on the interval $[\tau, \tau_u]$. Note, that at the point where $t = \tau$, we have $\Delta_\tau = 0$. So

$$\begin{aligned} d\Delta_t &= b_t\Delta_t dt + z_t(\tau_u)h_t[g_t m_t(\tau) - h_t m_t(\tau_u)]dt + \\ &\quad + \varepsilon[z_t(\tau_u)h_t - z_t(\tau)g_t)d\bar{W}_t = \\ &= [b_t - z_t(\tau_u)h_t^2]\Delta_t dt + z_t(\tau_u)h_t(g_t - h_t)m_t(\tau)dt + \\ &\quad + \varepsilon[z_t(\tau_u)h_t - z_t(\tau)g_t]d\bar{W}_t, \end{aligned}$$

hence

$$\begin{aligned} \Delta_{\tau_u} &= \int_{\tau}^{\tau_u} \exp\left\{ \int_s^{\tau_u} [b_v - z_v(\tau_u)h_v^2] dv \right\} z_s(\tau_u) h_s \times \\ &\quad \times (g_s - h_s) m_s(\tau) ds + \varepsilon \int_{\tau}^{\tau_u} \exp\left\{ \int_s^{\tau_u} [b_v - z_v(\tau_u)h_v^2] dv \right\} \times \\ &\quad \times [z_s(\tau_u)h_s - z_s(\tau)g_s] d\bar{W}_s. \end{aligned} \tag{6.39}$$

Then we obtain the required inequality

$$\mathbf{E}\Delta_{\tau_u}^2 \le C_1(\tau_u - \tau)^2 + C_2\varepsilon^2(\tau_u - \tau) \le C_3\varepsilon^4(u^2 + u). \tag{6.40}$$

Let us now consider the difference $\delta_t = z_t(\tau_u) - z_t(\tau)$. First, we suppose that $t \in [\tau, \tau_u]$, then for δ_t, we obtain the equation

$$\begin{aligned} \frac{\partial}{\partial t}\delta_t &= 2b_t\delta_t - h_t^2(z_t(\tau_u) + z_t(\tau))\delta_t + \\ &\quad + z_t(\tau)^2(g_t^2 - h_t^2), \quad \delta_\tau = 0. \end{aligned}$$

Hence, δ_t has the representation

$$\delta_t = \int_\tau^t \exp\left\{\int_s^t [2b_v - h_v^2(z_v(\tau_u) + z_v(\tau))]dv\right\} \times \\ \times z_s(\tau)^2(g_s^2 - h_s^2)ds$$

and, from the boundedness of the expression under the integral sign, we have

$$|\delta_{\tau_u}| \leq C\,|\tau_u - \tau|.$$

Let $t \in [\tau_u, \tau]$. The equations for $z_t(\tau_u)$ and $z_t(\tau)$ coincide and the difference is in the initial values. So, for δ_t, we have

$$\frac{\partial}{\partial t}\delta_t = 2b_t\delta_t - g_t^2[z_t(\tau_u) + z_t(\tau)]\delta_t, \quad \delta_{\tau_u},$$

hence we can write

$$\sup_{0\leq t\leq T} |\delta_t| \leq \tilde{C}\,|\delta_{\tau_u}| \leq C_0\,\varepsilon^2 u.$$

Substituting this inequality in (6.37) and then in (6.35) we obtain

$$\mathbf{P}_\tau - \lim_{\varepsilon\to 0}\frac{1}{\varepsilon^2}\int_{\tau_u}^T g_t^2[m_t(\tau_u) - m_t(\tau)]^2 dt = 0. \tag{6.41}$$

This convergence is uniform on $\tau \in \mathbf{K}_\tau$ and $u_\varepsilon \to u$ because all the estimates are majored by the constants which do not depend on τ and u_ε.

So we have proved the uniform convergence

$$\mathbf{P}_\tau - \lim_{\varepsilon\to 0}\frac{1}{\varepsilon^2}\int_0^T [f_t(\tau_u)m_t(\tau_u) - f_t(\tau)m_t(\tau)]^2 dt = uv_\tau^2. \tag{6.42}$$

From the convergences (6.41) and (6.42), it follows that

$$\lim_{\varepsilon\to 0}\frac{1}{\varepsilon^2}\mathbf{E}_\tau\left(\int_{\tau_u}^T g_t[m_t(\tau_u) - m_t(\tau)]dW_t\right)^2 = 0$$

and

$$\mathcal{L}_\tau\left\{\frac{1}{\varepsilon}\int_\tau^{\tau_u}[h_t m_t(\tau_u) - y_t m_t(\tau)]dW_t\right\} \Longrightarrow \mathcal{L}_\tau\left(v_\tau \tilde{W}(u)\right).$$

To prove the last equation we have to use the representation of the stochastic integral in the form

$$\begin{aligned}
&\frac{1}{\varepsilon}\int_0^T [f_t(\tau_u)m_t(\tau_u) - f_t(\tau)m_t(\tau)]dW_t = \\
&= \frac{1}{\varepsilon}\int_\tau^{\tau_u}[h_t m_t(\tau_u) - g_t m_t(\tau)]dW_t + o(1) = \\
&= \frac{1}{\varepsilon}\int_\tau^{\tau_u} v_t dW_t + o(1) = \frac{W(\tau + u\varepsilon^2) - W(\tau)}{\varepsilon}v_\tau + o(1) = \\
&= \tilde{W}(u)v_\tau + o(1).
\end{aligned}$$

So we have the convergence of one-dimensional distributions. The proof of the convergence of multi-dimensional distributions is a direct generalization of the presented proof.

Lemma 6.2. *Under the made assumptions*

$$\sup_{\tau\in\mathbf{K}_\tau,|u_i|<R,i=1,2} |u_2-u_1|^{-2}\mathbf{E}_\tau|Z_\varepsilon^{\frac{1}{4}}(u_2)-Z_\varepsilon^{\frac{1}{4}}(u_1)|^{\frac{1}{4}} \le B(1+R^2).$$

Proof of this lemma is similar to the proof of Lemma 5.4 and so is not presented here.

Lemma 6.3. *Under the previously made assumptions, there are the constants $C>0$ and $\kappa>0$ such that for any compact $\mathbf{K}_\tau$*

$$\sup_{\tau\in\mathbf{K}_\tau} \mathbf{E}_\tau Z_\varepsilon^{\frac{1}{2}}(u) \le C\,e^{-\kappa|u|}.$$

Proof is also a particular case of Lemma 5.5. We obtain the corresponding inequalities using (6.33) and the technique of Lemma 2.3. Note, that for positive u

$$\begin{aligned}\int_0^T [f_t(\tau_u)m_t(\tau_u)-f_t(\tau)m_t(\tau)]^2dt &\ge\\ \ge \int_\tau^{\tau_u}[h_tm_t(\tau_u)-g_tm_t(\tau)]^2dt &\ge\\ \ge \int_\tau^{\tau_u} v_t^2dt - C\varepsilon \sup_{\tau\le t\le\tau_u}\left|\int_\tau^t G(s)d\bar{W}_s\right|&,\end{aligned}$$

where $G(\cdot)$ is a bounded function.

The properties of the likelihood ratio presented by Lemmas 6.1 – 6.3 allow us to make a reference to Theorem 1.4.

REMARK 6.6. If τ is a random variable with a positive continuous density $\pi(y), y\in(\tau_1,\tau_2)$, then for a wide class of loss functions, the Bayes estimator $\tilde{\tau}_\varepsilon$ is also uniformly consistent and the normed difference $\varepsilon^{-2}(\tilde{\tau}_\varepsilon-\tau)$ converges in distribution to the random variable ζ:

$$\tilde{\zeta}=\frac{\int_{-\infty}^{\infty}uZ(u)du}{\int_{-\infty}^{\infty}Z(u)du}.$$

This follows from the properties of the likelihood ratio established in Lemmas 6.1 – 6.3 and Theorem 1.5. Moreover, in this problem there exists a minimax lower bound on the risk of any estimator τ_ε^*:

$$\lim_{\delta\to0}\lim_{\varepsilon\to0}\sup_{|\tau-\tau_0|<\delta}\mathbf{E}_\tau l(\varepsilon^{-2}(\tau_\varepsilon^*-\tau))\ge\mathbf{E}l(\tilde{\zeta})$$

and for polynomial loss functions $l(\cdot)\in\mathbf{W}_p$, the Bayes estimators are local asymptotic minimax.

Disorder in the state equation. Let the observed component of the partially observed system be

$$dX_t = f_t Y_t dt + \varepsilon dW_t, \quad X_0 = 0, \quad 0 \le t \le T \tag{6.43}$$

and in the state equation at time $t = \tau$ be a switching

$$\begin{aligned} dY_t &= h_t Y_t dt + \varepsilon \sigma_t dv_t, \quad 0 \le t \le \tau, \quad Y_0 = y_0 \ne 0, \\ dY_t &= g_t Y_t dt + \varepsilon \sigma_t dv_t, \quad \tau < y \le T, \quad Y_\tau. \end{aligned} \tag{6.44}$$

Here the measurable, bounded functions f_t, g_t, h_t and σ_t are known, f_t is positive, and we have to estimate the unknown parameter $\tau \in (\tau_1, \tau_2) \subset [0, T]$ by the observations $\{X_t, 0 \le t \le T\}$. As before, we suppose that $h_t \ne g_t$ for all $t \in (\tau_1, \tau_2)$ and $\mathbf{K}_\tau$ is arbitrary compact in (τ_1, τ_2).

The system (6.44) could be written as

$$dY_t = r_t(\tau) Y_t dt + \varepsilon \sigma_t dv_t, \quad Y_0 = y_0, \quad 0 \le t \le T, \tag{6.45}$$

where

$$r_t(\tau) = h_t \chi_{\{t \le \tau\}} + g_t \chi_{\{t > \tau\}}.$$

Supposing that $v_t = h_t - g_t > 0$, we also introduce the following notations:

$$\begin{aligned} y_t(\tau) &= v_\tau x_\tau \exp\left\{ \int_\tau^t [g_s - f_s z_s(\tau)] ds \right\}, \\ \frac{dx_t}{dt} &= r_t(\tau) x_t, \quad x_0 = y_0, \\ \frac{\partial z_t(\tau)}{\partial t} &= 2 r_t(\tau) z_t(\tau) - f_t^2 z_t(\tau) + \sigma_t^2, \ z_0(\tau) = 0, \\ I(\tau) &= \int_\tau^T f_t^2 y_t^2(t) dt. \end{aligned} \tag{6.46}$$

Theorem 6.7. *Under previously made assumptions, the MLE $\hat{\theta}_\varepsilon$, as $\varepsilon \to 0$, is uniformly on $\tau \in \mathbf{K}_\tau$*

(a) consistent

$$\mathbf{P}_\tau - \lim_{\varepsilon \to 0} \hat{\tau}_\varepsilon = \tau;$$

(b) asymptotically normal

$$\mathcal{L}_\tau\{\varepsilon^{-1}(\hat{\tau}_\varepsilon - \tau)\} \Longrightarrow \mathcal{N}(0, I(\tau)^{-1});$$

(c) local asymptotic minimax for $l(\cdot) \in \mathbf{W}_p$.

Proof. We shall show that this disorder problem corresponds to the regular case with the LAN likelihood ratio, so the proof of this theorem consists of verifying the conditions of Theorem 2.2 for the observations (6.43).

Lemma 6.4. *The family of measures $\{\mathbf{P}_\tau^{(\varepsilon)}, \tau \in (\tau_1, \tau_2)\}$ is uniformly on $\tau \in \mathbf{K}_\tau$ LAN, i.e.*

$$\begin{aligned} Z_\varepsilon(u_\varepsilon) &\equiv \frac{d\mathbf{P}^{(\varepsilon)}_{\tau+\varepsilon u_\varepsilon}}{d\mathbf{P}^{(\varepsilon)}_\tau}(X) = \\ &= \exp\left\{u\Delta(\tau) - \frac{1}{2}I(\tau)u^2 + \psi_\varepsilon(\tau, u_\varepsilon, u, X)\right\}, \end{aligned}$$

where

$$\Delta(\tau) = \int_\tau^T Y_t(\tau) f_t dW_t \sim \mathcal{N}(0, I(\tau))$$

and uniformly on $\tau \in \mathbf{K}_\tau$

$$\mathbf{P}_\tau - \lim_{\varepsilon \to 0} \psi_\varepsilon(\tau, u_\varepsilon, u, X) = 0$$

for any bounded $u_\varepsilon \to u$.

Proof. By the "innovation theorem"

$$dX_t = f_t m_t(\tau)dt + \varepsilon d\bar{W}_t, \quad X_0 = 0, \quad 0 \le t \le T,$$

where $\bar{W}_t$ is an innovation Wiener process and $m_t(\tau)$ satisfies the equations of the Kalman filter:

$$dm_t(\tau) = r_t(\tau) m_t(\tau)dt + z_t(\tau) f_t (dX_t - f_t m_t(\tau)dt),$$

$$\frac{\partial}{\partial t} z_t(\tau) = 2r_t(\tau) z_t(\tau) - f_t^2 z_t(\tau)^2 + \sigma_t^2, \quad z_0(\tau) = 0.$$

So, for $t \in [0, \tau]$,

$$m_t(\tau) = y_0 + \int_0^t h_s m_s(\tau)ds + \int_0^t f_s z_s(\tau)(dX_s - f_s m_s(\tau)ds)$$

and, for $t \in [\tau, T]$,

$$\begin{aligned} m_t(\tau) &= y_0 + \int_0^\tau h_s m_s(\tau)ds + \int_\tau^t g_s m_s(\tau)ds + \\ &+ \int_0^t f_s z_s(\tau)[dX_s - f_s m_s(\tau)ds]. \end{aligned} \tag{6.47}$$

The function $z_t(\tau)$ satisfies the equations

$$\frac{\partial}{\partial t} z_t(\tau) = 2h_t z_t(\tau) - f_t^2 z_t(\tau)^2 + \sigma_t^2, \quad z_0(\tau) = 0, \ t \in [0, \tau]$$

and

$$\begin{aligned} z_t(\tau) &= 2\int_0^\tau h_s z_s(\tau)ds + 2\int_\tau^t g_s z_s(\tau)ds - \\ &- \int_0^t f_s^2 z_s(\tau)^2 ds + \int_0^t \sigma_s^2 ds, \quad t \in [\tau, T]. \end{aligned}$$

Differentiating the last pair of equations, we obtain

$$\dot{z}_t(\tau) = 0, \quad t \in [0, \tau],$$

$$\begin{aligned} \dot{z}_t(\tau) &= 2v_\tau z_\tau(\tau) + 2\int_0^t r_s(\tau)\dot{z}_s(\tau)ds - \\ &\quad - 2\int_0^t f_s^2 z_s(\tau)\dot{z}_s(\tau)ds, \end{aligned} \tag{6.48}$$

i.e. $\dot{z}_t(\tau)$ has at point $t = \tau$ the jump equal $2v_\tau z_\tau(\tau)$, so

$$\int_0^T [\dot{z}_t(\tau + v) - \dot{z}_t(\tau)]^2 f_t^2 dt \leq C\,|v|. \tag{6.49}$$

The stochastic integral

$$\dot{\eta}(\tau) = \int_0^T f_t \dot{z}_t(\tau) dW_t$$

is a Gaussian random process that has a continuous modification (according to (6.49)) and, hence, the limit transition in Lemma 1.1 is valid. Thus, the process $m_t(\tau)$ is derivable w.r.t. τ with probability 1 and, for the derivative $\dot{m}_t(\tau)$, we have

$$\begin{aligned} \dot{m}_t(\tau) &= v_\tau m_\tau(\tau) + \int_\tau^t g_s \dot{m}_s(\tau)ds + \\ &\quad + \varepsilon\int_\tau^t f_s \dot{z}_s(\tau)d\bar{W}_s - \int_\tau^t f_s z_s(\tau)\dot{m}_s(\tau)ds, \quad t \in [\tau, T]. \end{aligned} \tag{6.50}$$

Using standard arguments, we derive

$$\mathbf{P}_\tau - \lim_{\varepsilon\to 0} \sup_\tau \sup_{0\leq t\leq T} |m_t(\tau) - x_t(\tau)| = 0$$

and

$$\mathbf{P}_\tau - \lim_{\varepsilon\to 0} \sup_\tau \sup_{\tau\leq t\leq T} |\dot{m}_t(\tau) - y_t(\tau)| = 0.$$

The proof of these convergences follows directly from the representations (6.47), (6.50) and Lemma 1.11.

So we have proved the convergence

$$\mathbf{P}_\tau - \lim_{\varepsilon\to 0} \int_0^T f_t^2 \dot{m}_t(\tau)^2 dt = I(\tau)$$

and this convergence is uniform w.r.t. $\tau \in \mathbf{K}_\tau$.

Moreover, the Gaussian process $\dot{m}_t(\tau)$ is uniformly continuous in $L_2[0,T]$ on τ and on $X_t, 0 \leq t \leq T$. Hence, conditions III and IV are also satisfied. The positiveness of the Fisher information follows from the condition $y_0 \neq 0$ and the positiveness of f_t and v_t.

Lemma 6.5. *Under previously made assumptions*

$$\sup_{\tau\in\mathbf{K}_\tau, |u_i|<R} \mathbf{E}_\tau \left| Z_\varepsilon^{\frac{1}{2}}(u_2) - Z_\varepsilon^{\frac{1}{2}}(u_1) \right|^2 \leq C|u_2 - u_1|^2.$$

Proof. Denote $\Delta_t = f_t[m_t(\tau + \varepsilon u_2) - m_t(\tau + \varepsilon u_1)]$, then, by the Itô formula, the process $H_T = [Z_\varepsilon(u_2)Z_\varepsilon^{-1}(u_1)]^{\frac{1}{2}}$ admits the representation

$$H_T = 1 - \frac{1}{8\varepsilon^2}\int_0^T H_t\Delta_t^2 dt + \frac{1}{2\varepsilon}\int_0^T H_t\Delta_t d\bar{W}_t,$$

hence

$$\begin{aligned}
\mathbf{E}_\tau \left| Z_\varepsilon^{\frac{1}{2}}(u_2) - Z_\varepsilon^{\frac{1}{2}}(u_1)\right|^2 &= 2 - 2\mathbf{E}_{\tau_1} H_T = \\
&= \frac{1}{4\varepsilon^2}\mathbf{E}_{\tau_1}\int_0^T \left(\frac{d\mathbf{P}_{\tau_2}^{(\varepsilon)}}{d\mathbf{P}_{\tau_1}^{(\varepsilon)}}\right)_t \Delta_t^2 dt = \\
&= \frac{1}{4\varepsilon^2}\mathbf{E}_{\tau_2}\int_0^T f_t^2\left(\int_{\tau_1}^{\tau_2} \dot{m}_t(\tau + v)dv\right)^2 dt \leq \\
&\leq \frac{(u_2 - u_1)}{4\varepsilon}\int_{\tau_1}^{\tau_2}\left(\int_{\tau+v}^T f_t^2 \mathbf{E}\dot{m}_t(\tau + v)^2 dt\right) dv \leq \\
&\leq C|u_2 - u_1|^2
\end{aligned}$$

because $\dot{m}_t(\tau)$ is a Gaussian process of bounded variance.

Lemma 6.6. *For any* $\mathbf{K}_\tau$

$$\sup_{\tau\in\mathbf{K}_\tau} \mathbf{E}_\tau Z_\varepsilon^{\frac{1}{2}}(u) \leq C\exp\{-\kappa u^2\}.$$

Proof. Denote $\Delta_t = f_t[m_t(\tau + u) - m_t(\tau)]$. Using (6.47), for Δ_t we obtain the equation ($u \geq 0$)

$$\begin{aligned}
d\Delta_t &= [r_t(\tau + u) - z_t(\tau + u)f_t^2]\Delta_t dt + \\
&\quad + [r_t(\tau + u) - r_t(\tau)]m_t(\tau)dt + \\
&\quad + \varepsilon[r_t(\tau + u) - r_t(\tau)]f_t d\bar{W}_t, \quad \Delta_\tau = 0, \quad t \in [\tau, T]. \qquad (6.51)
\end{aligned}$$

For $t \in [0, \tau]$, we have, of course $\Delta_t = 0$ because the equations for $m_t(\tau)$ and $m_t(\tau + u)$ coincide. From the representation (6.47), we obtain the inequality

$$\sup_{\tau\leq t\leq T}|m_t(\tau) - x_t(\tau)| \leq C\,\varepsilon \sup_{\tau\leq t\leq T}\left|\int_\tau^t z_s(\tau)f_s d\bar{W}_s\right|$$

and from (6.51), the estimate

$$\begin{aligned}
\sup_{\tau\leq t\leq T}|\Delta_t - \Delta_t^\circ| &\leq C_1\varepsilon \sup_{\tau\leq t\leq T}|m_t(\tau) - x_t(\tau)| + \\
&\quad + C_2\varepsilon \sup_{\tau\leq t\leq T}\left|\int_\tau^t [z_s(\tau + u) - z_s(\tau)]f_s d\bar{W}_s\right|.
\end{aligned}$$

Denote

$$R_t(u) = \int_\tau^t [z_s(\tau + u) - z_\tau(\tau)]^2 f_s^2 ds, \quad R_\tau(u) = 0.$$

Then there exists such a Wiener process $\tilde{W}(r)$ that

$$\sup_{\tau \leq t \leq T} \left| \int_{\tau}^{t} [z_s(\tau + u) - z_s(\tau)] f_s d\bar{W}_s \right| = \sup_{0 \leq r \leq R_T(u)} |\tilde{W}(r)|$$

and, by Lemma 1.14,

$$\mathbf{E} \exp \left\{ \gamma \sup_{0 \leq r \leq R(u)} |\tilde{W}(r)| \right\} \leq C \exp\{\gamma^2 C |u|^2\}.$$

So we can choose such a $\gamma > 0$ in the proof of Lemma 2.3 which provides the finiteness of the corresponding mathematical expectation.

It only remains to refer to Theorem 2.2. The Bayes estimators are also consistent, asymptotically normal with the same parameters and LAM in this problem.

Chapter 7

Minimum Distance Estimation

We introduce a new class of estimators – minimum distance estimators – and describe their properties in regular and nonstandard situations. These estimators, in the regular case of Hilbert metrics, are consistent and asymptotically normal. In nonstandard situations, their behavior is similar to the behavior of the MLE. We find that in certain circumstances these estimators are local asymptotic minimax (asymptotically optimal) and they are better than the MLE and BE. In the case of L_1-norm and sup-norm the limit distributions of the estimators are non-Gaussian, but the limit (as $T \to 0$) distributions of these (limit) as $\varepsilon \to 0$ random variables are Gaussian.

7.1 Definitions and examples of the MDE

Let the observed diffusion-type process be

$$dX_t = S_t(\theta, X)dt + \varepsilon dW_t, \quad X_0 = x_0, \quad 0 \leq t \leq T, \tag{7.1}$$

where $\{S_t(\theta, X), 0 \leq t \leq T\} \in \mathcal{P}$ is some smooth on θ and X functional of the "past"and θ is an unknown finite-dimensional parameter, $\theta \in \Theta \subset \mathcal{R}^d$.

The maximum likelihood approach solves this problem: in the regular case, it provides a consistent, asymptotically normal, and LAM estimator but this approach has several disadvantages. To find a MLE in the nonlinear case, the statistician has to calculate the Itô stochastic integral several times and if, for example, the noise process is not a process of infinite variation, then the result differs from what is wanted. The optimality of the MLE was defined using Hajek-Le Cam's bound (Theorem 2.1)

$$\lim_{\delta \to 0} \lim_{\varepsilon \to 0} \sup_{|\theta - \theta_0| < \delta} \int l(\varepsilon^{-1} I(\theta_0)^{\frac{1}{2}}(\theta_\varepsilon - \theta)) d\mathbf{P}_\theta^{(\varepsilon)} \geq \mathbf{E} l(\xi), \tag{7.2}$$

where sup is taken over the set of probability measures $\{\mathbf{P}_\theta^{(\varepsilon)}, |\theta - \theta_0| < \delta\}$. If we consider a similar inequality with sup over a wider (nonparametric) vicinity of the measure $\mathbf{P}_{\theta_0}^{(\varepsilon)}$, then the right-hand side could increase and other estimator can arise as asymptotically optimal (minimax) [102].

In this chapter, we consider an approach concerned with the construction of minimal distance estimators which also provides the possibility of consistent, asymptotically normal and local asymptotic minimax (with nonparametric vicinity) estimation of an unknown parameter. The calculation of these estimators (except one) does not concerned with the calculation of the Itô integral.

One of the ways to introduce the minimum distance estimators in the classical i.i.d. situation is to define the empirical distribution function

$$\hat{F}_n(x) = \frac{1}{n}\sum_{j=1}^{n} \chi_{\{X_j \leq x\}},$$

which is (by the Glivenko-Cantelli theorem)a uniformly on x consistent estimator of the unknown distribution function $F(\theta_0, x)$:

$$\sup_x |\hat{F}_n(x) - F(\theta_0, x)| \to 0.$$

The difference $W_n(x) = \sqrt{n}(\hat{F}_n(x) - F(\theta_0, x))$ weakly converges to the Brownian bridge $W^\circ(F(\theta_0, x))$. Consider $\hat{F}_n(\cdot)$ and $F(\theta, \cdot)$ as elements of a Banach space $\mathcal{B}$ with the norm $\|\cdot\|_\mathcal{B}$ and introduce the minimum distance estimator θ_n^* as

$$\theta_n^* = \arg\inf_{\theta\in\Theta} \|\hat{F}_n(\cdot) - F(\theta, \cdot)\|_\mathcal{B}. \tag{7.3}$$

Using the above-mentioned convergencies, in the regular case it is possible to prove the consistence

$$\theta_n^* \longrightarrow \arg\inf_{\theta\in\Theta} \|F(\theta_0, \cdot) - F(\theta, \cdot)\|_\mathcal{B} = \theta_0$$

and asymptotic normality (in the case of Hilbert space $\mathcal{B}$)

$$\begin{aligned}
\sqrt{n}(\theta_n^* - \theta_0) &= \arg\inf_u \left\| \hat{F}_n(\cdot) - F(\theta_0, \cdot) + F(\theta_0, \cdot) - F\left(\theta_0 + \frac{u}{\sqrt{n}}, \cdot\right) \right\|_\mathcal{B} \Longrightarrow \\
&\Longrightarrow \arg\inf_u \|W^\circ(F(\cdot)) - (u, \dot{F}(\theta_0, \cdot))\|_\mathcal{B} = \\
&= \langle \dot{F}(\theta_0, \cdot), \dot{F}(\theta_0, \cdot)^T\rangle^{-1} \langle W^\circ(F(\cdot)), \dot{F}(\theta_0, \cdot)\rangle,
\end{aligned}$$

where $\langle f, g\rangle$ is an inner product in the Hilbert space $\mathcal{B}$, and T is transposing.

As was mentioned in Chapter 4, there exists a structural similarity between i.i.d. problems with the empirical distribution function

$$\hat{F}_n(x) = F(\theta_0, 0) + \int_0^x f(\theta_0, y)dy + \frac{1}{\sqrt{n}} W_n(x) \tag{7.4}$$

and our problem with the observations (7.1) which could be written as

$$X_t = x_0 + \int_0^t S_v(\theta_0, X)dv + \varepsilon W_t. \tag{7.5}$$

This analogy suggest defining the minimum distance estimator (MDE) θ_ε^* as

$$\theta_\varepsilon^* = \arg\inf_{\theta\in\Theta} |X - x(\theta)|_\mathcal{B}, \tag{7.6}$$

where $\mathcal{B}$ is a Banach space. We suppose, of course, that the trajectory $\{X_t, 0 \le t \le T\}$ and all the solutions $x(\theta) = \{x_t(\theta), 0 \le t \le T\}, \theta \in \Theta$ of the equation

$$\frac{dx_t}{dt} = S_t(\theta, x), \quad x_0, \quad 0 \le t \le T$$

belong to $\mathcal{B}$. We have the similar convergencies:

$$\sup_{0 \le t \le T} |X_t - x_t(\theta_0)| \longrightarrow 0$$

and

$$\varepsilon^{-1}(X_t - x_t(\theta_0)) \Longrightarrow x_t^{(1)},$$

where $x_t^{(1)}, 0 \le t \le T$ is a Gaussian process. So we can expect that, in the regular case, the above-defined MDE θ_ε^* is consistent and asymptotically normal. These properties of θ_ε^* will be established in the next section.

We consider the Banach spaces $\mathcal{C}_T$ (continuous on $[0, T]$ functions), $L_1[0, T]$ (integrable on $[0, T]$ functions) and a Hilbert space $L_2(\mu)$ of square integrable functions with some finite on $[0, T]$ measure $\mu(\cdot)$, i.e.

$$\|f\|^2_{L_2(\mu)} = \int_0^T f(t)^2 \mu(dt).$$

The disadvantage of definition (7.6) is that even in the simplest linear case $S_t(\theta, X) \equiv \theta f_t(X)$ the function $x(\theta)$ is nonlinear w.r.t. θ and the MDE θ_ε^* could not be directly calculated. So we propose another MDE θ_ε^{**} which is preferable from the point of view of calculation. We introduce a family of stochastic processes

$$\hat{X}_t(\theta) = x_0 + \int_0^t S_v(\theta, X) dv, \quad 0 \le t \le T, \quad \theta \in \Theta, \tag{7.7}$$

where the argument X of $S_v(\theta, \cdot)$ is our observations $X_s, 0 \le s \le v$. Recall, that

$$x_t(\theta) = x_0 + \int_0^t S_v(\theta, x) dv.$$

Define the MDE θ_ε^{**} by the equality

$$\theta_\varepsilon^{**} = \arg \inf_{\theta \in \Theta} |X - \hat{X}(\theta)|_{\mathcal{B}}. \tag{7.8}$$

Then, in the case of $\mathcal{B} = L_2(\mu)$ and linear trend $S_t(\theta, X) = (\theta, f_t(X))$, we can calculate this estimator in the exact form:

$$\theta_\varepsilon^{**} = \langle F(X), F(X)^T \rangle^{-1} \langle X - x_0, F(X) \rangle, \tag{7.9}$$

where $F_t(X) = \int_0^t f_s(X) ds$.

We can also consider the case when the measure $\mu(\cdot)$ depends on θ and, for example, $(\mu(dt) = \mu_t(\theta) dt)$

$$\theta_\varepsilon^* = \arg \inf_{\theta \in \Theta} \int_0^T |X_t - x_t(\theta)|^2 \mu_t(\theta) dt$$

(analog of the weighted Cramér-von Mises estimator).

The discrete variant of least squares estimators can be obtained using the discrete measure μ. We can consider the sequence $\{t_{0,\varepsilon}, t_{1,\varepsilon}, \ldots, t_{N(\varepsilon)}\}$ of "experiments" with $\mu_\varepsilon(t_{k,\varepsilon}) > 0$ and $N(\varepsilon) \to \infty$ as $\varepsilon \to 0$, and define the estimator

$$\theta_\varepsilon^* = \arg \inf_{\theta \in \bar{\Theta}} \sum_{k=0}^{N(\varepsilon)} |X_{t_{k,\varepsilon}} - x_{t_{k,\varepsilon}}(\theta)|^2 \mu_\varepsilon(t_{k,\varepsilon}).$$

The estimator of the minimum χ^2 type could be defined as

$$\theta_\varepsilon^* = \arg \inf_{\theta \in \bar{\Theta}} \sum_{k=0}^{N(\varepsilon)-1} \frac{[Y_k - y_k(\theta)]^2}{y_k(\theta)},$$

where

$$Y_t = X_{t_{k+1,\varepsilon}} - X_{t_{k,\varepsilon}} \quad \text{and} \quad y_k(\theta) = \int_{t_{k,\varepsilon}}^{t_{k+1,\varepsilon}} S_t(\theta, x)dt.$$

In regular cases, these estimators are also consistent and asymptotically normal.

We introduce the last minimum distance estimator using the nonparametric estimator of the trend:

$$\hat{S}_t = \frac{1}{\varphi_\varepsilon} \int_0^T G\left(\frac{t-\tau}{\varphi_\varepsilon}\right) dX_\tau$$

with the kernel $G(\cdot)$ and normalizing function φ_ε taken from Section 4.1. We define

$$\theta_\varepsilon^{***} = \arg \inf_{\theta \in \bar{\Theta}} \int_0^T [\hat{S}_t - S_t(\theta, x(\theta))]^2 \mu(dt) \tag{7.10}$$

or

$$\theta_\varepsilon^{***} = \arg \inf_{\theta \in \bar{\Theta}} \int_0^T [\hat{S}_t - S_t(\theta, X)]^2 \mu(dt).$$

Another possibility of introducing the MDE is to consider other metrics. We shall also study the MDE defined by *sup*-norm and L_1-norm. The functions $\{X_t, 0 \le t \le T\}$ and $\{x_t(\theta), 0 \le t \le T\}, \theta \in \Theta$ are the elements of C_T and $L_1(\mu)$ spaces, ($\mu(\cdot)$ is a finite measure on $[0,T]$). So we can define the two corresponding MDE's as

$$\breve{\theta}_\varepsilon = \arg \inf_{\theta \in \bar{\Theta}} \sup_{0 \le t \le T} |X_t - x_t(\theta)| \tag{7.11}$$

and

$$\theta_\varepsilon^o = \arg \inf_{\theta \in \bar{\Theta}} \int_0^T |X_t - x_t(\theta)| \mu(dt). \tag{7.12}$$

The variety of the minimum distance estimators is defined by the variety of the problems to solve. These problems could be characterized as "fitting the models". Every MDE solves the problem of corresponding choice of parametrical model. For example, $\breve{\theta}_\varepsilon$ and θ_ε^o allow us to find models closer in those metrics to the true model, estimator (7.10) allows us to find the model with the closest trend, and so on. The estimator (7.8) could even be used in the situation when a reasonable limit model does not exist (see Section 7.3 below).

A more general definition of the MDE can be given as follows. Let $Y(\theta,\cdot)$ be a family of measurable mappings from the space of observations $\mathcal{C}_T$ in a Banach space $\mathcal{B}$ with norm $\|\cdot\|_\mathcal{B}$ such that

$$\mathbf{P}_{\theta_0}-\lim_{\varepsilon\to 0}\|Y(\theta,\cdot)\|_\mathcal{B}=\|y(\theta,\theta_0,\cdot)\|_\mathcal{B},$$

where $y(\theta,\theta_0,\cdot)$ is a deterministic function such that

$$\theta_0=\arg\inf_{\theta\in\Theta}\|y(\theta,\theta_0,\cdot)\|_\mathcal{B}.$$

Define the MDE

$$\theta_\varepsilon^*=\arg\inf_{\theta\in\bar\Theta}\|Y(\theta,\cdot)\|_\mathcal{B}.$$

So the minimum-distance approach is a particular case of the minimum contrast approach (see [23]). For example, the MLE

$$\hat\theta_\varepsilon=\arg\sup_{\theta\in\Theta}\frac{d\mathbf{P}_\theta^{(\varepsilon)}}{d\mathbf{P}^{(\varepsilon)}}(X)=\arg\sup_{\theta\in\bar\Theta}\left\{\int_0^T S_t(\theta,X)dX_t-\frac{1}{2}\int_0^T S(\theta,X_t)^2dt\right\}$$

can be defined as MDE

$$\theta_\varepsilon^*=\arg\inf_{\theta\in\bar\Theta}\exp\left\{\frac{1}{2}\int_0^T S_t(\theta,X_t)^2dt-\int_0^T S_t(\theta,X)dX_t\right\}=\hat\theta_\varepsilon$$

with $\mathcal{B}=\mathbf{R}^1$ and

$$y(\theta,\theta_0)=\exp\left\{\frac{1}{2}\int_0^T[S_t(\theta,x)-S_t(\theta_0,x_t)]^2dt\right\}.$$

The asymptotic behavior of the MDE was studied by P. Millar [101], [102] within a rather general framework. To formulate his result, we need several conditions. Let $\mathcal{B}$ be a separable Hilbert space with norm $\|\cdot\|$ and inner product $\langle\cdot,\cdot\rangle$. For each $\theta\in\Theta$, (Θ is an open subset of $\mathbf{R}^d$) and $\varepsilon\in(0,1]$, suppose that $Y_\varepsilon(\theta,X)$ is a measurable mapping from $\mathcal{C}_T$ to $\mathcal{B}$. Define the random variable θ_ε^* by the equation

$$\inf_{\theta\in\bar\Theta}\|Y_\varepsilon(\theta,X)\|=\|Y_\varepsilon(\theta_\varepsilon^*,X)\|,$$

i.e., θ_ε^* any point in $\bar\Theta$ which achieves this inf.

Now fix $\theta_0\in\Theta$. There are three basic assumptions that must hold at θ_0. The first is the *hypothesis of identifiability :*

(I) *for every $\delta>0$, c, there exists $\gamma>0$ such that*

$$\mathbf{P}_{\theta_0}^{(\varepsilon)}\left\{\inf_{|\theta-\theta_0|>c}\|Y_\varepsilon(\theta,X)-Y_\varepsilon(\theta_0,X)\|>\delta\right\}\geq 1-\gamma$$

for all $\varepsilon\in(0,1]$.

The second assumption is the *hypothesis of convergence* :

(C) *there exists a function* $\Phi_\varepsilon \to 0$ *such that* $\Phi_\varepsilon^{-1} Y_\varepsilon(\theta_0, X)$ *converges in distribution to a random variable* ξ *($\mathcal{B}$-valued).*

The third assumption is the *hypothesis of differentiability:*

(D) *there exists* $\dot{y}(\theta_0) = (\dot{y}^{(1)}, \dots, \dot{y}^{(d)}), \dot{y}^{(j)} \in \mathcal{B}$ *such that*

$$Y_\varepsilon(\theta, X) = Y_\varepsilon(\theta_0, X) + (\dot{y}(\theta_0), \theta - \theta_0) + |\theta - \theta_0| \psi_\varepsilon(|\theta - \theta_0|, X),$$

where $(\cdot, \cdot)$ *and* $|\cdot|$ *are the inner product and norm in* $\mathbf{R}^d$ *and* $\psi_\varepsilon(\cdot, X)$ *is an increasing function on the line such that for any* $\delta > 0, c > 0$, *there exists a* $\gamma > 0$ *such that*

$$\mathbf{P}_{\varepsilon_0}^{(\varepsilon)} \left\{ \sup_{|\theta| < \gamma} |\psi_\varepsilon(|\theta|, X)| \le c \right\} \ge 1 - \delta$$

for all $\varepsilon \in (0, 1]$. *The vector* $\dot{y}(\theta_0)$ *is nonrandom and non-singular: there exists* $\kappa_0 > 0$ *such that*

$$\lambda^T J(\theta_0) \lambda \ge \kappa_0 |\lambda|^2,$$

where the vector $\lambda \in \mathbf{R}^d$ *and the matrix* $J(\theta_0) = \langle \dot{y}(\theta_0), \dot{y}(\theta_0)^T \rangle$.

Theorem 7.1. *If the hypotheses (I), (C) and (D) hold, then with probability approaching to 1, the MDE* θ_ε^* *exists and is unique. Moreover,*

$$\mathcal{L}_{\theta_0} \left\{ \Phi_\varepsilon^{-1} (\theta_\varepsilon^* - \theta_0) \right\} \Longrightarrow \mathcal{L}_{\theta_0}\{\zeta\},$$

where

$$\zeta = J^{-1}(\theta_0) \langle \dot{y}(\theta_0), \xi \rangle.$$

Proof can be found in [101] and [102].

7.2 Consistence and limit distributions

We consider the problem of parameter estimation by observations of the diffusion-type process

$$dX_t = S_t(\theta, X) dt + \varepsilon dW_t, \quad X_0 = x_0, \quad 0 \le t \le T, \tag{7.13}$$

where $\theta \in \Theta \subset \mathbf{R}^d$, Θ is an open bounded set and we suppose that the trend coefficient $\{S_t(\theta, X), 0 \le t \le T\}$ satisfies the conditions $\mathcal{L}$, providing the existence and uniqueness of the solution of this equation for all $\theta \in \Theta$ and the equivalence of all measures $\{\mathbf{P}_\theta^{(\varepsilon)}, \theta \in \Theta\}$ induced by the process (7.13) in $(\mathcal{C}_T, \mathcal{B}_T)$. Moreover, we recall that under these conditions we have, with $\mathbf{P}_\theta^{(\varepsilon)}$ probability 1,

$$\sup_{0 \le t \le T} |X_t - x_t(\theta)| \le C\varepsilon \sup_{0 \le t \le T} |W_t|, \tag{7.14}$$

where $x_t(\theta) = x_t$ is a solution of the nonperturbated system

$$\frac{dx_t}{dt} = S_t(\theta, x), \quad x_0, \quad 0 \le t \le T. \tag{7.15}$$

Below, we would like to prove the consistence of the all the MDEs θ_ε^*, θ_ε^{**}, θ_ε^{***}, $\breve{\theta}_\varepsilon$, θ_ε°. Note, that for all metrics $\|\cdot\|$ introduced, we can write

$$\|X_t - x_t(\theta)\| \le C\varepsilon \sup_{0\le t\le T} |W_t|,$$

where the constant C depends on the chosen metric.

First, we consider three metrics of the spaces $L_2(\mu)$, $L_1(\mu)$, and C_T (estimators $\theta_\varepsilon^*, \theta_\varepsilon^\circ, \breve{\theta}_\varepsilon$) and we put $\|\cdot\|^*$ to denote any of them.

We introduce the function

$$g_{\theta_0}(\nu) = \inf_{|\theta-\theta_0|>\nu} \|x(\theta) - x(\theta_0)\|^*$$

and

$$g(\nu) = \inf_{\theta_0\in\Theta} g_{\theta_0}(\nu)$$

and denote θ_ε as one of these three estimators.

Theorem 7.2. *For all (small) $\nu > 0$, let the function $g(\nu) > 0$, then*

$$\sup_{\theta_0\in\Theta} \mathbf{P}_{\theta_0}^{(\varepsilon)}\{|\theta_\varepsilon - \theta_0| \ge \nu\} \le 2\exp\left\{-\frac{g(\nu)^2}{2TC_*^2\varepsilon^2}\right\}.$$

Proof. Note that for the MDE, we have, by definition,

$$\{\omega : |\theta_\varepsilon - \theta_0| \ge \nu\} = \left\{\omega : \inf_{|\theta-\theta_0|<\nu} \|X - x(\theta)\|^* > \inf_{|\theta-\theta_0|\ge\nu} \|X - x(\theta)\|^*\right\}$$

and by the inequality of the triangle for the norm

$$\begin{aligned}\|X - x(\theta)\|^* &\le \|X - x(\theta_0)\|^* + \|x(\theta) - x(\theta_0)\|^*,\\ \|X - x(\theta)\|^* &\ge \|x(\theta) - x(\theta_0)\|^* - \|X - x(\theta_0)\|^*.\end{aligned}$$

Using these inequalities, the estimate (7.14), and the obvious equality

$$\inf_{|\theta-\theta_0|<\nu} \|x(\theta) - x(\theta_0)\| = 0,$$

we can write

$$\begin{aligned}&\mathbf{P}_{\theta_0}^{(\varepsilon)}\{|\theta_\varepsilon - \theta_0| \ge \nu\} =\\ &= \mathbf{P}_{\theta_0}^{(\varepsilon)}\left\{\inf_{|\theta-\theta_0|<\nu} \|X - x(\theta)\|^* > \inf_{|\theta-\theta_0|\ge\nu} \|X - x(\theta)\|^*\right\} \le\\ &\le \mathbf{P}_{\theta_0}^{(\varepsilon)}\left\{\|X - x(\theta_0)\|^* > \inf_{|\theta-\theta_0|\ge\nu} \|x(\theta) - x(\theta_0)\|^* - \|X - x(\theta_0)\|^*\right\} \le\\ &\le \mathbf{P}\left\{C_*\varepsilon \sup_{0\le t\le T} |W_t| > g(\nu)\right\} \le 2\exp\left\{-\frac{g(\nu)^2}{2TC_*\varepsilon^2}\right\}\end{aligned}$$

with the corresponding constant C_*.

So we have established the exponential consistency of these three estimators.

Consider now θ_ε^{**} defined by (7.8). Denote

$$g^{**}(\nu) = \inf_{\theta_0\in\Theta}\inf_{|\theta-\theta_0|>\nu}\|\hat{x}(\theta) - x(\theta_0)\|,$$

where the function $\hat{x}_t(\theta) = \hat{x}_t$ is a solution of the equation

$$\frac{d\hat{x}_t}{dt} = S_t(\theta, x(\theta_0)), \quad \hat{x}_0 = x_0, \quad 0 \le t \le T. \tag{7.16}$$

Here $x(\theta_0) = \{x_t(\theta_0), 0 \le t \le T\}$ is a solution of (7.15) as $\theta = \theta_0$.

Theorem 7.3. *Let the function $g^{**}(\nu) > 0$ for $\nu > 0$, then*

$$\sup_{\theta_0\in\Theta}\mathbf{P}_{\theta_0}^{(\varepsilon)}\{|\theta_\varepsilon^{**} - \theta_0| \ge \nu\} \le 2\exp\left\{-\frac{g^{**}(\nu)^2}{2TC_{**}^2\varepsilon^2}\right\}$$

with the constant $C_{**} > 0$.

Proof is similar that presented above. Note that

$$X_t - \hat{X}_t(\theta_0) = \varepsilon W_t, \quad 0 \le t \le T$$

and

$$\inf_{|\theta-\theta_0|<\nu}\|\hat{X}(\theta) - \hat{X}(\theta_0)\| = 0,$$

so we can write

$$\begin{aligned}
&\mathbf{P}_{\theta_0}^{(\varepsilon)}\{|\theta_\varepsilon^{**} - \theta_0| \ge \nu\} = \\
&= \mathbf{P}_{\theta_0}^{(\varepsilon)}\{\inf_{|\theta-\theta_0|<\nu}\|X - \hat{X}(\theta)\| > \inf_{|\theta-\theta_0|\ge\nu}\|X - \hat{X}(\theta)\|\} \le \\
&\le \mathbf{P}_{\theta_0}^{(\varepsilon)}\{\inf_{|\theta-\theta_0|<\nu}(\|X - \hat{X}(\theta_0)\| + \|\hat{X}(\theta_0) - x(\theta_0)\| + \\
&\quad +\|\hat{X}(\theta) - \hat{x}(\theta)\| + \|x(\hat{\theta}) - x(\theta_0)\|) > \\
&\quad > \inf_{|\theta-\theta_0|\ge\nu}(\|\hat{x}(\theta) - x(\theta_0)\| - \|X - \hat{X}(\theta_0)\| - \\
&\quad -\|\hat{X}(\theta_0) - x(\theta_0)\| - \|\hat{X}(\theta) - \hat{x}(\theta)\|)\} \le \\
&\le \mathbf{P}\{C_{**}\varepsilon\sup_{0\le t\le T}|W_t| > g^{**}(\nu)\} \le 2\exp\left\{-\frac{g^{**}(\nu)^2}{2C_{**}^2T\varepsilon^2}\right\}
\end{aligned}$$

with $C_{**} > 0$ defined by the inequality (7.14).

Finally, we consider θ_ε^{***} defined by (7.10).The corresponding function

$$g^{***}(\nu) = \inf_{|\theta-\theta_0|>\nu}\|S(\theta, x(\theta)) - S(\theta_0, x(\theta_0))\|$$

and remember that the kernel $G(\cdot)$ satisfies the conditions

$$\int_A^B G(u)du = 1, \quad G(u) = 0,\ u \notin [A, B]. \tag{7.17}$$

The estimator $\hat{S}_t$ under condition $\mathcal{L}$ is uniformly on $t \in [c_\varepsilon, d_\varepsilon] \to [0, T]$ consistent as $\varepsilon \to 0$ and $\varepsilon^2 \varphi_\varepsilon^{-1} \to 0$ (Theorem 4.1). Below, we suppose that the Hölder condition

$$|S_t(\theta, x) - S_s(\theta, x)| \leq L_\alpha |t - s|^\alpha, \quad \alpha > 0$$

is also satisfied and supp $\mu \subset (0, T)$. So, there exists $\tilde{\varepsilon} > 0$ such that for all $\varepsilon < \tilde{\varepsilon}$, we have supp $\mu \subset [c_\varepsilon, d_\varepsilon]$. Without loss of generality, we suppose that below $\varepsilon < \tilde{\varepsilon}$.

Theorem 7.4. *Let $g^{***}(\nu) > 0$ for $\nu > 0$, then under previously made assumptions*

$$\sup_{\theta_0 \in \Theta} \mathbf{P}_{\theta_0}^{(\varepsilon)} \{|\theta_\varepsilon^{***} - \theta_0| > \nu\} \leq K \exp\left\{-\kappa \frac{g^{***}(\nu)^2}{\varepsilon^2} \varphi_\varepsilon\right\}$$

with some constants $K > 0$ and $\kappa > 0$.

Proof. First, consider the difference

$$|\hat{S}_t - S_t(\theta_0, x(\theta_0))| = \left|\frac{1}{\varphi_\varepsilon} \int_0^T G\left(\frac{t-\tau}{\varphi_\varepsilon}\right) dX_\tau - S_t(\theta_0, x(\theta_0))\right| \leq$$

$$\leq \left|\frac{1}{\varphi_\varepsilon} \int_0^T G\left(\frac{t-\tau}{\varphi_\varepsilon}\right) S_\tau(\theta, X) d\tau - S_t(\theta_0, x(\theta_0))\right| + \left|\frac{\varepsilon}{\varphi_\varepsilon} \int_0^T G\left(\frac{t-\tau}{\varphi_\varepsilon}\right) dW_\tau\right| \leq$$

$$\leq \left|\int G(u)[S_{t-u\varphi_\varepsilon}(\theta_0, X) - S_{t-u\varphi_\varepsilon}(\theta_0, x(\theta_0))] du\right| +$$

$$+ \left|\int G(u)[S_{t-u\varphi_\varepsilon}(\theta_0, x(\theta_0)) - S_t(\theta_0, x(\theta_0))] du\right| + \frac{\varepsilon}{\varphi_\varepsilon^{\frac{1}{2}}} \left|\int G(u) d\bar{W}(u)\right| \leq$$

$$\leq C_1 \varepsilon \sup_{0 \leq t \leq T} |W_t| + C_2 \varphi_\varepsilon^\alpha \int |uG(u)| du + \frac{\varepsilon}{\varphi_\varepsilon^{\frac{1}{2}}} |xi_\varepsilon(t)|,$$

$$\xi_\varepsilon(t) = \frac{1}{\sqrt{\varphi_\varepsilon}} \int_0^T G\left(\frac{t-\tau}{\varphi_\varepsilon}\right) dW_\tau,$$

where we have used (7.14) and changed the variable $u = (t - \tau)\varphi_\varepsilon^{-1}$. Here $\bar{W}(u) = \varphi_\varepsilon^{-\frac{1}{2}} (W_t - W_{t-u\varphi_\varepsilon})$ is a Wiener process and $\xi_\varepsilon(t)$ is a Gaussian random variable with parameters $(0, \sigma^2)$, $\sigma^2 = \int G(u)^2 du$.

So, we can write

$$\mathbf{P}_{\theta_0}^{(\varepsilon)} \{|\theta_\varepsilon^{***} - \theta_0| > \nu\} =$$

$$= \mathbf{P}_{\theta_0}^{(\varepsilon)} \{\inf_{|\theta - \theta_0| < \nu} \|\hat{S} - S(\theta, x(\theta))\| > \inf_{|\theta - \theta_0| > \nu} \|\hat{S} - S(\theta, x(\theta))\|\} \leq$$

$$\leq \mathbf{P}_{\theta_0}^{(\varepsilon)} \{\inf_{|\theta - \theta_0| < \nu} (\|\hat{S} - S(\theta_0, x(\theta_0))\| + \|S(\theta_0, x(\theta_0)) - S(\theta, x(\theta))\|) >$$

$$> \inf_{|\theta - \theta_0| > \nu} (\|S(\theta, x(\theta)) - S(\theta_0, x(\theta_0))\| - \|\hat{S} - S(\theta_0, x(\theta_0))\|)\} \leq$$

$$\leq \mathbf{P}_{\theta_0}^{(\varepsilon)} \left\{\|\hat{S}(X) - S(\theta_0, x(\theta_0))\|^2 \geq \frac{g^{***}(\nu)^2}{2}\right\} \leq$$

$$\leq \mathbf{P} \left\{3C_1^2 \varepsilon^2 \sup_{0 \leq t \leq T} |W_t|^2 \mu^* > \frac{g^{***}(\nu)^2}{4} - 3C_2^2 \varphi_\varepsilon^{2\alpha} \left(\int |uG(u)| du\right)^2 \mu^*\right\} +$$

$$+ \mathbf{P}_{\theta_0}^{(\varepsilon)} \left\{\frac{\varepsilon^2}{\varphi_\varepsilon} \int_0^T \xi_\varepsilon^2(t) d\mu_t > \frac{g^{***}(\nu)^2}{4}\right\}.$$

Put

$$\varepsilon_0 = \sup\left\{\varepsilon : \varphi_\varepsilon^{2\alpha}\mu^* 3C_2^2 \int |uG(u)|^2 du = \frac{g^{***}(\nu)^2}{8}\right\}.$$

Hence, for $\varepsilon < \varepsilon_0$, the first probability

$$\mathbf{P}\left\{3C_1^2\varepsilon^2\mu^* \sup_{0\le t\le T}|W_t|^2 > \frac{g^{***}(\nu)^2}{4}\right\} \le$$

$$\le \mathbf{P}\left\{\sup_{0\le t\le T}|W_t| > \gamma_1\frac{g^{***}(\nu)}{\varepsilon}\right\} \le 2\exp\left\{-\gamma_1^2\frac{g^{***}(\nu)^2}{2T\varepsilon^2}\right\},$$

and for the second probability, using the Jensen inequality

$$\exp\left(\int_0^T f_t^2 d\mu_t\right) \le \frac{1}{\mu^*}\int_0^T \exp(\mu^* f_t^2)d\mu_t$$

and the property of $\xi_\varepsilon(t)$, we can write

$$\mathbf{P}_{\theta_0}^{(\varepsilon)}\left\{\int_0^T \xi_\varepsilon^2(t)d\mu(t) > \frac{g^{***}(\nu)^2\varphi_\varepsilon}{4\varepsilon^2}\right\} \le$$

$$\le \exp\left\{-\gamma_2\frac{g^{***}(\nu)^2\varphi_\varepsilon}{4\varepsilon^2}\right\}\mathbf{E}\exp\left\{\int_0^T \xi_\varepsilon^2(t)d\mu(t)\right\}$$

$$\le \frac{1}{\mu^*}\exp\left\{-\gamma_2\frac{g^{***}(\nu)^2\varphi_\varepsilon}{4\varepsilon^2}\right\}\int_0^T \mathbf{E}\exp(\gamma_2\mu^*\xi_t^2(t))d\mu(t).$$

So we can choose

$$\gamma_2 < \frac{1}{2\mu^*\sigma^2}$$

and obtain the finiteness of the last mathematical expectation.

We can now write

$$\mathbf{P}_{\theta_0}^{(\varepsilon)}\{|\theta_\varepsilon^{***} - \theta_0| > \nu\} \le 2\left\{-\gamma_1\frac{g^{***}(\nu)^2}{2T\varepsilon^2}\right\} + \tilde{K}\exp\left\{-\gamma_2\frac{g^{***}(\nu)^2\varphi_\varepsilon}{4\varepsilon^2}\right\},$$

and now choose ε_1 such that, for $\varepsilon < \varepsilon_1$

$$\mathbf{P}_{\theta_0}^{(\varepsilon)}\{|\theta_\varepsilon^{***} - \theta_0| > \nu\} \le K\exp\left\{-\gamma_2\frac{g^{***}(\nu)^2\varphi_\varepsilon}{4\varepsilon^2}\right\}.$$

REMARK 7.1 As follows from Theorems 7.1 and 7.2, the rate of convergence depends on the corresponding function $g(\nu)$ and it is sometimes possible to choose a good metric for some families of "signals". For example, if we have a family of narrow high functions $x_t(\theta), 0 \le t \le T$, then the *sup*-norm difference should be more than $L_2(\mu)$, and so on.

Below, we establish the asymptotic normality of the above-considered consistent MDEs. As before, the observed process is

$$dX_t = S_t(\theta, X)dt + \varepsilon dW_t, \quad X_0 = x_0, \quad 0 \le t \le T, \tag{7.18}$$

where $\theta \in \Theta \subset \mathbf{R}^d$, Θ is an open bounded set. We suppose that the trend coefficient is of the integral type

$$S_t(\theta, X) = V(\theta, t, X) + \int_0^t K(\theta, t, s, X_s)ds, \tag{7.19}$$

where the functions $V(\theta, t, x)$ and $K(\theta, t, s, x)$ satisfy the conditions $\mathcal{L}$. Then, Equation (7.18) has a unique solution and we have the estimate (Lemma 1.13)

$$\sup_{0\le t\le T} |X_t - x_t| \le C\varepsilon \sup_{0\le t\le T} |W_t|,$$

where $x_t = x_t(\theta)$ is a solution of the integro-differential equation

$$\frac{dx_t}{dt} = V(\theta, t, x_t) + \int_0^t K(\theta, t, s, x_s)ds, \quad x_0.$$

MDE θ_ε^*. We have a family of solutions $\{x_t(\theta), 0 \le t \le T, \theta \in \Theta\}$ and we define the MDE θ_ε^* and the function $g_{\theta_0}^*(\nu) = \inf_{|\theta-\theta_0|>\nu} \|x(\theta) - x(\theta_0)\|$.

Denote $x_t^{(1)} = x_t^{(1)}(\theta_0)$ as the Gaussian process, satisfying the linear equation (see Lemma 2.12)

$$\begin{aligned} dx_t^{(1)} &= [v'(\theta_0, t, x_t(\theta_0))x_t^{(1)} + \int_0^t K'(\theta, t, s, x_s(\theta_0))x_s^{(1)}ds]dt + \\ &\quad + dW_t, \quad x_0^{(1)} = 0, \; 0 \le t \le T \end{aligned} \tag{7.20}$$

and $\dot{x}_t = \dot{x}_t(\theta_0)$ as the vector-function of the derivatives of $x_t(\theta)$ w.r.t. θ at the point θ_0, $\dot{x}_t$ satisfies the system of equations

$$\begin{aligned} \frac{d\dot{x}_t(\theta_0)}{dt} &= [V'(\theta_0, t, x_t)\dot{x}_t(\theta_0) + \dot{V}(\theta_0, t, x_t) + \\ &\quad + \int_0^t (\dot{K}(\theta_0, t, s, x_s) + K'(\theta_0, t, s, x_s)\dot{x}_s(\theta_0)]ds, \quad \dot{x}_0(\theta_0) = 0. \end{aligned} \tag{7.21}$$

Here, as before, the prime corresponds to differentiating on x and the point – differentiating on θ, i.e.

$$\begin{aligned} V'(\theta, t, x_t) &= \frac{\partial}{\partial x} V(\theta, t, x)\bigg|_{x=x_t}, \\ \dot{V}(\theta_0, t, x_t(\theta_0)) &= \left(\frac{\partial}{\partial \theta^{(1)}} V(\theta, t, x), \ldots, \frac{\partial}{\partial \theta^{(d)}} V(\theta, t, x)\right)\bigg|_{\theta=\theta_0, x=x_t(\theta_0)}. \end{aligned}$$

Introduce also the matrix

$$J(\theta) = \int_0^T \dot{x}_t(\theta)\dot{x}_t^T(\theta)d\mu_t$$

and the Gaussian vector

$$\xi = J(\theta_0)^{-1} \int_0^T x_t^{(1)}(\theta_0)\dot{x}_t^T(\theta_0)d\mu_t.$$

Conditions:

I. *Stochastic process $X_t, 0 \le t \le T$ is differentiable on ε at the point $\varepsilon = 0$ in the sense*

$$\mathbf{P}_{\theta_0} - \lim_{\varepsilon \to 0} \|\varepsilon^{-1}(X - x) - x^{(1)}\| = 0,$$

where $x_t^{(1)}, 0 \le t \le T$ is the Gaussian process (7.20) with bounded coefficients $V'(\cdot)$ and $K'(\cdot)$.

II. *The function $x_t(\theta), 0 \le t \le T$ is derivable on θ at the point θ_0 in $L_2(\mu)$, i.e.*

$$\|x(\theta_0 + h) - x(\theta_0) - (h, \dot{x}(\theta_0))\| = o(|h|)$$

and the matrix $J(\theta_0)$ is positive definite.

Theorem 7.5. *Let the function $g^*(\nu) > 0$ for $\nu > 0$ and conditions I, II be fulfilled, then the MDE θ_ε^* is consistent and asymptotically normal*

$$\mathcal{L}_{\theta_0}\{\varepsilon^{-1}(\hat{\theta}_\varepsilon - \theta_0)\} \Longrightarrow \mathcal{L}\{\xi\} = \mathcal{N}(0, \sigma^2),$$

where the matrix

$$\sigma^2 = J(\theta_0)^{-1} \int_0^T \int_0^T \dot{x}_t(\theta_0)\dot{x}_s(\theta_0)^T \mathbf{E} x_t^{(1)} x_s^{(1)} d\mu_t d\mu_s J(\theta_0)^{-1}.$$

Proof. We check the conditions (I), (C), (D) of the Theorem 7.1 The *identifiability* condition follows from $g(\nu) > 0$ for $\nu > 0$, and *convergence* and *differentiability* from conditions I and II respectively.

Strengthening the conditions of regularity, we obtain below the rate of convergence in probability $\varepsilon^{-1}(\theta_\varepsilon^* - \theta_0) \to \xi$. To do this, we denote $H_x(L, a)$ as a class of functions satisfying the Hölder condition of order a, i.e.

$$H_x(L, a) = \{f(\cdot) : \ |f(x) - f(y)| \le L|x - y|^a\}.$$

Theorem 7.6. *Let the functions $V(\theta, t, x)$ and $K(\theta, t, s, x)$ have continuous bounded derivatives on x and θ of the first order such that*

$$V'(\theta, t, x), K'(\theta, t, s, x) \in H_x(L, a), \quad \dot{V}(\theta, t, x_t), \dot{K}(\theta, t, s, x_s) \in H_\theta(L, b), \qquad (7.22)$$

where $a, b \in (0, 1]$, are constants and L does not depend on θ. Suppose also that $g_{\theta_0}^*(\nu)$ is positive uniformly on $\theta_0 \in \mathbf{K}$ and $J(\theta_0)$ is positive definite uniformly on $\theta_0 \in \mathbf{K}$, i.e.

$$\inf_{\theta_0 \in \mathbf{K}} g_{\theta_0}(\nu) > 0, \qquad \inf_{\theta_0 \in \mathbf{K}} \inf_{|l|=1} (J(\theta_0) l, l) > 0. \qquad (7.23)$$

($\mathbf{K}$ is arbitrary compact in Θ). Then

$$\mathbf{P}_{\theta_0} - \lim_{\varepsilon \to 0} \varepsilon^{-1}(\theta_\varepsilon^* - \theta_0) = \xi$$

and, for any $\nu > 0$, $\mathbf{K}$ and $\varepsilon < \varepsilon_0(\nu)$

$$\sup_{\theta_0 \in \mathbf{K}} \mathbf{P}_{\theta_0}^{(\varepsilon)} \{ |\varepsilon^{-1}(\theta_\varepsilon^* - \theta_0) - \xi| > \nu \} \leq 8 \exp\{-\kappa_1 \varepsilon^{-\gamma_1}\} \tag{7.24}$$

with some positive constants $\kappa_1 = \kappa_1(\nu)$ and γ_1.

Moreover, uniformly on $\theta \in \mathbf{K}$ for any $p > 0$

$$\lim_{\varepsilon \to 0} \mathbf{E}_{\theta_0} |\varepsilon^{-1}(\theta_\varepsilon^* - \theta_0)|^p = \mathbf{E}_{\theta_0} |\xi|^p.$$

Proof. First, we change the variables:

$$u = \varepsilon^{-1}(\theta - \theta_0), \quad u \in U_{\theta_0,\varepsilon} = \{u : \ \theta_0 + \varepsilon u \in \Theta\},$$

$$u_\varepsilon^* = \varepsilon^{-1}(\theta_\varepsilon^* - \theta_0) = \arg \inf_{u \in U_{\theta_0,\varepsilon}} \|\tilde{X}^{(1)} - \varepsilon^{-1}(x(\theta_0 + \varepsilon u) - x(\theta_0))\|,$$

where $\tilde{X}_t^{(1)} = \varepsilon^{-1}(X_t - x_t(\theta_0))$, and localize the problem introducing a "good" set

$$\mathcal{A}_1 = \{\omega : \ \|X - x(\theta_0)\| < \inf_{|u| > \lambda_\varepsilon, u \in U_{\theta_0,\varepsilon}} \|X - x(\theta_0 + \varepsilon u)\|\},$$

where $\lambda_\varepsilon = \varepsilon^{-\delta}, \delta \in (0,1)$. Then, for $\omega \in \mathcal{A}_1$, we have $|\theta_\varepsilon^* - \theta_0| < \varepsilon\lambda_\varepsilon$ and $|u_\varepsilon^*| < \lambda_\varepsilon$. Differentiating the integral

$$\int_0^T [X_t - x_t(\theta_0 + \varepsilon u)]^2 d\mu_t$$

on u, we obtain a system of equations ($\dot{x}_t$ is a vector)

$$\int_0^T [X_t - x_t(\theta_0 + \varepsilon u)] \dot{x}_t(\theta_0 + \varepsilon u) d\mu_t = 0 \tag{7.25}$$

and, for $\omega \in \mathcal{A}_1$, the "MDE" u_ε^* is one of the solutions of this system satisfying $|u_\varepsilon^*| < \lambda_\varepsilon$.

Using the Taylor formula $x_t(\theta_0 + \varepsilon u) = x_t(\theta_0) + \varepsilon(u, \dot{x}_t(\theta_\varepsilon))$, where $\theta_\varepsilon = \theta_0 + \gamma_t \varepsilon u, \gamma_t \in [0,1)$, we can rewrite (7.25) as

$$J(\theta_\varepsilon, \theta_\varepsilon^*) u_\varepsilon^* = \int_0^T \tilde{X}_t^{(1)} \dot{x}_t(\theta_\varepsilon^*) d\mu_t \tag{7.26}$$

with the matrix

$$J(\theta_1, \theta_2) = \int_0^T \dot{x}_t(\theta_1) \dot{x}_t(\theta_2)^T d\mu_t.$$

To estimate the difference $|u_\varepsilon^* - \xi|$, we first consider the process $l_t = |\varepsilon^{-1}(X_t - x_t(\theta_0)) - x_t^{(1)}(\theta_0)|$. Condition (7.22) allows us to write

$$l_t \ \leq \ \int_0^t \|[V(\theta, s, X_s) - V(\theta, s, x_s)]\varepsilon^{-1} - V'(\theta, s, x_s) x_s^{(1)}| ds +$$

$$+\int_0^t\int_0^s |[K(\theta,s,v,X_v)-K(\theta,s,v,X_v)]\varepsilon^{-1}-K'(\theta,s,v,X_v)X_v^{(1)}|dvds \leq$$

$$\leq \int_0^t |V'(\theta,s,x_s)|l_s ds+\int_0^t\int_0^s |K'(\theta,s,v,X_v)|l_v dvds +$$

$$+\int_0^t |V'(\theta,s,\tilde{X}_s)-V'(\theta,s,x_s)||x_s^{(1)}|ds+$$

$$+\int_0^t\int_0^s |K'(\theta,s,v,\tilde{X}_v)-K'(\theta,s,v,x_v)||x_v^{(1)}|dvds \leq$$

$$\leq C_1\int_0^t l_s ds + C_2\int_0^t\int_0^s l_v dvds + C_3\varepsilon^a \sup_{0\leq t\leq T}|W_t|^{1+a}, \tag{7.27}$$

where $\tilde{X}_v \in (x_v, X_v)$ and we have used the estimate

$$\sup_{0\leq t\leq T}|x_t^{(1)}| \leq C \sup_{0\leq t\leq T}|W_t|$$

obtained from (7.20) with the help of Lemma 1.11.

This lemma allows us also to derive the estimate

$$\sup_{\theta_0\in\mathbf{K}}\sup_{0\leq t\leq T}|\varepsilon^{-1}(X_t-x_t(\theta_0))-x_t^{(1)}(\theta_0)| \leq C\varepsilon^a \sup_{0\leq t\leq T}|W_t|^{1+a}. \tag{7.28}$$

Further, from Equation (7.21) and using the same lemma and conditions (7.22), we obtain

$$\sup_{0\leq t\leq T}|\dot{x}_t(\theta)-\dot{x}_t(\theta_0)| \leq C|\theta-\theta_0|^b,$$

where the constant C can be chosen without depending on $\theta_0 \in \mathbf{K}$.

Equation (7.25) can be rewritten in the form

$$u_\varepsilon^* = J^{-1}(\theta_0)\int_0^T \tilde{X}_t^{(1)}\dot{x}_t(\theta_\varepsilon^*)d\mu_t - J^{-1}(\theta_0)(J(\theta_\varepsilon,\theta_\varepsilon^*)-J(\theta))u_\varepsilon^*.$$

Conditions (7.22) and (7.23) allow us to derive the estimate

$$|J^{-1}(\theta_0)(J(\theta_\varepsilon,\theta_\varepsilon^*)-J(\theta_0))u_\varepsilon^*| \leq CL^{1+b}\varepsilon^b.$$

Hence

$$\begin{aligned}
|u_\varepsilon^*-\xi| &\leq C_0\int_0^T |\tilde{X}_t^{(1)}\dot{x}_t(\theta_\varepsilon^*)-x_t^{(1)}(\theta_0)\dot{x}_t(\theta_0)|d\mu_t + CL^{1+b}\varepsilon^b \leq \\
&\leq C_1\int_0^T |\tilde{X}_t^{(1)}-x_t^{(1)}(\theta_0)||\dot{x}_t(\theta_0)|d\mu_t + \\
&\quad + C_2\int_0^T |\dot{x}_t(\theta_\varepsilon^*)-\dot{x}_t(\theta_0)||x_t^{(1)}(\theta_0)|d\mu_t + C_3L^{1+b}\varepsilon^b \leq \\
&\leq C\varepsilon^a\sup_{0\leq t\leq T}|W_t|^{1+a} + C_3L^b\varepsilon^b\sup_{0\leq t\leq T}|W_t| + C_4L^{1+b}\varepsilon^b = \\
&= C\varepsilon^a\sup_{0\leq t\leq T}|W_t|^{1+a} + C_3\varepsilon^{b-\delta}\sup_{0\leq t\leq T}|W_t| + C_4\varepsilon^{b-\delta(1+b)}.
\end{aligned} \tag{7.29}$$

Put $\delta = \frac{b}{2(1+b)}$ and denote $r = [a \wedge (b-\delta)]$. Then the last expression in (7.29) is not greater than

$$C\varepsilon^{\frac{r}{2}}\left(\varepsilon^{a-\frac{r}{2}} \sup_{0\le t\le T} |W_t|^{1+a} + \varepsilon^{b-\delta-\frac{r}{2}} \sup_{0\le t\le T} + \varepsilon^{\frac{b-r}{2}}\right).$$

So, if we also introduce the sets

$$\begin{aligned}\mathcal{A}_2 &= \left\{\omega : \sup_{0\le t\le T} |W_t| < \varepsilon^{-\frac{2a-r}{2(b+1)}}\right\},\\ \mathcal{A}_3 &= \left\{\omega : \sup_{0\le t\le T} |W_t| < \varepsilon^{-(b-\delta-\frac{r}{2})}\right\},\end{aligned}$$

and denoting

$$\gamma = \min\left(\frac{2a-r}{2(b+1)}, \left(b-\delta-\frac{r}{2}\right)\right)$$

on $\mathcal{A} = \mathcal{A}_1 \bigcap \mathcal{A}_2 \bigcap \mathcal{A}_3$, we can write

$$|u_\varepsilon^* - \xi| \le C\varepsilon^{\frac{r}{2}} \tag{7.30}$$

with some constant $C > 0$. It only remains to estimate the probability of the complement of $\mathcal{A}$

$$\mathbf{P}_{\theta_0}^{(\varepsilon)}(\mathcal{A}^C) \le \sum_{j=1}^{3} \mathbf{P}_{\theta_0}^{(\varepsilon)}(\mathcal{A}_j^C).$$

We begin with $\mathbf{P}_{\theta_0}^{(\varepsilon)}(\mathcal{A}_1^C)$. Below, we use the inequality of Schwarz

$$\begin{aligned}&\mathbf{P}_{\theta_0}^{(\varepsilon)}\left\{\inf_{|u|\ge\lambda_\varepsilon} \|\varepsilon^{-1}(X - x(\theta_0)) - \varepsilon^{-1}(x(\theta_0+\varepsilon u) - x(\theta_0)\| < \|\varepsilon^{-1}(X - x(\theta_0))\|\right\} \le\\ &\le \mathbf{P}_{\theta_0}^{(\varepsilon)}\left\{\inf_{|u|>\lambda_\varepsilon} (\varepsilon^{-2}\|x(\theta_0+\varepsilon u) - x(\theta_0)\| - \right.\\ &\qquad \left. -2\,\varepsilon^{-2}\int_0^T (X_t - x_t(\theta_0))(x_t(\theta_0+\varepsilon u) - x_t(\theta_0))d\mu_t < 0\right\} \le\\ &\le \mathbf{P}_{\theta_0}^{(\varepsilon)}\left\{\inf_{|u|\ge\lambda_\varepsilon} [(J(\theta_\varepsilon)u,u) - 2\varepsilon^{-1}(J(\theta_\varepsilon)u,u)^{\frac{1}{2}}\|X - x(\theta_0)\|] > 0\right\} \le\\ &\le \mathbf{P}_{\theta_0}^{(\varepsilon)}\left\{\varepsilon^{-1}\|X - x(\theta_0)\| \ge \frac{\kappa}{2}\lambda_\varepsilon\right\}.\end{aligned}$$

We obtain the last inequality after division on $(J(\theta_\varepsilon)u,u)^{\frac{1}{2}}$ and substitution

$$\inf_{|u|>\lambda_\varepsilon} (J(\theta_\varepsilon)u,u) \ge \kappa\lambda_\varepsilon^2,$$

which follows from (7.23) with some $\kappa > 0$. By Lemma 1.13, we have with $\mathbf{P}_{\theta_0}^{(\varepsilon)}$ probability 1

$$\sup_{0\le t\le T} \varepsilon^{-1}|X_t - x_t(\theta_0)| \le C \sup_{0\le t\le T} |W_t|.$$

Thus

$$\mathbf{P}_{\theta_0}^{(\varepsilon)}(\mathcal{A}_1^C) \leq \mathbf{P}\left\{\sup_{0\leq t\leq T}|W_t| > \frac{\kappa}{2C\mu^*}\varepsilon^{-\delta}\right\} \leq 4\exp\{-\kappa_0\varepsilon^{-2\delta}\}$$

with a positive constant κ_0.

We also have

$$\mathbf{P}_{\theta_0}^{(\varepsilon)}(\mathcal{A}_2^C) \leq +\mathbf{P}_{\theta_0}^{(\varepsilon)}(\mathcal{A}_3^C)4\exp\left\{-\frac{1}{2T}\varepsilon^{-2\gamma}\right\}.$$

Hence

$$\mathbf{P}_{\theta_0}^{(\varepsilon)}(\mathcal{A}^C) \leq 8\exp\{-\kappa_1\varepsilon^{-\gamma_1}\},$$

where $\gamma_1 = \min(2\delta, 2\gamma)$.

This allows us to write

$$\begin{aligned}\mathbf{P}_{\theta_0}^{(\varepsilon)}\{|(\theta_\varepsilon^* - \theta_0)\varepsilon^{-1} - \xi| > \nu\} &= \\ &= \mathbf{P}_{\theta_0}^{(\varepsilon)}\{|u_\varepsilon^* - \xi| > \nu, \mathcal{A}\} + \mathbf{P}_{\theta_0}^{(\varepsilon)}\{|u_\varepsilon^* - \xi| > \nu, \mathcal{A}^C\} \leq \\ &\leq \mathbf{P}_{\theta_0}^{(\varepsilon)}\{C\varepsilon^{\frac{r}{2}} > \nu, \mathcal{A}\} + \mathbf{P}_{\theta_0}^{(\varepsilon)}(\mathcal{A}^C) \leq 8\exp\{-\kappa_1\varepsilon^{-\gamma_1}\}\end{aligned}$$

for $\varepsilon < \varepsilon_0 = (\frac{\nu}{C})^{\frac{2}{r}}$. Here C is the constant from (7.30).

For small $|u|$, we have

$$\|x(\theta+u) - x(\theta)\|^2 = (J(\theta)u, u)(1 + o(1)),$$

so by conditions (7.23), we can find a positive τ such that

$$\|x(\theta+u) - x(\theta)\|^2 \geq \tau|u|^2 \tag{7.31}$$

for all $u \in U = \{u : \theta + u \in \Theta\}$. Hence, by Theorem 7.2

$$\begin{aligned}\sup_{\theta_0\in\mathbf{K}} \mathbf{P}_{\theta_0}^{(\varepsilon)}\{\varepsilon^{-1}|\theta_\varepsilon^* - \theta_0| > x\} &\leq 2\exp\left\{-\frac{1}{2TC^2\varepsilon^2}g^*(\varepsilon x)\right\} \leq \\ &\leq 2\exp\left\{-\frac{\tau}{2TC^2}x^2\right\}.\end{aligned}$$

The uniform integrability of $|u_\varepsilon^*|^p$ follows from the following inequalities:

$$\begin{aligned}\mathbf{E}_{\theta_0}|u_\varepsilon^*|^p &= \int_0^\infty x^p d\mathbf{P}_{\theta_0}^{(\varepsilon)}\{|u_\varepsilon^*| < x\} = -\int_0^\infty x^p d\mathbf{P}_{\theta_0}^{(\varepsilon)}\{|u_\varepsilon^*| > x\} = \\ &= -x^p\mathbf{P}_{\theta_0}^{(\varepsilon)}\{|u_\varepsilon^*| > x\}\Big|_0^\infty + p\int_0^\infty x^{p-1}\mathbf{P}_{\theta_0}^{(\varepsilon)}\{|u_\varepsilon| > x\}dx \leq \\ &\leq 2p\int_0^\infty x^{p-1}\exp\left\{-\frac{\tau}{2TC^2}x^2\right\}dx < \infty.\end{aligned}$$

Using these inequalities, we can immediately obtain the upper and lower bounds on the deviations from the Gaussian law, because on $\mathcal{A}$

$$\xi - C\varepsilon^{\frac{r}{2}} < u_\varepsilon^* < \xi + C\varepsilon^{\frac{r}{2}}.$$

Hence

$$\begin{aligned}\mathbf{P}_{\theta_0}^{(\varepsilon)}\{u_\varepsilon^* < x, \mathcal{A}\} &\leq \mathbf{P}(\xi < x + C\varepsilon^{\frac{r}{2}}),\\ \mathbf{P}_{\theta_0}^{(\varepsilon)}\{u_\varepsilon^* < x, \mathcal{A}\} &\geq \mathbf{P}(\xi < x - C\varepsilon^{\frac{r}{2}}) - \mathbf{P}_{\theta_0}^{(\varepsilon)}(\mathcal{A}^C),\end{aligned}$$

where ξ is Gaussian.

REMARK 7.2 This behavior of the MDE could also be obtained with the help of Theorem 1.4. Indeed, if we introduce a stochastic processes

$$Z_\varepsilon(u) = \exp\left\{-\frac{\alpha}{\varepsilon^2}\int_0^T |X_t - x_t(\theta_0 + \varepsilon u)|^2 d\mu_t\right\}$$

and

$$Z(u) = \exp\left\{-\alpha \int_0^T |x_t^{(1)}(\theta_0) - (u, \dot{x}_t(\theta_0))|^2 d\mu_t\right\},$$

where α is a positive constant, then we can easily establish the uniform on $\theta_0 \in \mathbf{K}$ convergence of the vector $\{Z_\varepsilon(u_1), \ldots, Z_\varepsilon(u_n)\}$ to the vector $\{Z(u_1), \ldots, Z(u_n)\}$ for any integer n. Moreover, for the derivative $\partial Z_\varepsilon(u)/\partial u$, we have

$$\frac{\partial Z_\varepsilon(u)}{\partial u} = -2\frac{\alpha}{\varepsilon} Z_\varepsilon(u) \int_0^T [X_t - x_t(\theta_0 + \varepsilon u)]\dot{x}_t(\theta_0 + \varepsilon u) d\mu_t.$$

Hence (below $u_v = u_1 + (u_2 - u_1)v$ and $\theta_v = \theta_0 + \varepsilon u_v$, $v \in [0,1]$)

$$\begin{aligned}&\mathbf{E}_{\theta_0}|Z_\varepsilon(u_2) - Z_\varepsilon(u_1)|^{2m} =\\ &= \mathbf{E}_{\theta_0}\left|\int_0^1 Z_\varepsilon(u_v)\left\{-\frac{2\alpha}{\varepsilon}\int_0^T [X_t - x_t(\theta_v)](u_2 - u_1, \dot{x}_t(\theta_v))\right\} d\mu_t\right|^{2m} dv \leq\\ &\leq \left(\frac{2\alpha}{\varepsilon}\right)^{2m} \int_0^1 \mathbf{E}_{\theta_0}\left|Z_\varepsilon(u_v)\int_0^T [X_t - x_t(\theta_v)](u_2 - u_1, \dot{x}_t(\theta_v))]d\mu_t\right|^{2m} dv \leq\\ &\leq 2^{4m-1}\frac{\alpha^{2m}}{\varepsilon^{2m}} T^{2m-1}\int_0^1\int_0^T (\mathbf{E}_{\theta_0}|X_t - x_t(\theta_0)|^{2m} + |x_t(\theta_v) - x_t(\theta_0)|^{2m}) \times\\ &\qquad \times |(u_2 - u_1, \dot{x}_t(\theta_v))|^{2m} d\mu_t dv \leq C_1(|u_2 - u_1|^{2m} + |u_2 - u_1|^{4m}),\end{aligned}$$

where we have used the estimates:

$$\mathbf{E}_{\theta_0}|X_t - x_t(\theta_0)|^{2m} \leq C\varepsilon^{2m},$$

$$|x_t(\theta_0 + \varepsilon u_v) - x_t(\theta_0)| \leq C|u_2 - u_1|\varepsilon.$$

So we have proved the inequality

$$\sup_{|u_1|<R, |u_2|<R} |u_2 - u_1|^{-2m}\mathbf{E}_\theta|Z_\varepsilon(u_2) - Z_\varepsilon(u_1)|^{2m} \leq C(1 + R^{2m}).$$

The last condition (1.77) could be also checked directly, because

$$\mathbf{E}_{\theta_0}\exp\left\{-\frac{\alpha}{\varepsilon^2}\int_0^T |X_t - x_t(\theta_0 + \varepsilon u)|^2 d\mu_t\right\} \leq$$

$$\leq \exp\left\{-\frac{\alpha}{\varepsilon^2}\|x(\theta_0+\varepsilon u)-x(\theta_0)\|^2\right\}\times$$
$$\times\,\mathbf{E}_{\theta_0}\exp\left\{\frac{2\alpha}{\varepsilon^2}\int_0^T|X_t-x_t(\theta_0)||x_t(\theta_0+\varepsilon u)-x_t(\theta_0)|d\mu_t\right\}\leq$$
$$\leq \exp\left\{-\alpha\kappa|u|^2\right\}\mathbf{E}_{\theta_0}\exp\left\{2\alpha C|u|\sup_{0\leq t\leq T}|W_t|\right\}\leq$$
$$\leq \exp\left\{-\alpha\kappa|u|^2+c_*\alpha^2u^2\right\}$$

with the corresponding constant $c_* > 0$. Above, we used the inequalities (7.31) and (1.50).

Now we choose

$$\alpha=\frac{\kappa}{2c_*}$$

and obtain

$$\sup_{\theta_0\in\mathbf{K}}\mathbf{E}_{\theta_0}Z_\varepsilon(u)\leq e^{-\alpha\kappa|u|^2}.$$

So for $Z_\varepsilon(\cdot)$ all the conditions of Theorem 1.4 are fulfilled and for the MDE, we have uniformly on $\theta_0\in\mathbf{K}$

$$\mathbf{P}_{\theta_0}^{(\varepsilon)}\{\varepsilon^{-1}(\theta_\varepsilon^*-\theta_0)<x\}=$$
$$=\mathbf{P}_{\theta_0}^{(\varepsilon)}\{\inf_{u<x}\|X-x(\theta_0+\varepsilon u)\|<\inf_{u>x}\|X-x(\theta_0+\varepsilon u)\|\}=$$
$$=\mathbf{P}_{\theta_0}^{(\varepsilon)}\{\sup_{u<x}Z_\varepsilon(u)>\sup_{u>x}Z_\varepsilon(u)\}\longrightarrow\mathbf{P}\{\sup_{u<x}Z(u)>\sup_{u>x}Z(u)\}=$$
$$=\mathbf{P}\{J^{-1}(\theta_0)\langle x^{(1)}(\theta_0),\dot{x}(\theta_0)\rangle<x\},$$

for any vector $x\in\mathbf{R}^d$.

MDE θ_ε^{**}. Now consider the estimator θ_ε^{**} defined by (7.7), (7.8). Remember that $\hat{X}_t(\theta)$ converges to the function $\hat{x}_t(\theta)$ satisfying (7.16). The corresponding matrix

$$J^*(\theta)=\int_0^T\dot{x}_t^*(\theta)\dot{x}_t^*(\theta)^Td\mu_t$$

and the vector $\dot{x}_t^*(\theta)$ of derivatives $\hat{x}_t(\theta)$ w.r.t. θ is equal to

$$\dot{x}_t^*(\theta)=\int_0^t\dot{V}(\theta,s,x_s(\theta_0))ds+\int_0^t\int_0^s\dot{K}(\theta,s,v,x_v(\theta_0))dvds.$$

Introduce also the Gaussian vector

$$\zeta=J^*(\theta_0)^{-1}\int_0^TW_t\dot{x}_t^*(\theta_0)d\mu_t$$

with zero-mean and the covariance matrix

$$\mathbf{E}\zeta\zeta^T=J^*(\theta_0)^{-1}\int_0^T\int_0^T(t\wedge s)\dot{x}_t^*(\theta_0)\dot{x}_s^*(\theta_0)^Td\mu_td\mu_sJ(\theta_0)^{-1}.$$

Proposition 7.1. *Let the functions $V(\theta,t,x), K(\theta,t,s,x)$ have continuous bounded derivatives on x and the functions $V(\theta,t,x_t(\theta_0)), K(\theta,t,s,x_s(\theta_0))$ have continuous bounded derivatives on θ on θ belonging to $H_\theta(L,b)$, the matrix $J^*(\theta)$ is positive definite uniformly on $\theta \in \mathbf{K}$, and the function $g^{**}(\nu)$ is positive for $\nu > 0$, then uniformly on $\theta \in \mathbf{K}$*

$$\mathbf{P}_\theta - \lim_{\varepsilon \to 0} \varepsilon^{-1}(\theta_\varepsilon^{**} - \theta) = \zeta$$

and the exponential-type estimate like (7.24) is also valid.

Proof is not presented because it differs only slightly from the proof of Theorem 7.5.

MDE θ_ε^{***} defined by (7.10) has another asymptotics. Recall that $\varphi_\varepsilon = \varepsilon^{\frac{2}{3}}$ and that the kernel $G(\cdot)$ satisfies (7.17).

Introduce the notations:

(a) vector

$$\begin{aligned}\dot{S}_t(\theta, x(\theta)) &= \dot{V}(\theta,t,x_t(\theta)) + V'(\theta,t,x_t(\theta))\dot{x}_t(\theta) + \\ &+ \int_0^t [\dot{K}(\theta,t,s,x_s(\theta)) + K'(\theta,t,s,x_s(\theta))\dot{x}_s(\theta)]ds;\end{aligned}$$

(b) matrix

$$\tilde{J}(\theta) = \int_0^T \dot{S}_t(\theta,x(\theta))\dot{S}_t(\theta,x(\theta))^T d\mu_t;$$

(c) vector

$$m(\theta) = -\sigma \tilde{J}(\theta)^{-1} \int_0^T \dot{S}_t(\theta,x(\theta))\frac{\partial}{\partial t}S_t(\theta,x(\theta))d\mu_t,$$

where $\sigma = \int_A^B uG(u)du$.

Theorem 7.7. *Let the functions $V(\theta,t,x)$, $K(\theta,t,s,x)$ have continuous bounded derivatives on x and t, the functions $V(\theta,t,x_t(\theta))$, $K(\theta,t,s,x_s(\theta))$ have continuous bounded derivatives on θ belonging to $H_\theta(L,a)$, the matrix $\tilde{J}(\theta)$ is positive definite uniformly on $\theta \in \mathbf{K}$, and $g^{***}(\nu) > 0$ for $\nu > 0$. Then if the measure $\mu(\cdot)$ is absolutely continuous, we have the convergence*

$$\mathbf{P}_{\theta_0} - \lim_{\varepsilon \to 0} \varepsilon^{-\frac{2}{3}}(\theta_\varepsilon^{***} - \theta_0) = m(\theta_0)$$

uniformly on $\theta_0 \in \mathbf{K}$ and if $\mu(\cdot)$ has a discrete component $\mu_d(\cdot)$, the limit is

$$\mathbf{P}_{\theta_0} - \lim_{\varepsilon \to 0} \varepsilon^{-\frac{2}{3}}(\theta_\varepsilon^{***} - \theta_0) = \zeta^*,$$

where $\mathcal{L}_{\theta_0}\{\zeta^*\} = \mathcal{N}(m(\theta_0), d^2(\theta_0))$ with

$$d^2(\theta_0) = \tilde{J}^{-1}(\theta_0) \sum_l \dot{S}_{t_l}(\theta_0, x(\theta_0))\dot{S}_{t_l}(\theta_0,x(\theta_0))^T \mu_\alpha(t_l)^2 \int G(v)^2 dv.$$

Proof. First note that according to Theorem 4.2, the normalizing function φ_ε can be chosen as $\varphi_\varepsilon = \varepsilon^{\frac{2}{3}}$. The demonstration of this theorem is similar to that of Theorem 7.5. We localize the problem introducing the "good" set $\mathcal{A}_1$, providing the solution $|u_\varepsilon^{***}| < L_\varepsilon = \varepsilon^{-\delta}, \delta \in (0, \frac{1}{3})$. Then repeating the calculation from Theorem 4.1 (see Remark 4.2), we write

$$\begin{aligned}
&\varepsilon^{-\frac{2}{3}}[\hat{S}_t(X) - S_t(\theta, x(\theta))] = \\
&= \varepsilon^{-\frac{2}{3}}[\hat{S}_t(X) - S_t(\theta_0, x(\theta_0))] - [S_t(\theta_0 + \varepsilon^{-\frac{2}{3}}u, x(\theta_0 + \varepsilon^{-\frac{2}{3}}u)) - S_t(\theta_0, x(\theta_0))] = \\
&= \varepsilon^{-\frac{2}{3}} \int G(v)[S_{t-v\varphi_\varepsilon}(\theta_0, x(\theta_0)) - S_t(\theta_0, x(\theta_0))]dv + \\
&+\varepsilon^{-\frac{2}{3}} \int G(v)[S_{t-v\varphi_\varepsilon}(\theta_0, X) - S_{t-v\varphi_\varepsilon}(\theta_0, x(\theta_0))]dv + \\
&+\varepsilon^{-\frac{1}{3}} \int_0^T G\left(\frac{t-\tau}{\varphi_\varepsilon}\right) dW_\tau = -\frac{\partial}{\partial t} S_t(\theta_0, x(\theta_0))\sigma + \xi_\varepsilon(t) + o(1).
\end{aligned}$$

The random variables $\xi_\varepsilon(t)$ for all $t \in (0, T)$ are Gaussian with the parameters

$$\begin{aligned}
\mathbf{E}\xi_\varepsilon(t) &= \mathbf{E}\varepsilon^{-\frac{1}{3}} \int_0^T G\left(\frac{t-\tau}{\varphi_\varepsilon}\right) dW_\tau = 0, \\
\mathbf{E}\xi_\varepsilon(t)\xi_\varepsilon(s) &= \mathbf{E}\varepsilon^{-\frac{2}{3}} \int_0^T G\left(\frac{t-\tau}{\varphi_\varepsilon}\right) G\left(\frac{s-\tau}{\varphi_\varepsilon}\right) d\tau = \\
&= \int_A^B G(v) G\left(v + \frac{s-t}{\varphi_\varepsilon}\right) dv \to 0
\end{aligned}$$

if $s \neq t$. Hence

$$\begin{aligned}
&\mathbf{E}_{\theta_0}(u_\varepsilon^{***} - m(\theta_0))^2 = \\
&\quad = \tilde{J}^{-1}(\theta_0) \int_0^T \int_0^T \dot{S}_t(\theta_0, x(\theta_0)) \dot{S}_s(\theta_0, x(\theta_0))^T \mathbf{E}\xi_\varepsilon(t)\xi_\varepsilon(s) d\mu_t d\mu_s \tilde{J}^{-1}(\theta_0) + \\
&\quad +o(1) = o(1)
\end{aligned}$$

in the case of absolutely continuous $\mu(\cdot)$ and if $\mu(\cdot)$ has a discrete component, then

$$\begin{aligned}
&\int_0^T \int_0^T \dot{S}_t(\theta_0, x(\theta_0)) \dot{S}_s(\theta_0, x(\theta_0))^T \mathbf{E}\xi_\varepsilon(t)\xi_\varepsilon(s) d\mu_t d\mu_s = \\
&\quad = \sum_l \dot{S}_{t_l}(\theta_0, x(\theta_0)) \dot{S}_{t_l}(\theta_0, x(\theta_0))^T \mu_d(t_l)^2 \int G(v)^2 dv
\end{aligned}$$

and the random variable u_ε^{***} is asymptotically nondegenerate (Gaussian).

REMARK 7.3 We can always have $m(\theta_0) = 0$ if the kernel $G(\cdot)$ satisfies the condition

$$\int_A^B uG(u)du = 0.$$

$\breve{\theta}_\varepsilon$**-estimator**. Suppose that the conditions of Theorem 7.5 are fulfilled with $\|\cdot\|^*$ as *sup*-metric. Define the random vector

$$\begin{aligned}
\breve{\xi} &= \arg \inf_{u \in \mathbf{R}^d} \sup_{0 \le t \le T} |x_t^{(1)} - (u, \dot{x}_t(\theta))| = \\
&= \arg \inf_{u \in \mathbf{R}^d} |x_{\tau(u)}^{(1)} - (u, \dot{x}_{\tau(u)}(\theta))|, \qquad (7.32)
\end{aligned}$$

where $\tau(u)$ is instant when *sup* is achieved. Suppose also that $\breve{\xi}$ is a well-defined unique solution of this minimax problem.

REMARK 7.4 In the one-dimensional case ($d = 1$), the uniqueness of $\breve{\xi}$ follows from the condition $\dot{x}_t(\theta) \neq 0$. Indeed, denote

$$\varphi_t(u) = |x_t^{(1)} - u\dot{x}_t(\theta)|, \quad \varphi(u) = \varphi_{\tau(u)}(u), \quad \varphi(\breve{\xi}) = \inf_u \varphi(u).$$

The norm $\varphi(u)$ is a convex continuous function of u and $\varphi(u) \to \infty$ as $|u| \to \infty$, so the *inf* on u exists and could be one point or an interval. Suppose there are at least two points u_1 and $u_2 \neq u_1$ such that $\varphi(u_1) = \varphi(u_2) = \varphi(\breve{\xi})$. The line (on u) $\varphi_{\tau(u_1)}(u)$ is not parallel to the axis $u = 0$, so we can have

$$\varphi(u_1) = \varphi_{\tau(u_1)}(u_1) < \varphi_{\tau(u_1)}(u_2) \leq \sup_{0\leq t\leq T} \varphi_t(u_2) = \varphi(u_2).$$

This contradiction proves the uniqueness.

Denote the matrix-function

$$I_t(\theta) = \dot{x}_t(\theta)\dot{x}_t(\theta)^T, \quad 0 \leq t \leq T$$

and $\|f\|_s$ the *sup*-norm in $\mathcal{C}_T$. We also need the condition

$$\inf_{\theta\in\mathbf{K}} \sup_{0\leq t\leq T} \inf_{|l|=1} (I_t(\theta)l, l) > 0. \tag{7.33}$$

Recall also that the function

$$\breve{g}(\nu) = \inf_{\theta_0\in\mathbf{K}} \inf_{|\theta-\theta_0|>\nu} \|x(\theta) - x(\theta_0)\|_s > 0 \tag{7.34}$$

for $\nu > 0$ (one of the conditions of Theorem 7.5). These two last conditions provide the inequality

$$\inf_{\theta\in\mathbf{K}} \|x(\theta + u) - x(\theta)\|_s^2 \geq \kappa|u|^2.$$

According to Theorem 7.1, we can put $\nu = \nu_\varepsilon = \varepsilon\lambda_\varepsilon \to 0$, where $\lambda_\varepsilon \to \infty$ and obtain

$$\sup_{\theta\in\mathbf{K}} \mathbf{P}_{\theta_0}^{(\varepsilon)}\{|\breve{\theta}_\varepsilon - \theta| > \nu_\varepsilon\} \leq \sup_{\theta\in\mathbf{K}} \mathbf{P}_{\theta_0}^{(\varepsilon)}\{|\breve{u}_\varepsilon| > \lambda_\varepsilon\} \leq \exp\{-\kappa^*\lambda_\varepsilon^2\},$$

where $\breve{u}_\varepsilon = \varepsilon^{-1}(\breve{\theta}_\varepsilon - \theta)$ and $\kappa^* = \kappa(2TC_*^2)^{-1}$.

So we can choose the set

$$\mathcal{A}_1 = \left\{\omega : \inf_{|\theta-\theta_0|<\nu_\varepsilon} \|X - x(\theta)\|_s < \inf_{|\theta-\theta_0|>\nu_\varepsilon} \|X - x(\theta)\|_s\right\}$$

as "good" and in this set have the necessary approximation:

$$\sup_{|u|<\lambda_\varepsilon} \|\varepsilon^{-1}(X - x(\theta + \varepsilon u)) - x^{(1)} + (u, \dot{x}(\theta))\|_s \to 0.$$

Hence

$$\sup_{|u|<\lambda_\varepsilon} \left|\varepsilon^{-1}\|X - x(\theta+\varepsilon u)\|_s - \|x^{(1)} - (u,\dot{x}(\theta))\|_s\right| \to 0.$$

and

$$\begin{aligned}\breve{u}_\varepsilon &= \arg\inf_{|u|<\lambda_\varepsilon} \|\varepsilon^{-1}(X - x(\theta+\varepsilon u))\|_s \to \\ &\to \arg\inf_u \|x^{(1)} - (u,\dot{x}(\theta))\|_s = \breve{\xi}.\end{aligned}$$

So we have the following result.

Proposition 7.2. *Suppose that the functions $V(\cdot)$ and $K(\cdot)$ are as smooth as was required in Theorem 7.5, the conditions (7.33), (7.34) are fulfilled, and $\breve{\xi}$ is a unique solution of the problem (7.32), then, uniformly on $\theta \in \mathbf{K}$,*

$$\mathbf{P}_\theta - \lim_{\varepsilon\to 0} \varepsilon^{-1}(\breve{\theta}_\varepsilon - \theta) = \breve{\xi}.$$

θ^o_ε**-estimator** is investigated under similar conditions.

REMARK 7.5 Suppose that the observed process (7.18), (7.19) has the functions $V(\cdot)$ and $K(\cdot)$ as smooth as in Theorem 7.5, the function

$$g^o(\nu) = \inf_{\theta\in\mathbf{K}} \inf_{|\theta-\theta_0|>\nu} \|x(\theta) - x(\theta_0)\|_L$$

is positive for $\nu > 0$ (here $\|\cdot\|_L$ - $L_1(\mu)$-norm), the norm

$$\inf_{\theta\in\mathbf{K}} \|(\dot{x}(\theta), h)\|_l \geq \kappa|h|,$$

and the random vector

$$\xi^o = \arg\inf_{u\in\mathbf{R}^d} \|x^{(1)}(\theta) - (u,\dot{x}(\theta))\|_L$$

is unique with probability 1. Then by a similar way as in Proposition 7.2 it is possible to show that the random variables $\varepsilon^{-1}(\theta^o_\varepsilon - \theta)$ converge to ξ^o as $\varepsilon \to 0$.

We return to these last two problems in the next section in the problem of parameter estimation for the Ornstein-Uhlenbeck process.

7.3 Linear systems

We return to the problems of parameter estimation of Section 2.3 and describe the properties of the MDE for the same models. Then we compare the asymptotics ($T \to \infty$) of the limit variances of different estimators. The first asymptotic ($\varepsilon \to 0$) corresponds, as before, the "small noise" and the "large signal".

Example 7.1. Let the observed Ornstein-Uhlenbeck process be

$$dX_t = \theta X_t dt + \varepsilon dW_t, \quad X_0 = x_0 \neq 0, \quad 0 \leq t \leq T \tag{7.35}$$

with an unknown parameter $\theta \in (\alpha,\beta)$. Denote $x_t(\theta) = x_0 e^{\theta t}$ as the limit ($\varepsilon \to 0$) solution of this equation and consider the

θ_ε^* and θ_ε^{**} estimators. The conditions of Theorem 7.5 are fulfilled and the MDEs θ_ε^* and θ_ε^{**} are consistent and asymptotically normal:

$$\mathcal{L}_\theta\{\varepsilon^{-1}(\theta_\varepsilon^* - \theta)\} \Longrightarrow \mathcal{N}(0, \sigma_*(\theta)^2),$$

$$\mathcal{L}_\theta\{\varepsilon^{-1}(\theta_\varepsilon^{**} - \theta)\} \Longrightarrow \mathcal{N}(0, \sigma_{**}(\theta)^2),$$

where the limit variances are

if $\theta = 0$

$$\sigma_*(0)^2 = \left(x_0 \int_0^T t^2 d\mu_t\right)^{-2} \int_0^T \int_0^T ts(t \wedge s) d\mu_t d\mu_s, \tag{7.36}$$

$$\sigma_{**}(0)^2 = (x_0 \mu([0,T]))^{-2} \int_0^T \left[\int_0^t s d\mu_s + t\mu([t,T])\right] d\mu_t; \tag{7.37}$$

if $\theta \neq 0$

$$\begin{aligned} \sigma_*(\theta)^2 &= \left(x_0 \int_0^T t^2 e^{2\theta t} d\mu_t\right)^{-2} \times \\ &\times \frac{1}{2\theta} \int_0^T \int_0^T ts e^{2\theta(t+s)} [1 - e^{-2\theta(t \wedge s)}] d\mu_s d\mu_t, \end{aligned} \tag{7.38}$$

$$\begin{aligned} \sigma_{**}(\theta)^2 &= \theta^2 \left(x_0 \int_0^T (e^{\theta t} - 1)^2 d\mu_t\right)^{-2} \times \\ &\times \int_0^T \int_0^T (t \wedge s)(e^{\theta t} - 1)(e^{\theta s} - 1) d\mu_s d\mu_t. \end{aligned} \tag{7.39}$$

These values of variances can be obtained directly from the representations of the limit variables ξ_* and ξ_{**}. Particularly, if $\theta \neq 0$,

$$\xi_* = \left(\int_0^T x_0 t^2 e^{2\theta t} d\mu_t\right)^{-1} \int_0^T x_t^{(1)}(\theta) t e^{\theta t} d\mu_t,$$

$$\xi_{**} = \left(x_0 \int_0^T (e^{\theta t} - 1)^2 d\mu_t\right)^{-1} \int_0^T W_t (e^{\theta t} - 1) d\mu_t,$$

where $x_t^{(1)}(\theta) = x_t^{(1)}$ is a solution of the equation

$$dx_t^{(1)} = \theta x_t^{(1)} dt + dW_t, \quad x_0^{(1)} = 0, \quad 0 \le t \le T. \tag{7.40}$$

To compare these estimators, we put $d\mu_t = dt$ and calculate the integrals (7.36) – (7.39). The expressions of the obtained results are rather cumbersome, so we do not show them here, but show their asymptotics of large T (more precisely, large θT, if $\theta \neq 0$ and large T if $\theta = 0$).
Asymptotics of the variances are:

If $\theta = 0$ then as $T \to \infty$

$$\begin{aligned} \sigma_*(0)^2 &\simeq 6/5\, x_0^{-2} T^{-1}, \\ \sigma_{**}(0)^2 &\simeq 5/6\, x_0^{-2} T. \end{aligned} \tag{7.41}$$

If $\theta < 0$, then

$$\begin{aligned} \sigma_*(\theta)^2 &\simeq -5/2\, \theta x_0^{-2}, \\ \sigma_{**}(\theta)^2 &\simeq 1/4\, \theta^2 4 x_0^{-2} T. \end{aligned} \tag{7.42}$$

If $\theta > 0$, then

$$\begin{aligned} \sigma_*(\theta)^2 &\simeq 1/2\, \theta^{-1} x_0^{-2} T^{-2}, \\ \sigma_{**}(\theta)^2 &\simeq 4\theta^2 x_0^{-2} T e^{-2\theta T}. \end{aligned} \tag{7.43}$$

Recall, that the MLE has the following asymptotics of its variance

$$\sigma^2(\theta) = \begin{cases} -2\theta x_0^{-2}, & \text{if } \theta < 0, \\ x_0^{-2} T^{-1}, & \text{if } \theta = 0, \\ 2\theta x_0^{-2} e^{-2\theta T}, & \text{if } \theta > 0. \end{cases} \tag{7.44}$$

So this estimator is always better than the above-mentioned MDEs, but as we shall show in Section 7.5, the MDE are sometimes LAM and the MLE are not.

$\breve{\theta}_\varepsilon$ - estimator. The limit ($\varepsilon \to 0$) random variable $\breve{\xi} = \breve{\xi}_T$ of the difference $\varepsilon^{-1}(\breve{\theta}_\varepsilon - \theta)$ is defined by the equation

$$\breve{\xi}_T = \arg\inf_u \sup_{0 \le t \le T} \left| x_t^{(1)} - u x_0 t e^{\theta t} \right|.$$

Below, we suppose that $x_0 \neq 0$ and $\theta \in (\alpha, \beta), \alpha > 0$.

Theorem 7.8.

$$\mathcal{L}_\theta\{\breve{\xi}_T T \sqrt{\theta}\} \Longrightarrow \mathcal{N}\left(0, \frac{1}{2x_0^2}\right)$$

as $T \to \infty$.

Proof. The process (7.40) is represented as the stochastic integral

$$x_t^{(1)} = e^{\theta t} \int_0^t e^{-\theta s} dW_s = e^{\theta t} \tilde{W}\left(\int_0^t e^{-2\theta s} ds\right) = \frac{W(\tau)}{\sqrt{2\theta(1-\tau)}}, \tag{7.45}$$

where $\tilde{W}(\cdot)$ and $W(\cdot)$ are Wiener processes and we denote

$$\tau = 1 - e^{-2\theta t} \le T^* = 1 - e^{-2\theta T} < 1. \tag{7.46}$$

Hence, we can rewrite $\breve{\xi}_T$ as

$$\begin{aligned}\breve{\xi}_T &= \arg\inf_u \sup_{0\le\tau\le T^*} \frac{1}{\sqrt{1-\tau}}\left|\frac{W(\tau)}{\sqrt{2\theta}} - \frac{ux_0}{\theta}\ln\frac{1}{\sqrt{1-\tau}}\right| = \\ &= \frac{\theta}{2}x_0^{-1}\arg\inf_v \sup_{0\le\tau\le T^*} h(\tau)|W(\tau) - v\ln h(\tau)|,\end{aligned}$$

where we put $h(\tau) = (1-\tau)^{-\frac{1}{2}}$ and $v = ux_0\sqrt{2}\theta^{-\frac{1}{2}}$.

Note that the random variable

$$\zeta_{T^*} = \arg\inf_v \sup_{0\le\tau\le T^*} h(\tau)|W(\tau) - v\ln h(\tau)|$$

depends on one parameter T^* only and $\breve{\xi}_T$ depends on three parameters (θ, x_0, T). The asymptotics of $\breve{\xi}_T$ as $T \to \infty$ corresponds to $T^* \to 1$ (we obtain the same limit if $\theta \to \infty$).

Lemma 7.1.

$$\mathbf{P} - \lim_{T^*\to 1} \zeta_{T^*} = 0.$$

Proof. Denote

$$\begin{aligned}\tau_u &= \arg \sup_{0\le\tau\le T^*} h(\tau)|W(\tau) - u\ln h(\tau)|, \\ \tau_u^* &= \arg \sup_{0\le\tau\le T^*} |W(\tau) - u\ln h(\tau)|.\end{aligned}$$

The increasing multiplier $h(\tau)$ can only shift the point of maximum at the right, so

$$\tau_u^* \le \tau_u \le T^*.$$

Put

$$u^* = W(T^*)\left(\ln\frac{1}{\sqrt{1-T^*}}\right)^{-1}$$

and introduce the set

$$A_\nu = \{u : |u - u^*| \ge \nu\}$$

with $\nu > 0$.

We have

$$\begin{aligned}&\min_{u\in A_\nu} \sup_{0\le\tau\le T^*} |u\ln h(\tau) - W(\tau)| = \\ &= \min_{u\in A_\nu} \sup_{0\le\tau\le T^*} \left|(u-u^*)\ln h(\tau) - (W(\tau) - W(\tau^*)) - \right. \\ &\quad \left. - W(T^*)\left(\frac{\ln h(\tau)}{\ln h(T^*)} - 1\right)\right| \ge \\ &\ge \min_{u\in A_\nu} |u - u^*| \sup_{0\le\tau\le T^*} \ln h(\tau) - \sup_{0\le\tau\le T^*} |W(\tau) - W(T^*)| - \\ &\quad -|W(T^*)| \sup_{0\le\tau\le T^*} \left|\frac{\ln h(\tau)}{\ln h(T^*)} - 1\right| \ge \\ &\ge \nu\ln\frac{1}{\sqrt{1-T^*}} - 3\sup_{0\le\tau\le T^*} |W(\tau)| \to +\infty\end{aligned}$$

as $T^* \to 1$. At the same time for $u \in A_\nu^C$, we have

$$\begin{aligned}
&\inf_{|u-u^*|<\nu} \sup_{0\le\tau\le T^*} |W(\tau) - u\ln h(\tau)| \le \\
&\le \sup_{0\le\tau\le T^*} |W(\tau) - u^*\ln h(\tau)| = \\
&= \sup_{0\le\tau\le T^*} \left| W(\tau) - W(T^*) + W(T^*)\left(\frac{\ln h(\tau)}{\ln h(T^*)} - 1\right)\right| \le \\
&\le \sup_{0\le\tau\le T^*} \left(|W(\tau) - W(T^*)| + |W(T^*)|\left(\frac{\ln h(\tau)}{\ln h(T^*)} - 1\right)\right) \le \\
&\le 3 \sup_{0\le\tau\le T^*} |W(\tau)| < \infty.
\end{aligned}$$

So the random variable

$$\xi^* = \arg\inf_u |W(\tau_u^*) - u\ln h(\tau_u^*)|$$

belongs to the interval $(u^* - \nu, u^* + \nu)$ with probability tending to 1, i.e.

$$\lim_{T^*\to 1} \mathbf{P}\{|\xi^* - u^*| > \nu\} = 0$$

for any $\nu > 0$. Moreover, it follows that for any u $\tau_u^* \to 1$ as $T^* \to 1$.

We have similar inequalities for

$$\zeta_{T^*} = \arg\inf_v \sup_{0\le\tau\le T^*} h(T^*)h(\tau)|W(\tau) - v\ln h(\tau)|$$

because

$$\min_{u\in A_\nu} \sup_{0\le\tau\le T^*} h(T^*)^{-1}h(\tau)|W(\tau) - W(T^*)\frac{\ln h(\tau)}{\ln h(T^*)} + (u^* - u)\ln h(\tau)| \to +\infty.$$

Hence, we have the same convergence for ζ_{T^*}: for any $\nu > 0$

$$\lim_{T^*\to 1} \mathbf{P}\{|\zeta_{T^*} - u^*| > \nu\} = 0. \tag{7.47}$$

But $u^* \to 0$.

Consider now the asymptotics of the random variable

$$\eta_{T^*} = \zeta_{T^*} \ln h(T^*).$$

Put

$$\begin{aligned}
\nu &= \nu_{T^*} = [\ln h(T^*)]^{-\frac{1}{2}} \to 0, \\
\lambda_{T^*} &= \frac{1}{2}[\ln h(T^*)]^{\frac{1}{2}} \to +\infty
\end{aligned}$$

and note that (7.47) is also valid for $\nu = \nu_{T^*}$. So

$$\mathbf{P}\{|\zeta_{T^*} - u^*| > \nu_{T^*}\} = \mathbf{P}\{|\eta_{T^*} - W(T^*)| > 2\lambda_{T^*}\} \to 0.$$

Let us consider the behavior of η_{T^*} on the set $|\eta_{T^*}| < \lambda_{T^*}$. So

$$\begin{aligned}\eta_{T^*} &= \arg\inf_{|v|<\lambda_{T^*}} \sup_{0\le\tau\le T^*} h(\tau)\left|W(\tau) - v\frac{\ln(1-\tau)}{\ln(1-T^*)}\right| = \\ &= \arg\inf_{|v|<\lambda_{T^*}} h(\tau_v)\left|W(\tau_v) - v\frac{\ln(1-\tau_v)}{\ln(1-T^*)}\right|.\end{aligned}$$

Introduce the set

$$\mathcal{B}_\Delta = \{v: \ |W(\tau_v) - v| > \Delta\}.$$

Then on this set

$$\begin{aligned}&h(T^*)^{-1}h(\tau_v)\left|W(\tau_v) - v\frac{\ln(1-\tau_v)}{\ln(1-T^*)}\right| \ge \\ &\ge h(T^*)^{-1}h(\tau_v)\Delta\frac{\ln(1-\tau_v)}{\ln(1-T^*)} - \frac{h(\tau_v)}{h(T^*)^{-1}}|W(\tau_v)|\left(1 - \frac{\ln(1-\tau_v)}{\ln(1-T^*)}\right) \ge \frac{\Delta}{2}\end{aligned}$$

because $\tau_v \to T^* \to 1$. On the complement of this set, we have at the point $v = W(\tau_v)$

$$h(T^*)^{-1}h(\tau_v)\left|W(\tau_v) - W(\tau_v)\frac{\ln(1-\tau_v)}{\ln(1-T^*)}\right| \to 0.$$

Hence, as $T^* \to 1$

$$|\eta_{T^*} - W(1)| \le |\eta_{T^*} - W(\tau_v)| + |W(\tau_v) - W(1)| \le 2\Delta.$$

So we have proved that

$$\check{\xi}_T x_0\sqrt{\frac{2}{\theta}}\ln h(T^*) \Longrightarrow W(1)$$

and finally

$$\mathcal{L}\{\check{\xi}_T T\sqrt{\theta}\} \Longrightarrow \mathcal{L}\left\{\frac{W(1)}{x_0\sqrt{2}}\right\}.$$

θ_ε^o **- estimator**. If we put $d\mu_t = dt$ in the definition of θ_ε^o estimator (7.12), then

$$\theta_\varepsilon^o = \arg\inf_{\theta\in\Theta}\int_0^T |X_t - x_t(\theta)|dt$$

and according to Remark 7.4, the normed difference $\varepsilon^{-1}(\theta_\varepsilon^o - \theta)$ converges to the random variable $\xi^o = \xi_T^o$

$$\xi_T^o = \arg\inf_u \int_0^T |x_t^{(1)} - ux_0te^{\theta t}|dt.$$

We are interested in its asymptotics as $T \to \infty$ under the same assumptions as before: $x_0 \neq 0$, $\theta \in (\alpha, \beta)$, $\alpha > 0$.

Theorem 7.9.

$$\mathcal{L}_\theta\{\xi_T^o\sqrt{\theta}\} \Longrightarrow \mathcal{N}\left(0, \frac{1}{2x_0^2}\right).$$

Proof. The process

$$z_0(u) = \int_0^T |x_t^{(1)} - u\dot{x}_t(\theta_0)|dt, \quad u \in \mathbf{R}$$

has a unique minimum with probability 1. Indeed,

$$\begin{aligned}\frac{\partial z_0(u)}{\partial u} &= -\int_0^T \text{sign}\,\{x_t^{(1)} - u\dot{x}_t(\theta_0)\}\dot{x}_t(\theta_0)dt = \int_{A_t(u)} \dot{x}_t(\theta_0)dt - \int_{\bar{A}_t(u)} \dot{x}_t(\theta_0)dt = \\ &= \int_0^T \dot{x}_t(\theta_0)dt - 2\int_{\bar{A}_t(u)} \dot{x}_t(\theta_0)dt,\end{aligned}$$

where $A_t(u) = \{t \in [0,T]: \; x_t^{(1)} < u\dot{x}_t(\theta_0)\}$ and $\bar{A}_t(u) = [0,T] \setminus A_t(u)$. Note that the function

$$F_T(u) = \int_{\bar{A}_t(u)} \dot{x}_t(\theta_0)dt$$

is positive, strictly monotone and continuous on u. Hence, the equation

$$F_t(u) = \frac{1}{2}\int_0^T \dot{x}_t(\theta_0)dt$$

has a unique solution u^*.

As in the preceding proof, we change the variables (7.45), (7.46) and put

$$v = \frac{ux_0}{\sqrt{2\theta}} \ln\left(\frac{1}{1-T^*}\right).$$

Then we can write

$$\begin{aligned}\xi_T^o &= \arg\inf_u \|x^{(1)} - u\dot{x}\| = \\ &= \arg\inf_u \int_0^T \left|W(\tau) - \frac{ux_0}{\sqrt{2\theta}}\ln\left(\frac{1}{1-\tau}\right)\right| \frac{d\tau}{(1-\tau)^{\frac{3}{2}}} = \\ &= \frac{\sqrt{2\theta}}{x_0 \ln\left(\frac{1}{1-T^*}\right)} \arg\inf_v \int_0^{T^*} \left|W(\tau) - v\frac{\ln\left(\frac{1}{1-\tau}\right)}{\ln\left(\frac{1}{1-T^*}\right)}\right| (1-\tau)^{-\frac{3}{2}}d\tau = \\ &= \frac{\sqrt{2\theta}}{x_0 \ln\left(\frac{1}{1-T^*}\right)}\zeta_{T^*}^o.\end{aligned}$$

Note also that the random variable $\zeta_{T^*}^o$ depends on one parameter T^*. We study the asymptotics of $\zeta_{T^*}^o$ as $T^* \to 1$.

Denote

$$J(v) = \int_0^T \left|W(\tau) - v\frac{\ln\left(\frac{1}{1-\tau}\right)}{\ln\left(\frac{1}{1-T^*}\right)}\right| \frac{d\tau}{(1-\tau)^{\frac{3}{2}}}$$

and introduce the set

$$V_* = \{v : \; |v - W(1)| > \nu\},$$

where $\nu > 0$.

We shall prove that

$$\text{(a)} \qquad \mathbf{P}_\theta - \limsup_{T^* \to 1} J(W(1))\sqrt{1-T^*}\ln\left(\frac{1}{1-T^*}\right) < \infty,$$

$$\text{(b)} \qquad \mathbf{P}_\theta - \lim_{T^* \to 1} \inf_{v \in V_*} J(v)\sqrt{1-T^*}\ln\left(\frac{1}{1-T^*}\right) = \infty.$$

Then from these two limits it follows that for any $\nu > 0$

$$\zeta^o_{T^*} \in (W(1) - \nu, W(1) + \nu)$$

with probability tending to 1. Hence

$$\mathcal{L}\{\zeta^o_{T^*}\} \Longrightarrow \mathcal{L}\{W(1)\} = \mathcal{N}(0,1).$$

We begin with (a):

$$\begin{aligned} J(W(1)) &= \int_0^{T^*} \left| W(\tau) - W(1)\frac{\ln\left(\frac{1}{1-\tau}\right)}{\ln\left(\frac{1}{1-T^*}\right)} \right| \frac{d\tau}{(1-\tau)^{\frac{3}{2}}} \leq \\ &\leq \int_0^{T^*} \frac{|W(\tau) - W(1)|}{(1-\tau)^{\frac{3}{2}}} d\tau + \frac{|W(1)|}{\ln\left(\frac{1}{1-T^*}\right)} \int_0^{T^*} \frac{\ln\left(\frac{1-\tau}{1-T^*}\right)}{(1-\tau)^{\frac{3}{2}}} d\tau = \\ &= I_1 + \frac{|W(1)|}{\ln\left(\frac{1}{1-T^*}\right)} I_2. \end{aligned}$$

The mathematical expectation of the first integral is

$$\mathbf{E}I_1 = \sqrt{\frac{2}{\pi}} \int_0^{T^*} \frac{(1-\tau)^{\frac{1}{2}}}{(1-\tau)^{\frac{3}{2}}} d\tau = \sqrt{\frac{2}{\pi}} \ln\left(\frac{1}{1-T^*}\right).$$

In the second integral we change the variables

$$x = \frac{1-\tau}{1-T^*},$$

then we can write

$$I_2 = \frac{1}{\sqrt{1-T^*}} \int_1^{\frac{1}{1-T^*}} \frac{\ln x}{x^{\frac{3}{2}}} dx.$$

Hence, the random variables

$$J(W(1))\sqrt{1-T^*}\ln\left(\frac{1}{1-T^*}\right)$$

are bounded in probability and (a) is proved.

Consider (b):

$$\begin{aligned} J(v) &= \int_0^{T^*} \left| W(\tau) - W(1)\frac{\ln\left(\frac{1}{1-\tau}\right)}{\ln\left(\frac{1}{1-T^*}\right)} + \right. \\ &\quad \left. + (W(1)-v)\frac{\ln\left(\frac{1}{1-\tau}\right)}{\ln\left(\frac{1}{1-T^*}\right)} \right| \frac{d\tau}{(1-\tau)^{\frac{3}{2}}} \geq \\ &\geq |W(1)-v| \int_0^{T^*} \frac{\ln\left(\frac{1}{1-\tau}\right)}{\ln\left(\frac{1}{1-T^*}\right)} \frac{d\tau}{(1-\tau)^{\frac{3}{2}}} - \\ &\quad - \int_0^{T^*} \left| W(\tau) - W(1)\frac{\ln\left(\frac{1}{1-\tau}\right)}{\ln\left(\frac{1}{1-T^*}\right)} \right| \frac{d\tau}{(1-\tau)^{\frac{3}{2}}} \geq \\ &\geq \Delta \int_0^{T^*} \frac{d\tau}{(1-\tau)^{\frac{3}{2}}} - \Delta \int_0^{T^*} \left(1 - \frac{\ln\left(\frac{1}{1-\tau}\right)}{\ln\left(\frac{1}{1-T^*}\right)}\right) \frac{d\tau}{(1-\tau)^{\frac{3}{2}}} - J(W(1)) = \\ &= 2\Delta\left(\frac{1}{\sqrt{1-T^*}} - 1\right) - J(W(1)) - \frac{\Delta}{\ln\left(\frac{1}{1-T^*}\right)} \int_0^{T^*} \ln\left(\frac{1-\tau}{1-T^*}\right) \frac{d\tau}{(1-\tau)^{\frac{3}{2}}}. \end{aligned}$$

Hence

$$\begin{aligned} &\inf_{v \in V_*} J(v)\sqrt{1-T^*} \ln\left(\frac{1}{1-T^*}\right) \geq \\ &\geq 2\Delta(1-\sqrt{1-T^*}) \ln\left(\frac{1}{1-T^*}\right) - \Delta \int_0^{\frac{1}{1-T^*}} \frac{\ln x}{x^{\frac{3}{2}}} dx - \\ &\quad -\sqrt{1-T^*} \ln\left(\frac{1}{1-T^*}\right) J(W(1)) \geq \\ &\geq \Delta \ln\left(\frac{1}{1-T^*}\right) \end{aligned}$$

for sufficiently large T. So Theorem 7.8 is proved.

REMARK 7.6 As follows from Theorems 7.7 and 7.8, the MDEs $\check{\theta}_\varepsilon$ and θ_ε^o have limit ($\varepsilon \to 0$) random variables $\check{\xi}_T$ and ξ_T^o which have the same asymptotics as large T. These two variables have essentially different definitions and this result is not evident. Technically, it is explained by the fact that their asymptotics is defined in two cases by the value of the Wiener process in the final point.

Example 7.2. If the observed process is

$$dX_t = \theta\left(\int_0^t X_s ds\right) dt + \varepsilon dW_t, \quad X_0 = x_0 \neq 0, \quad 0 \leq t \leq T$$

and the unknown parameter $\theta \in (\alpha, \beta) = \Theta$, $\beta < 0$, then the limit solution is

$$x_t = x_0 \cos \omega t, \quad 0 \leq t \leq T$$

with $\omega = \sqrt{-\theta}$. The MDE θ_ε^* is defined by the equality

$$\theta_\varepsilon^* = \arg\inf_{\theta\in\Theta} \int_0^T [X_t + x_0\sqrt{-\theta}\sin(\sqrt{-\theta}t)]^2 d\mu_t,$$

is consistent and asymptotically normal (Theorem 7.5):

$$\mathcal{L}_\theta\{\varepsilon^{-1}(\theta_\varepsilon^* - \theta)\} \Longrightarrow \{\xi^*\} = \mathcal{N}(0, \sigma_*(\theta)^2),$$

where

$$\xi^* = \left(\int_0^T \dot{x}_t(\theta)^2 d\mu_t\right)^{-1} \int_0^T x_t^{(1)}(\theta)\dot{x}_t(\theta) d\mu_t$$

with a Gaussian process

$$dx_t^{(1)} = \theta[\int_0^t x_s^{(1)} ds]dt + dW_t, \quad x_0^{(1)} = 0, \quad 0 \le t \le T$$

and

$$\dot{x}_t(\theta) = -\frac{x_0}{2\sqrt{-\theta}}\sin(\sqrt{-\theta}t) - \frac{x_0\sqrt{-\theta}t}{2\sqrt{-\theta}}\cos(\sqrt{-\theta}t).$$

The asymptotics of θ_ε^{**} could be also described.

Example 7.3. Let

$$dX_t = \frac{\theta}{\varepsilon}X_t dt + dW_t, \quad X_0 = x_0, \quad 0 \le t \le T, \tag{7.48}$$

where $\varepsilon \in (0,1]$, θ is an unknown parameter, $\theta \in (\alpha,\beta) = \Theta$, and the limit corresponds $\varepsilon \to 0$. Recall that this diffusion process is not a particular case of the model (7.1). The limit behavior of MDEs also depends of the values of θ and x_0 (as in Section 2.3). In this problem we could not use the MDE θ_ε^* because X_t has no reasonable limit function, but we can always calculate

$$\begin{aligned}\theta_\varepsilon^{**} &= \arg\inf_{\theta\in\Theta} \int_0^T [X_t - x_0 - \frac{\theta}{\varepsilon}\int_0^t X_s ds]^2 d\mu_t = \\ &= \varepsilon\frac{\int_0^T (X_t - x_0)\int_0^t X_s ds d\mu_t}{\int_0^T \left(\int_0^t X_s ds\right)^2 d\mu_t} = \theta + \varepsilon\frac{\int_0^T \left(W_t\int_0^t X_s ds\right) d\mu_t}{\int_0^T \left(\int_0^t X_s ds\right)^2 d\mu_t}\end{aligned} \tag{7.49}$$

if $\theta_\varepsilon^* \in (\alpha,\beta)$ with the corresponding definition $\theta_\varepsilon^* = \alpha$ and $\theta_\varepsilon^* = \beta$ if the last expression is out of (α,β).

Note that

$$\int_0^t X_s ds = \frac{\varepsilon}{\theta}(X_t - x_0 - W_t).$$

Hence, we can also write

$$\theta_\varepsilon^{**} = \theta + \theta\frac{\int_0^T W_t(X_t - x_0 - W_t)d\mu_t}{\int_0^T (X_t - x_0 - W_t)^2 d\mu_t}.$$

Below we put $d\mu_t = dt$.

Theorem 7.10.

(a) *If* $\theta > 0$ *and* $x_0 = 0$, *then*

$$\mathcal{L}_\theta\{(\theta_\varepsilon^{**} - \theta)\varepsilon^{\frac{1}{2}} e^{\frac{\theta T}{\varepsilon}}\} \Longrightarrow (2\theta)^{\frac{3}{2}}\sqrt{T}\zeta,$$

where ζ *has Cauchy distribution;*

(b) *if* $\theta > 0$ *and* $x_0 \neq 0$, *then*

$$\mathcal{L}_\theta\{(\theta_\varepsilon^{**} - \theta)\varepsilon e^{\frac{\theta T}{\varepsilon}}\} \Longrightarrow \mathcal{N}\left(0, \frac{4\theta^2 T}{x_0^2}\right);$$

(c) *if* $\theta < 0$ *and* $x_0 = 0$, *then*

$$\mathbf{P}_\theta - \lim_{\varepsilon\to 0} \theta_\varepsilon^{**} = 0;$$

(d) *if* $\theta < 0$ *and* $x_0 \neq 0$, *then*

$$\mathbf{P}_\theta - \lim_{\varepsilon\to 0} \theta_\varepsilon^{**} = x_0 \left(\int_0^T [x_0 + W_t]^2 dt\right)^{-1} \int_0^T [x_0 + W_t] dt.$$

Proof. (a) If $x_0 = 0$, then we have the representation

$$\begin{aligned} X_t e^{-\frac{\theta t}{\varepsilon}} &= \int_0^t e^{-\frac{\theta s}{\varepsilon}} dW_s = \sqrt{\varepsilon}\int_0^{\frac{t}{\varepsilon}} e^{-\theta s} d\bar{W}_s = \\ &= \sqrt{\varepsilon}\left(\int_0^\infty e^{-\theta s} d\bar{W}_s - \int_{\frac{t}{\varepsilon}}^\infty e^{-\theta s} d\bar{W}_s\right) = \sqrt{\varepsilon}\left(\xi - \tilde{W}\left(\int_{\frac{t}{\varepsilon}}^\infty e^{-2\theta s} ds\right)\right) = \\ &= \sqrt{\varepsilon}\left(\xi - \tilde{W}\left(\frac{e^{-\frac{2\theta t}{\varepsilon}}}{2\theta}\right)\right) = \sqrt{\varepsilon}\left(\xi - \frac{1}{\sqrt{2\theta}} e^{-\frac{\theta t}{\varepsilon}} \tilde{\tilde{W}}(1)\right), \end{aligned}$$

where we have changed the variables $s/\varepsilon \to s$ and introduced the corresponding Wiener processes $\bar{W}_s, \tilde{W}_s$ and $\tilde{\tilde{W}}_s$. Recall, that $W(a) = \sqrt{a}\bar{W}(1)$.

Hence

$$\begin{aligned} \int_0^T \left(\int_0^t X_s ds\right)^2 dt &= \frac{\varepsilon^2}{\theta^2}\int_0^T (X_t - W_t)^2 dt = \frac{\varepsilon^2}{\theta^2}\int_0^T e^{\frac{2\theta t}{\varepsilon}} \xi^2 dt(1 + o(1)) = \\ &= \frac{\varepsilon^4}{2\theta^3}\xi^2 e^{\frac{2\theta T}{\varepsilon}}(1 + o(1)). \end{aligned}$$

In a similar way, we have

$$\begin{aligned} \int_0^T W_t\left(\int_0^t X_s ds\right) dt &= \frac{\varepsilon}{\theta}\int_0^T W_t(X_t - W_t) dt = \\ &= \frac{\varepsilon^{\frac{3}{2}}}{\theta}\int_0^T W_t \xi e^{\frac{\theta t}{\varepsilon}} dt(1 + o(1)) = \end{aligned}$$

$$
\begin{aligned}
&= \frac{\varepsilon^{\frac{5}{2}}}{\theta^2}\xi\int_0^T W_t de^{\frac{\theta t}{\varepsilon}}(1+o(1)) = \\
&= \frac{\varepsilon^{\frac{5}{2}}}{\theta^2}\xi\left(W_T e^{\frac{\theta T}{\varepsilon}} - \int_0^T e^{\frac{\theta t}{\varepsilon}}dW_t\right)(1+o(1)) = \\
&= \frac{\varepsilon^{\frac{5}{2}}}{\theta^2}\xi W_T e^{\frac{\theta T}{\varepsilon}}(1+o(1)).
\end{aligned}
$$

Note also that the random variables ξ and W_T are asymptotically independent because

$$
\xi = \lim_{\varepsilon\to 0}\frac{1}{\sqrt{\varepsilon}}\int_0^t e^{-\frac{\theta s}{\varepsilon}}dW_s = \lim_{\varepsilon\to 0}\xi_\varepsilon
$$

and

$$
\begin{aligned}
\mathbf{E}W_T\xi_\varepsilon &= \mathbf{E}\left(\int_0^T dW_s\frac{1}{\sqrt{\varepsilon}}\int_0^t e^{-\frac{\theta s}{\varepsilon}}dW_s\right) = \\
&= \frac{1}{\sqrt{\varepsilon}}\int_0^t e^{-\frac{\theta s}{\varepsilon}}ds = \sqrt{\varepsilon}\int_0^{\frac{t}{\varepsilon}} e^{-\theta s}ds \to 0.
\end{aligned}
$$

Finally we obtain (by distribution)

$$
\begin{aligned}
\theta_\varepsilon^{**} &= \theta + 2\theta\sqrt{\varepsilon}\frac{W_T}{\xi}e^{-\frac{\theta T}{\varepsilon}}(1+o(1)) = \\
&= \theta + \sqrt{\varepsilon}(2\theta)^{\frac{3}{2}}\sqrt{T}\exp\left\{-\frac{\theta T}{\varepsilon}\right\}\frac{W^*}{\xi^*}(1+o(1)),
\end{aligned}
$$

where W^* and ξ^* are independent $\mathcal{N}(0,1)$ random variables.

(b) The solution of (7.48) is

$$
\begin{aligned}
X_t &= x_0 e^{\frac{\theta t}{\varepsilon}} + e^{\frac{\theta t}{\varepsilon}}\int_0^t e^{-\frac{\theta s}{\varepsilon}}dW_s = \\
&= x_0 e^{\frac{\theta t}{\varepsilon}} + e^{\frac{\theta t}{\varepsilon}}\bar{W}\left(\int_0^t e^{-\frac{2\theta s}{\varepsilon}}ds\right) = \\
&= x_0 e^{\frac{\theta t}{\varepsilon}} + \sqrt{\frac{\varepsilon}{2\theta}}e^{\frac{\theta t}{\varepsilon}}\bar{W}(1-e^{-\frac{2\theta t}{\varepsilon}}) = \\
&= x_0 e^{\frac{\theta t}{\varepsilon}}(1+O(\sqrt{\varepsilon})).
\end{aligned}
$$

Hence

$$
\begin{aligned}
\int_0^T\left(\int_0^t X_s ds\right)^2 dt &= \frac{\varepsilon^2}{\theta^2}\int_0^T (X_t - x_0 - W_t)^2 dt = \\
&= \frac{\varepsilon^2}{\theta^2}\int_0^T x_0^2 e^{\frac{2\theta t}{\varepsilon}}(1+o(1))dt = \frac{x_0^2\varepsilon^3}{2\theta^3}e^{\frac{2\theta T}{\varepsilon}}(1+o(1))
\end{aligned}
$$

and

$$
\int_0^T W_t\int_0^t X_s ds dt = \frac{\varepsilon}{\theta}\int_0^T W_t(X_t - x_0 - W_t)dt =
$$

$$
\begin{aligned}
&= \frac{x_0\varepsilon}{\theta}\int_0^T W_t e^{\frac{\theta t}{\varepsilon}}dt(1+o(1)) = \frac{x_0^2\varepsilon^2}{\theta^2}\int_0^T W_t de^{\frac{\theta t}{\varepsilon}}(1+o(1)) = \\
&= \frac{x_0\varepsilon^2}{\theta^2}\left(W_T e^{\frac{\theta T}{\varepsilon}} - \int_0^T e^{\frac{\theta t}{\varepsilon}}dW_t\right)(1+o(1)) = \\
&= \frac{x_0\varepsilon^2}{\theta^2}W_T e^{\frac{\theta T}{\varepsilon}}(1+o(1)).
\end{aligned}
$$

So if we substitute this in (7.49), then we obtain

$$
e^{\frac{\theta T}{\varepsilon}}(\theta_\varepsilon^{**} - \theta) = \frac{2\theta}{x_0}W_T(1+o(1)).
$$

(c) If $\theta < 0$ then

$$
\begin{aligned}
\mathbf{E}X_t^2 &= x_0^2\exp\left\{\frac{2\theta t}{\varepsilon}\right\} + \int_0^t \exp\left\{\frac{2\theta(t-s)}{\varepsilon}\right\}ds = \\
&= x_0^2\exp\left\{\frac{2\theta t}{\varepsilon}\right\} + \frac{\varepsilon}{2\theta}\left[1 - \exp\left(\frac{2\theta t}{\varepsilon}\right)\right] \to 0
\end{aligned}
$$

as $\varepsilon \to 0$. Therefore, the process X_t converges to zero uniformly on $t \in [a,T]$ for any $a \in (0,T)$. We can choose a as being as close to zero as we want in order to prove the convergencies:

$$
\int_a^T (X_t - x_0 - W_t)^2 dt = \int_a^T (x_0 + W_t)^2 dt(1+o(1)),
$$

$$
\int_a^T W_t(X_t - x_0 - W_t)dt = -\int_a^T W_t(x_0 + W_t)dt(1+o(1)).
$$

These equalities allow us to write

$$
\begin{aligned}
\theta_\varepsilon^{**} &= \theta + \theta\frac{\int_0^T W_t(X_t - X_0 - W_t)dt}{\int_0^T (X_t - x_0 - W_t)^2 dt} = \\
&= x_0\left(\int_0^T (x_0 + W_t)^2 dt\right)^{-1}\int_0^T (x_0 + W_t)dt(1+o(1)) = \tilde{\xi}(1+o(1)).
\end{aligned}
$$

Hence,

$$
\mathbf{P}_\theta - \lim_{\varepsilon\to 0}\theta_\varepsilon^{**} = \tilde{\xi}
$$

and, particularly, if $x_0 = 0$, then for all θ, we have

$$
\mathbf{P}_\theta - \lim_{\varepsilon\to 0}\theta_\varepsilon^{**} = 0.
$$

7.4 Nonstandard situations and other problems

We shall approximately follow the same order of problems as in Chapter 2, so we begin with

Nondifferentiable trend. We consider the model

$$dX_t = X_{\theta t}dt + \varepsilon dW_t, \quad X_0 = x_0 > 0, \quad 0 \le t \le T, \tag{7.50}$$

where $\theta \in (\alpha, \beta), 0 \le \alpha < \beta < 1$. This is not a particular case of the model (7.18), (7.19) because it corresponds to the function $K(\theta, t, s, X_s) = X_s\ \delta(s - \theta t)$ with the Dirac function $\delta(\cdot)$. The limit equation is

$$\frac{dx_t}{dt} = x_{\theta t}, \quad x_0, \quad 0 \le t \le T$$

and the derivative $x_t^{(1)}$ satisfies the equation

$$dx_t^{(1)} = x_{\theta t}^{(1)}dt + dW_t, \quad x_0^{(1)} = 0, \quad 0 \le t \le T. \tag{7.51}$$

Note that the normed difference $\varepsilon^{-1}(X_t - x_t)$ satisfies the same equation (7.51), so the derivative $x_t^{(1)}$ of X_t w.r.t. ε at the point $\varepsilon = 0$ coincides with $\varepsilon^{-1}(X_t - x_t)$ for all $\varepsilon \in (0, 1]$.

Investigation of the MDE

$$\theta_\varepsilon^* = \arg \inf_{\theta \in \Theta} \|X - x(\theta)\|$$

has no additional difficulties in comparison with the model (7.51) because in Theorem 7.5 we do not require the smoothness of $S_t(\theta, X)$ but only $S_t(\theta, x)$. The function $x_{t\theta}$ is infinitely differentiable. Indeed, if we introduce a function $\dot{x}_t = \dot{x}_t(\theta)$ as a solution of the equation

$$\frac{d\dot{x}_t}{dt} = \dot{x}_{\theta t} + tx_{\theta^2 t}, \quad \dot{x}_0 = 0, \quad 0 \le t \le T \tag{7.52}$$

and denote

$$v_t(h) = \left| \frac{x_t(\theta + h) - x_t(\theta)}{h} - \dot{x}_t(\theta) \right|.$$

Then, from the equations of x_t for the values $\theta + h$ and h and (7.52), we have the inequalities

$$\begin{aligned} v_t(h) \ &= \ \left| \int_0^t \left[\frac{x_{s(\theta+h)}(\theta + h) - x_{s(\theta+h)}(\theta)}{h} - \dot{x}_{s(\theta+h)}(\theta) \right] ds + \right. \\ &\quad + \int_0^t [\dot{x}_{s(\theta+h)}(\theta) - \dot{x}_{s\theta}(\theta)] ds + \\ &\quad \left. + \int_0^t \left[\frac{x_{s(\theta+h)}(\theta + h) - x_{s\theta}(\theta)}{h} - s\dot{x}_{s\theta^2}(\theta) \right] ds \right| \le \\ &\le \ \frac{1}{\theta + h} \int_0^{t(\theta+h)} v_s(h) ds + \int_0^T |\dot{x}_{s(\theta+h)}(\theta) - \dot{x}_{s\theta}(\theta)| ds + \end{aligned}$$

$$+\int_0^T \left| \frac{x_{s(\theta+h)}(\theta+h) - x_{s\theta}(\theta)}{h} - s\dot{x}_{s\theta^2}(\theta) \right| ds \leq$$
$$\leq \frac{1}{\theta+h}\int_0^t v_s(h)ds + R_1(h) + R_2(h).$$

From Equation (7.52) we have the convergencies

$$R_1(h) \to 0, \quad R_2(h) \to 0$$

as $h \to 0$ and by Lemma 1.11 we have

$$\lim_{h\to 0} \sup_{0\leq t\leq T} v_t(h) = 0.$$

So the function $\dot{x}_t(\theta)$ is a first derivative of $x_t(\theta)$ w.r.t. θ. The higher derivatives can be obtained by a similar way.

The initial value $x_0 > 0$, so the solution x_t is an increasing function and the solution $\dot{x}_t$ of (7.52) is also a strictly increasing function. Hence, the quantity

$$J(\theta) = \int_0^T \dot{x}_t(\theta)^2 d\mu_t$$

is positive for all $\theta \in \Theta$. Repeating the proof of Theorem 7.5 shows that

$$\mathcal{L}_\theta\{\varepsilon^{-1}(\theta_\varepsilon^* - \theta)\} \Longrightarrow \mathcal{N}(0, J(\theta)^{-2}\mathcal{B}(\theta))$$

with

$$\mathcal{B}(\theta) = \int_0^T\int_0^T \dot{x}_t(\theta)\dot{x}_s(\theta)\mathbf{E}_\theta x_t^{(1)} x_s^{(1)} d\mu_s d\mu_t.$$

Denote the true value by θ_0. Then the MDE

$$\theta_\varepsilon^{**} = \arg\inf_{\theta\in\bar{\Theta}} \|X - \hat{X}(\theta)\|$$

with

$$\hat{X}_t(\theta) = x_0 + \int_0^t X_{\theta s}ds.$$

The difference $X_t - \hat{X}_t(\theta_0) = \varepsilon W_t$, so we need only to verify the differentiability of $\hat{X}_t(\theta)$ w.r.t. θ. Note that

$$\hat{X}_t(\theta+h) = x_0 + \int_0^t X_{(\theta+h)s}ds = x_0 + \frac{1}{\theta+h}\int_0^{(\theta+h)t} X_s ds,$$

where the process X_s satisfies (7.50). Hence

$$h^{-1}[\hat{X}_t(\theta+h) - \hat{X}_t(\theta)] =$$
$$= h^{-1}\left(\frac{1}{\theta+h}\int_0^{(\theta+h)t} X_s ds - \frac{1}{\theta+h}\int_0^{\theta t} X_s ds\right) + h^{-1}\left(\frac{1}{\theta+h} + \frac{1}{\theta}\right)\int_0^{\theta t} X_s ds =$$
$$= \frac{1}{(\theta+h)h}\int_{\theta t}^{(\theta+h)t} X_s ds - \frac{1}{\theta(\theta+h)}\int_0^{\theta t} X_s ds \longrightarrow \frac{t}{\theta}X_{\theta t} - \frac{1}{\theta^2}\int_0^{\theta t} X_s ds$$

as $h \to 0$. So if we put

$$\hat{\dot{x}}_t(\theta) = \frac{t}{\theta}x_{\theta t} - \frac{1}{\theta}\int_0^t x_{\theta s}ds = \frac{tx_{\theta t} - x_t + x_0}{\theta},$$

then with probability 1 we have

$$\|\hat{X}(\theta + h) - \hat{X}(\theta) - h\hat{\dot{x}}(\theta)\| = o(|h|).$$

Note also that the derivative on t

$$\frac{d}{dt}\dot{x}_t(\theta) = \frac{1}{\theta}\left(x_{\theta t} + t\frac{dx_{\theta t}}{dt} - \frac{dx_t}{dt}\right) = tx_{\theta^2 t}$$

for all $t > 0$ is positive and therefore $J_{**}(\theta) > 0$. So we can use Proposition 7.1 and establish the consistence and asymptotic normality of this estimator:

$$\mathcal{L}_\theta\{\varepsilon^{-1}(\theta_\varepsilon^{**} - \theta)\} \Longrightarrow \mathcal{L}\left\{J_{**}(\theta)^{-1}\int_0^T W_t\hat{\dot{x}}_t(\theta)d\mu_t\right\}.$$

The asymptotic behavior of θ_ε^{***} in this situation is described by Theorem 7.6.

Random initial value $x_0 = \eta$ in the observed process

$$dX_t = S_t(\theta, X)dt + \varepsilon dW_t, \quad 0 \le t \le T$$

makes the solution $x_t = x_t(\theta)$ of the limit system

$$\frac{dx_t}{dt} = S_t(\theta, x), \quad x_0 = \eta, \quad 0 \le t \le T$$

also random. Here $\theta \in \Theta \subset \mathbf{R}^d, \Theta$ is an open bounded set. Suppose that η is $\mathcal{F}_0$-measurable and its distribution does not depend on θ. Moreover, we assume that $\mathbf{E}\eta^2 < \infty$ and $S_t(\theta, X) = V(\theta, t, X_t) + \int_0^t K(\theta, t, s, X_s)ds$. We assume also that the functions $V(\theta, t, x)$ and $K(\theta, t, s, x)$ have two continuous bounded derivatives on θ and on x. The process

$$dx_t^{(1)} = \left[V'(\theta, t, x_t)x_t^{(1)} + \int_0^t K'(\theta, t, s, x_s)x_s^{(1)}ds\right]dt + dW_t, \quad x_0^{(1)} = 0, \quad 0 \le t \le T$$

is now conditionally Gaussian because x_t is random, so $\mathcal{L}\{x_t^{(1)}|\mathcal{F}_0\} = \mathcal{N}(0, \sigma_t^2)$.

To prove the convergence

$$\mathbf{P}_\theta - \lim_{\varepsilon \to 0}\varepsilon^{-1}(\theta_\varepsilon^* - \theta) = J_*(\theta)^{-1}\int_0^T x_t^{(1)}\dot{x}_t(\theta)d\mu_t$$

and exponential consistence of θ_ε^* using the same proofs as in Theorem 7.1 and 7.5, we need the additional conditions

$$\sup_{\theta \in \Theta}\|x(\theta)\| \le C,$$

$$\inf_{\theta_0 \in \mathbf{K}} \inf_{|\theta-\theta_0|>\nu} \|x(\theta) - x(\theta_0)\| \geq \kappa_1 |\theta - \theta_0|^{\delta},$$

$$\inf_{\theta \in \mathbf{K}} \inf_{|l|=1} (J_*(\theta)l, l) \geq \kappa_2$$

with some non random constants $\kappa_i > 0, i = 1, 2,\ \delta$ and C.

Other MDEs can be studied in similar way.

Misspecified model. As in Section 2.6, the observed process is

$$dX_t = b_t(X)dt + \varepsilon dW_t, \quad X_0 = x_0, \quad 0 \leq t \leq T, \tag{7.53}$$

but the trend $b(\cdot)$ is unknown to the observer who has to suppose that it is a member of the family (on θ) of the diffusion-type processes

$$dX_t = S_t(\theta, X)dt + \varepsilon dW_t, \quad X_0 = x_0, \quad 0 \leq t \leq T, \tag{7.54}$$

where $\theta \in \Theta \subset \mathbf{R}^d$ (Θ is an open bounded set), and then has to construct the MDE

$$\theta^*_\varepsilon = \arg \inf_{\theta \in \Theta} \|X - x(\theta)\|.$$

Here $x_t(\theta) = x_t$ is a solution of the limit equation for Equation (7.54)

$$\frac{dx_t}{dt} = S_t(\theta, x), \quad x_0, \quad 0 \leq t \leq T.$$

The observed process converges, of course, to the function

$$\frac{dy_t}{dt} = b_t(y), \quad y_0 = x_0, \quad 0 \leq t \leq T.$$

We describe the asymptotics of θ^*_ε under the condition that the point

$$\theta^* = \arg \inf_{\theta \in \Theta} \|y - x(\theta)\|$$

is unique and belongs to Θ.

Denote

$$g(\nu) = \inf_{|\theta-\theta^*|>\nu} (\|y - x(\theta)\|^2 - \|y - x(\theta^*)\|^2)$$

and $\mathbf{P}_b^{(\varepsilon)}$ is the measure induced by (7.53) in $(\mathcal{C}_T, \mathcal{B}_T)$.

Theorem 7.11. *Let the conditions $\mathcal{L}$ be fulfilled for the trend $b_t(X)$ and for the function $S_t(\theta, x)$ and the function $g(\nu) > 0$ for $\nu > 0$, then*

$$\mathbf{P}_b^{(\varepsilon)}\{|\theta^*_\varepsilon - \theta^*| > \nu\} \leq 4 \exp\left\{-\frac{\kappa^2 g(\nu)^2}{2T\varepsilon^2}\right\}$$

for $\nu < \nu_0$ with some constants $\kappa > 0$ and $\nu_0 > 0$.

Proof is similar to the proof of Theorem 7.2. Below, we denote A, B, C, D as constants.

$$\mathbf{P}_b^{(\varepsilon)}\{|\theta_\varepsilon^* - \theta^*| > \nu\} = \mathbf{P}_b^{(\varepsilon)}\{\inf_{|\theta-\theta^*|<\nu} \|X - x(\theta)\| > \inf_{|\theta-\theta^*|>\nu} \|X - x(\theta)\|\} =$$

$$= \mathbf{P}_b^{(\varepsilon)}\left\{\inf_{|\theta-\theta^*|<\nu} \left(\|X - x(\theta)\|^2 - \|y - x(\theta^*)\|^2\right) > \right.$$

$$\left. > \inf_{|\theta-\theta^*|>\nu} \left(\|X - x(\theta)\| - \|y - x(\theta^*)\|^2\right)\right\} \le$$

$$\le \mathbf{P}_b^{(\varepsilon)}\{\inf_{|\theta-\theta^*|<\nu} < X - y + x(\theta^*) - x(\theta), X - x(\theta) + y - x(\theta) > >$$

$$> \inf_{|\theta-\theta^*|>\nu} (\|y - x(\theta)\|^2 - \|y - x(\theta^*)\|^2 - 2 < y - x(\theta), X - y >)\} \le$$

$$\le \mathbf{P}_b^{(\varepsilon)}\{\inf_{|\theta-\theta^*|<\nu} (\|X - y + x(\theta^*) - x(\theta)\| \cdot \|X - x(\theta) + y - x(\theta^*)\| >$$

$$> g(\nu) - C\varepsilon \sup_{0\le t\le T} |W_t|\} \le$$

$$\le \mathbf{P}^{(\varepsilon)}\{A\varepsilon \sup_{0\le t\le T} |W_t|(B + C\varepsilon \sup_{0\le t\le T} |W_t| + D\varepsilon \sup_{0\le t\le T} |W_t| > g(\nu)\} \le$$

$$\le \mathbf{P}\left\{\sup_{0\le t\le T} |W_t| \ge \frac{\kappa g(\nu)}{\varepsilon}\right\} \le 4\exp\left\{-\frac{\kappa^2 g(\nu)^2}{2T\varepsilon^2}\right\}$$

with some $\kappa > 0$.

Recall that for $\eta = \sup_{0\le t\le T} |W_t|$ we have

$$\mathbf{P}\{A\varepsilon\eta(B + C\varepsilon\eta) + d\varepsilon\eta > g\} =$$

$$= \mathbf{P}\left(\eta + C\varepsilon\eta^2 > \frac{\gamma g}{\varepsilon}\right) = \mathbf{P}\left(\eta > \frac{-1 + \sqrt{1 + 4C\gamma g}}{2C\varepsilon}\right) \le$$

$$\le 4\exp\left\{-\frac{1}{2T}\frac{(-1 + \sqrt{1 + 4C\gamma g})^2}{4C^2\varepsilon^2}\right\}$$

and

$$\sqrt{1 + \alpha} - 1 = \frac{\alpha}{2\sqrt{1 + \tilde{\alpha}}} \ge \frac{\alpha}{2\sqrt{2}}$$

for the small values of α ($|\alpha| < 1$), i.e. $4C\gamma g(\nu_0) < 1$.

To prove the asymptotic normality we, as usual, concretize the model. Suppose that

$$\begin{aligned} b_t(\theta, X) &= V(t, X_t) + \int_0^t K(t, s, X_s)ds, \\ S_t(\theta, x) &= V(\theta, t, x_t) + \int_0^t K(\theta, t, s, x_s)ds \end{aligned}$$

and introduce the conditions

I. *The functions $V(t, x)$ and $K(t, s, x)$ have two continuous bounded derivatives on x.*

II. The functions $V(\theta, t, x)$ and $K(\theta, t, s, x)$ have two continuous bounded derivatives on θ and x on the set $\Theta \otimes \mathbf{Y}$, where $\mathbf{Y}$ is a set of values of the function $y_t, 0 \le t \le T$, so $\mathbf{Y} = [\min_{0 \le t \le T} y_t, \max_{0 \le t \le T} y_t]$.

III. The function $g(\nu) > 0$ for $\nu > 0$, the point $\theta^* \in \Theta$, and the matrix

$$J(\theta^*) = \langle \dot{x}(\theta^*), \dot{x}(\theta^*)^T \rangle - \langle y - x(\theta^*), \ddot{x}(\theta^*) \rangle$$

is positive definite.

Here T denotes the transposition and $\ddot{x}(\theta^*)$ is a matrix of the second derivatives on θ.

Denote $y_t^{(1)}$ as the solution of the equation

$$dy_t^{(1)} = [V_x'(t, y_t) y_t^{(1)} + \int_0^t K_x'(t, s, y_s) y_s^{(1)} ds] dt + dW_t, \quad y_0^{(1)} = 0$$

and introduce a random vector

$$\xi = J(\theta^*)^{-1} \langle y^{(1)}, \dot{x}(\theta^*) \rangle.$$

Theorem 7.12. *Let condition I-III be fulfilled, then*

$$\mathbf{P}_b - \lim_{\varepsilon \to 0} \varepsilon^{-1} (\theta_\varepsilon^* - \theta^*) = \xi.$$

Proof is similar to the proof of Theorem 7.5, so we only outline the main steps.

Introduce the set

$$\mathcal{A}_1 = \{\omega : \ |\theta_\varepsilon^* - \theta^*| < \nu_\varepsilon\},$$

where $\nu_\varepsilon = \varepsilon \lambda_\varepsilon$, $\lambda_\varepsilon \to \infty$ and $\nu_\varepsilon \to 0$ as $\varepsilon \to 0$. Moreover, we require that $g(\nu_\varepsilon)^2 \varepsilon^{-1} \to \infty$.

The probability $\mathbf{P}_b^{(\varepsilon)}(\mathcal{A}_1^C)$ is already estimated in Theorem 7.10. For $\omega \in \mathcal{A}_1$, we represent $\theta_\varepsilon^* = \theta^* + \varepsilon u_\varepsilon^*$, where u_ε^* is one of the solutions of the system of equations

$$\langle X - y + y - x(\theta^*) + x(\theta^*) - x(\theta^* + \varepsilon u), \dot{x}(\theta^* + \varepsilon u) \rangle = 0,$$

which we rewrite as

$$\begin{aligned} \langle x - y, \dot{x}(\theta^*) \rangle \ &+ \ \langle y - x(\theta^*), \dot{x}(\theta^*) \rangle - \varepsilon \langle u \dot{x}(\theta^*), \dot{x}(\theta^*) \rangle + \\ &+ \ \langle y - x(\theta^*), \ddot{x}(\theta^*) u \rangle + \varepsilon o(1) = 0. \end{aligned}$$

The point θ^* is a point of minimum of $\|y - x(\theta)\|$, so the first derivative

$$\langle y - x(\theta^*), \dot{x}(\theta^*) \rangle = 0.$$

This equality allows us to represent u_ε^* in the form

$$u_\varepsilon^* = J(\theta^*)^{-1}\varepsilon^{-1}\langle X - y, \dot{x}(\theta^*)\rangle + o(1)$$

and to pass to the limit $\varepsilon \to 0$. The detailed proof, of course, contains all the elements of the proof of Theorem 7.5.

The minimum distance approach can be applied to the problems like the following. We are given a diffusion-type process (7.53) with the unknown $b(\cdot)$ and a family of functions $\{(f_t(\theta), 0 \le t \le T), \theta \in \Theta\}$. Suppose that we are going to approximate the nonperturbed solution $\{y_t, 0 \le t \le T\}$ by a member of this family in the following sense:

$$\inf_{\theta \in \Theta} \|y - f(\theta)\| = \|y - f(\theta^*)\|,$$

then, we can introduce the MDE

$$\theta_\varepsilon^* = \arg \inf_{\theta \in \Theta} \|X - f(\theta)\|$$

and consider $\{f_t(\theta_\varepsilon^*), 0 \le t \le T\}$ as an estimator of the desired approximation. The properties of the MDE θ_ε^* described above allow us to study this approximation.

Nonconsistent estimation. Now consider the situation with a nonidentifiable trend, i.e. there exists a point $\theta^* \in \Theta$ such that

$$\|x(\theta_0) - x(\theta^*)\| = 0$$

and $\theta^* \neq \theta_0$. The observed process is

$$dX_t = S_t(\theta_0, X)dt + \varepsilon dW_t, \quad X_0 = x_0, \quad 0 \le t \le T$$

and the MDE

$$\theta_\varepsilon^* = \arg \inf_{\theta \in \Theta} \|X - x(\theta)\|.$$

As before, $x_t(\theta) = x_t$ is a solution of the equation

$$\frac{dx_t}{dt} = S_t(\theta, x), \quad x_0, \quad 0 \le t \le T$$

and we have a family (on θ) of such functions.

Let $\Theta = \Theta_0 \bigcup \Theta^*, \theta_0 \in \Theta_0, \theta^* \in \Theta^*$, and $\Theta_0 \bigcap \Theta^* = \emptyset$ is such a partition of Θ, that the following conditions are fulfilled:

$$\begin{aligned} g_0(\nu) &= \inf_{|\theta - \theta_0| > \nu, \theta \in \Theta_0} \|x(\theta) - x(\theta_0)\| > 0, \\ g^*(\nu) &= \inf_{|\theta - \theta^*| > \nu, \theta \in \Theta^*} \|x(\theta) - x(\theta^*)\| > 0 \end{aligned}$$

for $\nu > 0$.

Introduce the "good" sets

$$\mathcal{A}_0 = \left\{ \inf_{|\theta-\theta_0|<\nu,\theta\in\Theta_0} \|X - x(\theta)\| < \inf_{|\theta-\theta_0|>\nu,\theta\in\Theta_0} \|X - x(\theta)\| \right\},$$
$$\mathcal{A}^* = \left\{ \inf_{|\theta-\theta^*|<\nu,\theta\in\Theta^*} \|X - x(\theta)\| < \inf_{|\theta-\theta^*|>\nu,\theta\in\Theta^*} \|X - x(\theta)\| \right\}$$

and $\mathcal{A} = \mathcal{A}_0 \bigcup \mathcal{A}^*$. Then, for $\omega \in \mathcal{A}$, we have $|\theta_\varepsilon^* - \theta_0| < \nu$ or $|\theta_\varepsilon^* - \theta^*| < \nu$, i.e.

$$\theta_\varepsilon^* \in \Theta_{0,\nu} \bigcup \Theta_\nu^*; \quad \Theta_{0,\nu} = \{\theta : |\theta - \theta_0| < \nu\}, \quad \Theta_\nu^* = \{\theta : |\theta - \theta^*| < \nu\}.$$

So we have

$$\mathbf{P}_{\theta_0}^{(\varepsilon)}\{\min(|\theta_\varepsilon^* - \theta_0|, |\theta_\varepsilon^* - \theta^*|) > \nu\} = \mathbf{P}_{\theta_0}^{(\varepsilon)}\{\mathcal{A}^C\}$$

and, as follows from the proof of Theorem 7.5,

$$\mathbf{P}_{\theta_0}^{(\varepsilon)}\{\mathcal{A}^C\} \le 8\exp\left\{-\frac{g(\nu)^2}{2\kappa\varepsilon^2}\right\},$$

where $g(\nu) = \min(g_0(\nu), g^*(\nu))$, $\kappa = \max(\kappa_0, \kappa^*)$ in corresponding notations.

Let us now there are k points $\theta_0, \theta_1, \ldots, \theta_{k-1}$ such that

$$\|x(\theta_i) - x(\theta_0)\| = 0, \quad i = 1, \ldots, k-1.$$

We introduce a partition of Θ, i.e. $\Theta = \bigcup_{i=0}^{k-1} \Theta_i, \Theta_i \bigcap \Theta_j = \emptyset, i \neq j$ such that every $\theta_i \in \Theta_i$ with its nonempty vicinity.

Denote

$$g_i(\nu) = \inf_{|\theta-\theta_i|>\nu,\theta\in\Theta_i} \|x(\theta) - x(\theta_i)\|,$$
$$g(\nu) = \min_i g_i(\nu), \quad \Theta_{i,\nu} = \{\theta : |\theta - \theta_i| < \nu\}.$$

Using the similar arguments, the following theorem could be proved.

Theorem 7.13. *If $g(\nu) > 0$ for all $\nu > 0$, then*

$$\mathbf{P}_{\theta_0}^{(\varepsilon)}\left\{\theta_\varepsilon^* \in \bigcup_{i=0}^{k-1} \Theta_{i,\nu}\right\} \ge 1 - 4k\exp\left\{-\frac{g(\nu)^2}{2\kappa\varepsilon^2}\right\}$$

with the constant $\kappa > 0$.

We are interesting now in the probabilities of convergencies

$$\theta_\varepsilon^* \to \theta_i, \quad i = 0, 1, \ldots, k-1.$$

To find them we suppose that

$$S_t(\theta, X) = V(\theta, t, X_t) + \int_0^t K(\theta, t, s, X_s)ds$$

and preserve the notations $x_t^{(1)}(\theta_0)$ and $\dot{x}_t(\theta)$ for the derivative of X_t w.r.t. ε at the point $\varepsilon = 0$ and for the vector of derivatives of $x_t(\theta)$ w.r.t. θ. We also denote :

matrices :

$$J_i = \langle \dot{x}(\theta_i), \dot{x}(\theta_i)^T \rangle;$$

Gaussian vectors:

$$\xi_i = J_i^{-\frac{1}{2}} \langle x^{(1)}(\theta_0), \dot{x}(\theta_i) \rangle;$$

sets:

$$\mathcal{A}_i = \{\omega : \ |\xi_i| > \max_{j \neq i} |\xi_j|\};$$

probabilities:

$$p_i = \mathbf{P}\{\mathcal{A}_i\}, \quad i = 0, 1, \ldots, k-1$$

random vector

$$\theta_* = \sum_{i=0}^{k-1} \theta_i \chi_{\{\mathcal{A}_i\}}.$$

Theorem 7.14. *Let the functions $V(\theta, t, x)$, $K(\theta, t, s, x)$ have two continuous bounded derivatives w.r.t. x, the functions $V(\theta, t, x_t(\theta_0))$, $K(\theta, t, s, x_s(\theta_0))$ have two continuous bounded derivatives w.r.t. θ, the matrices $J_i, i = 0, 1, \ldots, k$, are positive definite, $\sum_{i=0}^{k-1} p_i = 1$, and $g(\nu) > 0$ for $\nu > 0$, then*

$$\mathbf{P}_{\theta_0} - \lim_{\varepsilon \to 0} \theta_\varepsilon^* = \theta_*.$$

Proof. Consider the random field

$$\eta_\varepsilon^{(i)}(u) = \varepsilon^{-2} \|X - x(\theta_i + \varepsilon u)\|^2, \quad u \in U_{i,\varepsilon} = \{u : \ \theta_i + \theta u \in \Theta_{i,\nu}\},$$

where $\nu = \nu_\varepsilon = \varepsilon \lambda_\varepsilon, \lambda_\varepsilon \to \infty, \varepsilon \lambda_\varepsilon \to 0$. Denote

$$\eta^{(i)}(u) = \|x^{(1)}(\theta_0) - (u, \dot{x}(\theta_0))\|^2, \quad u \in \mathbf{R}^d.$$

We have

$$\inf_u \eta^{(i)}(u) = \|x^{(1)}(\theta_0)\|^2 - |\xi_i|^2.$$

Recall several useful inequalities derived in Section 7.2

$$\begin{aligned}
\|X - x(\theta_0)\| &\leq C\varepsilon \sup_{0 \leq t \leq T} |W_t| \\
\|(X - x(\theta_0))\varepsilon^{-1} - x^{(1)}(\theta_0)\| &\leq C\varepsilon \sup_{0 \leq t \leq T} |W_t| \\
\|\varepsilon^{-1}(x(\theta_i + \varepsilon u) - x(\theta_i)) - (u, \dot{x}(\theta_i))\| &\leq C\varepsilon |u|^2 \\
\|x(\theta_i + \varepsilon u) - x(\theta_i)\| &\leq C\varepsilon |u|.
\end{aligned}$$

We have for some i:

$$
\begin{aligned}
&|\eta_\varepsilon^{(i)}(u) - \eta^{(i)}(u)| = \\
&= \|[(X - x(\theta_0))\varepsilon^{-1} - x^{(1)}(\theta_0) + (x(\theta_i) - x(\theta_i + \varepsilon u))\varepsilon^{-1} + (u, \dot{x}(\theta_i))] \times \\
&\quad \times [(X - x(\theta_0))\varepsilon^{-1} + x^{(1)}(\theta_0) + (x(\theta_i) - x(\theta_i + \varepsilon u))\varepsilon^{-1} - (u, \dot{x}(\theta_i)]\| \leq \\
&\leq (\|(X - x(\theta_0))\varepsilon^{-1} - x^{(1)}(\theta_0)\| + \|(x(\theta_i + \varepsilon u) - x(\theta_i))\varepsilon^{-1} - (u, \dot{x}(\theta_i))\|) \times \\
&\quad \times (\|(X - x(\theta_0))\varepsilon^{-1} + \|x^{(1)}(\theta_0)\| + \|x(\theta_i + \varepsilon u) - x(\theta_i)\|\varepsilon^{-1} + \|(u, \dot{x}(\theta_i))\|) \leq \\
&\leq (C\varepsilon \sup_{0 \leq t \leq T} |W_t| + C\varepsilon |u|^2)(C \sup_{0 \leq t \leq T} |W_t| + C|u|) \leq \\
&\leq C^* \varepsilon (\sup_{0 \leq t \leq T} |W_t|^2 + |u|^3)
\end{aligned}
$$

with some constant C^*. We choose that C^* does not depend on i.

Put

$$L_\varepsilon = (2C^*)^{-\frac{1}{3}} \varepsilon^{-\frac{1}{4}}$$

and introduce the set

$$\hat{\mathcal{A}} = \left\{ \omega : \sup_{0 \leq t \leq T} |W_t| < (2C^*)^{-\frac{1}{2}} \varepsilon^{-\frac{3}{8}} \right\}.$$

Then on this set we have the inequality

$$\max_{0 \leq i \leq k-1} \sup_{|u| < \lambda_\varepsilon} |\eta_\varepsilon^{(i)}(u) - \eta^{(i)}(u)| < \varepsilon^{\frac{1}{4}}$$

and, for each $i = 0, 1, \ldots, k-1$,

$$\eta^{(i)}(u) - \varepsilon^{\frac{1}{4}} < \eta_\varepsilon^{(i)}(u) < \eta^{(i)}(u) + \varepsilon^{\frac{1}{4}}.$$

So we can write

$$
\begin{aligned}
\|x^{(1)}(\theta_0)\|^2 - |\xi_i|^2 - \varepsilon^{\frac{1}{4}} &< \inf_{|u| < \lambda_\varepsilon} \eta_\varepsilon^{(\varepsilon)}(u) < \\
&< \|x^{(1)}(\theta_0)\|^2 - |\xi_i|^2 + \varepsilon^{\frac{1}{4}}
\end{aligned}
$$

on the set $\hat{\mathcal{A}} \bigcap \tilde{\mathcal{A}}$, where

$$\tilde{\mathcal{A}} = \left\{ \omega : \max_{0 \leq i \leq k-1} |\xi_i| < \lambda_\varepsilon \right\},$$

because we suppose here that the point

$$\hat{u}_i = \arg\inf_u \eta^{(i)}(u)$$

belongs to the set $\{u : |u| < \lambda_\varepsilon\}$.

Return to the original problem. Introduce the set

$$\mathcal{A}_0 = \bigcup_{i=0}^{k-1} \{\omega : \theta_\varepsilon^* \in \Theta_{i,\nu}\}$$

and note that, according to Theorem 7.12, its complement is exponentially small. Remember, that the complements of $\hat{\mathcal{A}}$ and $\tilde{\mathcal{A}}$ are also exponentially small (if we choose, for example, $\lambda_\varepsilon = \varepsilon^{-\frac{1}{2}}$).

Denote

$$\mathcal{A} = \mathcal{A}_0 \bigcap \hat{\mathcal{A}} \bigcap \tilde{\mathcal{A}}.$$

The probability of convergence $\theta_\varepsilon^* \to \theta_i$ could be calculated as follows:

$$\begin{aligned}
&\mathbf{P}_{\theta_0}^{(\varepsilon)}\{\theta_\varepsilon^* \in \Theta_{i,\nu}\} = \mathbf{P}_{\theta_0}^{(\varepsilon)}\{|\theta_\varepsilon^* - \theta_i| < \nu\} \leq \\
&\leq \mathbf{P}_{\theta_0}^{(\varepsilon)}\{\min_{|u|<\lambda_\varepsilon} \eta_\varepsilon^{(i)}(u) < \min_{j\neq i}\min_{|u|<\lambda_\varepsilon} \eta_\varepsilon^{(i)}(u), \mathcal{A}\} + \mathbf{P}_{\theta_0}^{(\varepsilon)}\{\mathcal{A}^C\} \leq \\
&\leq \mathbf{P}_{\theta_0}^{(\varepsilon)}\{\|x^{(1)}(\theta_0)\|^2 - |\xi_i|^2 - \varepsilon^{\frac{1}{4}} < \min(\|x^{(1)}(\theta_0)\|^2 - |\xi_j|^2 + \varepsilon^{\frac{1}{4}})\} + \\
&\quad + C\exp(-\kappa\varepsilon^{-\gamma}) = \\
&= \mathbf{P}_{\theta_0}^{(\varepsilon)}\{|\xi_i|^2 > \max_{j\neq i} |\xi_j|^2 - 2\varepsilon^{\frac{1}{4}}\} + C\exp(-\kappa\varepsilon^{-\gamma}) \to \\
&\to \mathbf{P}\{|\xi_i| > \max_{j\neq i} |\xi_j|\} = p_i,
\end{aligned}$$

and we obtain a similar low inequality for this probability.

The similar arguments allows to write

$$\begin{aligned}
&\mathbf{P}_{\theta_0}^{(\varepsilon)}\{|\theta_\varepsilon^* - \theta_*| < \nu\} = \sum_{i=0}^{k-1} \mathbf{P}_{\theta_0}^{(\varepsilon)}\{|\theta_\varepsilon^* - \theta_i| < \nu, \mathcal{A}_i\} \geq \\
&\geq \sum_{i=0}^{k-1} \mathbf{P}_{\theta_0}^{(\varepsilon)}\{\theta_\varepsilon^* \in \Theta_{i,\nu}, \mathcal{A}_i, \mathcal{A}\} - k\mathbf{P}_{\theta_0}^{(\varepsilon)}\{\mathcal{A}^C\} = \\
&= \sum_{i=0}^{k-1} \mathbf{P}_{\theta_0}^{(\varepsilon)}\{\theta_\varepsilon^* \in \Theta_{i,\nu}, \mathcal{A}|\mathcal{A}_i\}\mathbf{P}(\mathcal{A}) - k\mathbf{P}_{\theta_0}^{(\varepsilon)}\{\mathcal{A}^C\} \to \sum_{i=0}^{k-1} p_i = 1.
\end{aligned}$$

REMARK 7.7 If $\mathbf{P}\{|\xi_i| = |\xi_j|\} = 1$ for some i and j, then as in Section 2.7, we have to consider the next terms of the expansion of $\|X - x(\theta_0 + \varepsilon u)\|$ by the powers of ε, and so on.

Boundary of the parametric set. Let the unknown parameter $\theta \in [\alpha, \beta]$ and the true value $\theta_0 = \alpha$. So we observe the process

$$dX_t = [V(\alpha, t, X_t) + \int_0^t K(\alpha, t, s, X_s)ds]dt + \varepsilon dW_t, \quad X_0 = x_0, \quad 0 \leq t \leq T$$

and use the MDE θ_ε^* as an estimator of α.

We suppose that the functions $V(\alpha, t, x)$ and $K(\alpha, t, s, x)$ have two continuous bounded derivatives on x. The functions $V(\theta, t, x)$ and $K(\theta, t, s, x)$ in some right vicinity of the point α, have two continuous bounded derivatives on θ (at point α – two derivatives on the right). We also assume, that the function

$$g(\nu) = \inf_{|\theta-\alpha|>\nu} \|x(\theta) - x(\alpha)\| > 0$$

for any $\nu = \nu_\varepsilon > 0$ and

$$J(\alpha+) = \|\dot{x}(\alpha+)\|^2 > 0,$$

where $\dot{x}_t(\alpha+)$ is a right derivative at point α.

Put $\nu_\varepsilon = \varepsilon\lambda_\varepsilon$, where $\lambda_\varepsilon = \varepsilon^{-\frac{1}{4}} \to +\infty$ as $\varepsilon \to 0$ and introduce two processes:

$$\begin{aligned} \eta(u) &= \|x^{(1)}(\alpha) - u\dot{x}(\alpha+)\|^2, \quad u \in \mathbf{R}; \\ \eta_\varepsilon(u) &= \varepsilon^{-2}\|X - x(\alpha+\varepsilon u)\|^2, \quad u \in U_\varepsilon = \{u : \alpha + \varepsilon u \in [\alpha, \beta]\}. \end{aligned}$$

We can write

$$\begin{aligned} &|\eta_\varepsilon(u) - \eta(u)| = \\ &= |\langle \varepsilon^{-1}(X - x(\alpha)) - x^{(1)}(\alpha) + \\ &\quad +\varepsilon^{-1}(x(\alpha) - x(\alpha+\varepsilon u)) + u\dot{x}(\alpha+), \\ &\quad \varepsilon^{-1}(x - x(\alpha)) + x^{(1)}(\alpha) + \varepsilon^{-1}(x(\alpha) - x(\alpha+\varepsilon u)) - u\dot{x}(\alpha+)\rangle| \le \\ &\le (\|\varepsilon^{-1}(X - x(\alpha)) - x^{(1)}(\alpha)\| + \|\varepsilon^{-1}(x(\alpha) - x(\alpha+\varepsilon u)) + u\dot{x}(\alpha+)\|) \times \\ &\quad \times(\|\varepsilon^{-1}(X - x(\alpha))\| + \|x^{(1)}(\alpha)\| + \\ &\quad +\|\varepsilon^{-1}(x(\alpha) - x(\alpha+\varepsilon u))\| + \|u\dot{x}(\alpha+)\|) \le \\ &\le (C_1\varepsilon \sup_{0\le t\le T} |W_t|^2 + C_2|u|^2\varepsilon)(C_3 \sup_{0\le t\le T} |W_t| + C_4|u|) \end{aligned}$$

with the constants $C_i, i = 1, \ldots, 4$ (see the proof of Theorem 7.5). So if we introduce the set

$$\mathcal{A}_1 = \left\{\omega : \sup_{0\le t\le T} |W_t| < \varepsilon^{-\frac{1}{4}}\right\},$$

then on $\mathcal{A}_1$ we have the estimate

$$\sup_{|u|<\lambda_\varepsilon} |\eta_\varepsilon(u) - \eta(u)| \le C\varepsilon^{\frac{1}{4}} \tag{7.55}$$

with the corresponding constant $C > 0$.

Denote

$$\xi = J(\alpha+)^{-1}\langle x^{(1)}(\alpha), \dot{x}(\alpha+)\rangle.$$

The random variable

$$u_\varepsilon \equiv \arg \inf_{0\le u\le\lambda_\varepsilon} = \begin{cases} 0, & \text{if } \xi \le 0, \\ \xi, & \text{if } 0 < \xi < \lambda_\varepsilon, \\ \lambda_\varepsilon, & \text{if } \xi \ge \lambda_\varepsilon, \end{cases}$$

and, with probability 1, we have

$$u_\varepsilon \longrightarrow (\xi \vee 0)$$

as $\varepsilon \to 0$.

Introduce the set $\mathcal{A}_2 = \{\omega : |\theta^*_\varepsilon - \alpha| < \nu_\varepsilon\}$, then on $\mathcal{A} = \mathcal{A}_1 \bigcap \mathcal{A}_2$ we have $u^*_\varepsilon = \varepsilon^{-1}(\theta^*_\varepsilon - \alpha) < \lambda_\varepsilon$ and from the uniform convergence $\eta_\varepsilon(\cdot)$ to $\eta(\cdot)$ defined by

(7.55), we have

$$\lim_{\varepsilon\to 0}\mathbf{P}_\alpha^{(\varepsilon)}\{\varepsilon^{-1}(\theta_\varepsilon^*-\alpha)<x\} \doteq \mathbf{P}\{(0\vee\xi)<x\} = \begin{cases} 0, & \text{if } x<0,\\ \frac{1}{2}, & \text{if } x=0,\\ \Phi(\sigma(\alpha)x), & \text{if } x>0,\end{cases}$$

where $\Phi(y)$ is a Gaussian $(0,1)$ distribution function and

$$\sigma(\alpha)^2 = J(\alpha+)^{-2}\int_0^T\int_0^T \dot x_t(\alpha+)\dot x_s(\alpha+)\mathbf{E}x_t^{(1)}(\alpha)x_s^{(1)}(\alpha)d\mu_s d\mu_t.$$

Remember that the probabilities of $\mathcal{A}_1^C$ and $\mathcal{A}_2^C$ are exponentially small.

So the MDE θ_ε^* is also asymptotically "half normal"as is the MLE $\hat\theta_\varepsilon$ in Section 2.8. If the true value $\theta_0<\alpha$, then the behavior of θ_ε^* is also similar to the behavior of the MLE $\hat\theta_\varepsilon$ and is described above in the "misspecified model".

Expansions. We consider the same problem of estimation $\theta_0\in(\alpha,\beta)$ by observations

$$dX_t = S_t(\theta_0,X)dt+\varepsilon dW_t,\quad X_0=x_0,\quad 0\le t\le T \tag{7.56}$$

using the MDE θ_ε^*. As before, $\|\cdot\|$ is a norm in $L_2(\mu),\mu([0,T])<\infty, x_t=x_t(\theta)$ is a solution of the system

$$\frac{dx_t}{dt}=S_t(\theta,x),\quad x_0,\quad 0\le t\le T,$$

and the trend is

$$S_t(\theta,V)=V(\theta,t,X_t)+\int_0^t K(\theta,t,s,X_s)ds.$$

In regular case the estimator θ_ε^* is consistent and asymptotically normal (theorem 7.5), i.e.

$$\theta_\varepsilon^*=\theta_0+\varepsilon\xi(1+o(1)),$$

which we understand as a first two terms in the expansion of θ_ε^* by the powers of ε. In this section we obtain an expansion of θ_ε^* like the expansion of the MLE $\hat\theta_\varepsilon$ (3.4).

Introduce the notations:

$$\begin{aligned} g(\nu) &= \inf_{|\theta-\theta_0|>\nu}\|x(\theta)-x(\theta_0)\|,\\ J(\theta) &= \|\dot x(\theta)\|^2. \end{aligned}$$

Theorem 7.15. *Let the functions $V(\theta,t,x)$ and $K(\theta,t,s,x)$ have $k+2$ continuous bounded derivatives on θ and on x, the function $g(\nu)>0$ for all $\nu>0$ and $J(\theta_0)>0$,* *then there exist the random variables* $\xi_1,\ldots,\xi_k,\eta,\zeta$ *and a set* $\mathcal{A}$ *such that the MDE* θ^*_ε *admits the representation*

$$\theta^*_\varepsilon = \theta_0 + \left\{\sum_{j=1}^k \xi_j\varepsilon^j + \eta\varepsilon^{k+\frac{1}{2}}\right\}\chi_{\{\mathcal{A}\}} + \zeta\chi_{\{\mathcal{A}^C\}}, \tag{7.57}$$

where $|\eta| < 1$, *the random variables* $\xi_1,\cdot,\xi_k$ *do not depend on* ε, *and*

$$\begin{aligned}\mathbf{P}^{(\varepsilon)}_{\theta_0}\{\mathcal{A}^C\} &\le C_1\exp\{-\kappa_1\varepsilon^{-\gamma_1}\},\\ \mathbf{P}^{(\varepsilon)}_{\theta_0}\{|\zeta| > \varepsilon^\delta\} &\le C_2\exp\{-\kappa_2\varepsilon^{-\gamma_2}\}\end{aligned}$$

with some positive constants $C_i,\kappa_i,\gamma_i, i = 1,2$ *and* $\delta \in (0,\frac{1}{2})$.

Proof. We apply here the same method of "good sets" as in Chapter 3. So we introduce a first "good" set

$$\mathcal{A}_1 = \{\omega :\ |\theta^*_\varepsilon - \theta_0| < \nu_\varepsilon\},$$

where $\nu_\varepsilon = \varepsilon\lambda_\varepsilon \to 0, \lambda_\varepsilon \to \infty$ which localizes the problem, i.e. for $\omega \in \mathcal{A}_1$ the MDE $\theta^*_\varepsilon = \theta_0 + u^*_\varepsilon$, where u^*_ε is one of the solutions of the "minimum distance equation":

$$\langle x(\theta_0+u) - X, \dot{x}(\theta_0+u)\rangle = 0, \quad |u| < \nu_\varepsilon.$$

Denote this equation as

$$F(u,\varepsilon) = 0, \quad -\nu_\varepsilon < u < \nu_\varepsilon$$

and provide the uniqueness of its solution on some set $\mathcal{A}_2$ by the following way. The derivative

$$\begin{aligned}\frac{\partial F}{\partial u} &= J(\theta_0+u) + \langle x(\theta_0+u) - X, \ddot{x}(\theta_0+u)\rangle \ge\\ &\ge J(\theta_0) - C\nu_\varepsilon + \langle x(\theta_0+u) - x(\theta_0), \ddot{x}(\theta_0+u)\rangle + \langle x(\theta_0) - X, \ddot{x}(\theta_0+u)\rangle \ge\\ &\ge J(\theta_0) - c_1\nu_\varepsilon - c_2\varepsilon \sup_{0\le t\le T}|W_t|,\end{aligned}$$

where we have used the boundedness of the derivatives and Lemma 1.13. So on the set

$$\mathcal{A}_2 = \left\{\omega :\ \sup_{0\le t\le T}|W_t| < \frac{c_1}{c_2}\lambda_\varepsilon\right\}$$

for $\varepsilon < \varepsilon_0$ with ε_0 taken from

$$\varepsilon_0\lambda_{\varepsilon_0} \le \frac{1}{4c_1}J(\theta_0),$$

for the derivative, we obtain the inequality

$$\inf_{|u|<\nu_\varepsilon}\frac{\partial F}{\partial u}(u,\varepsilon) \ge \frac{1}{2}J(\theta_0) > 0.$$

Hence, on $\mathcal{A}_1 \bigcap \mathcal{A}_2$, the minimum distance equation has a unique solution $u^* = u^*(\varepsilon)$. We consider the implicitly defined function $u^* = u^*(\varepsilon)$ and calculate its Taylor formula

$$u^*(\varepsilon) = \sum_{j=1}^{k+1} \frac{u^{(j)}(0)}{j!}\varepsilon^j + \frac{u^{(k+1)}(\tilde{\varepsilon})}{(k+1)!}\varepsilon^{k+1}, \quad \tilde{\varepsilon} \in (0, \varepsilon_0).$$

The partial derivatives $F(u,\varepsilon)$ are

$$\begin{aligned}\frac{\partial^m F(u,\varepsilon)}{\partial \varepsilon^{l+1}\partial u^{m-l-1}} &= -\frac{\partial^{m-1}}{\partial \varepsilon^l \partial u^{m-l-1}}\langle X^{(1)}, \dot{x}(\theta_0+u)\rangle = \\ &= -\langle X^{(l+1)}, x^{(m-l)}(\theta_0+u)\rangle\end{aligned}$$

with $m = 1, \ldots, k+1; l = 0, 1, \ldots, m-1$ and

$$\begin{aligned}\frac{\partial^m F(u,\varepsilon)}{\partial u^m} &= \langle x(\theta_0+u) - X, x^{(m+1)}(\theta_0+u)\rangle + \\ &+ \sum_{j=1}^{m}\binom{m}{j}\langle x^{(j)}(\theta_0+u), \quad x^{(m-j+1)}(\theta_0+u)\rangle\end{aligned}$$

(Leibnitz formula). Here the processes $X_t^{(l)} = \frac{\partial^l}{\partial \varepsilon^l}X_t$ satisfy the system of equations

$$\begin{aligned}dX_t^{(1)} &= [V'(\theta,t,X_t)X_t^{(1)} + \int_0^t K'(\theta,t,s,X_s)X_s^{(1)}ds]dt + dW_t, \quad X_0^{(1)} = 0;\\ \frac{dX_t^{(2)}}{dt} &= V_{xx}^{(2)}(\theta,t,X_t)(X_t^{(1)})^2 + V'(\theta,t,X_t)X_t^{(2)} + \\ &+ \int_0^t [K_{xx}^{(2)}(\theta,t,s,X_s)(X_s^{(1)})^2 + K'(\theta,t,s,X_s)X_s^{(2)}]ds, \quad X_0^{(2)} = 0;\\ \ldots & \quad \ldots\ldots \\ \frac{dX_t^{(k)}}{dt} &= \sum_I C_I V_x^{(P_I)}(\theta,t,X_t)\prod_{j=1}^{k}(X_t^{(j)})^{i_j} + \\ &+ \sum_I \int_0^T C_I K_x^{(P_I)}(\theta,t,s,X_s)\prod_{j=1}^{k}(X_s^{(j)})^{i_j}ds, \quad X_0^{(k)} = 0, \end{aligned} \tag{7.58}$$

where the types of summation $I = \{i_1, \ldots, i_k\}$ and the constants P_I, C_I are the same as in (3.13).

So, if we also introduce the set

$$\mathcal{A}_3 = \{\omega : \sup_{0<\tilde{\varepsilon}<\varepsilon} |u^{(k+1)}(\tilde{\varepsilon})| < (k+1)!\varepsilon^{-\frac{1}{2}}\},$$

then on the set $\mathcal{A} = \bigcap_{j=1}^3 \mathcal{A}_j$ for $\varepsilon < \varepsilon_0$, we have the representation

$$\theta_\varepsilon^* = \theta_0 + \sum_{j=1}^{k}\xi_j\varepsilon^j + \eta\varepsilon^{k+\frac{1}{2}},$$

where $\xi_j = \frac{u^{(j)}(0)}{j!}$ and $|\eta| < 1$.

By Theorem 7.2

$$\mathbf{P}_{\theta_0}^{(\varepsilon)}\{\mathcal{A}_1^C\} \le 4\exp\left\{-\frac{g(\nu)^2}{T\varepsilon^2}\right\}.$$

We can also immediately write

$$\mathbf{P}\{\mathcal{A}_2^C\} \le 2\exp\left\{-\frac{C_1^2\lambda_\varepsilon^2}{2TC_2^2}\right\}.$$

The estimate of $\mathbf{P}_{\theta_0}^{(\varepsilon)}\{\mathcal{A}_3^C\}$ is reduced to estimates of the probabilities

$$\begin{aligned}\mathbf{P}_{\theta_0}^{(\varepsilon)}\{|\langle X^{(l+1)}, x^{(m-l)}(\theta_0+u)\rangle| > h\varepsilon^{-\delta_1}\} \le \\ \le \mathbf{P}_{\theta_0}^{(\varepsilon)}\{\sup_{0\le t\le T}|X_t^{(l+1)}| > h^*\varepsilon^{-\delta_1}\}\end{aligned}$$

and

$$\begin{aligned}\mathbf{P}_{\theta_0}^{(\varepsilon)}\{|\langle X - x(\theta_0+u), x^{(m+1)}(\theta_0+u)\rangle| > h_2\varepsilon^{-\delta_2}\} \le \\ \le \mathbf{P}\{\sup_{0\le t\le T}|W_t| > h_2^*\varepsilon^{-(\delta_2+1)}\}\end{aligned} \tag{7.59}$$

with the positive constants $h_1^*, h_2^*, \delta_1, \delta_2$.

The boundedness of the derivatives of $V(\cdot)$ and $K(\cdot)$ in the system (7.57) and Lemma 1.11 allow us to derive the estimates

$$\sup_{0\le t\le T}|X_t^{(j)}| \le N_j \sup_{0\le t\le T}|W_t|^j$$

similar to (3.24).

Hence, we can write

$$\begin{aligned}\mathbf{P}_{\theta_0}^{(\varepsilon)}\left\{\sup_{0\le t\le T}|X_t^{(l)}| > h^*\varepsilon^{-\delta_1}\right\} \le \\ \le \mathbf{P}\left\{\sup_{0\le t\le T}|W_t| > (h^*N_l^{-1})^{\frac{1}{l}}\varepsilon^{-\frac{\delta_1}{l}}\right\} \le \\ \le 2\exp\left\{-\tilde{\kappa}\varepsilon^{-\frac{2\delta_1}{l}}\right\}, \quad \tilde{\kappa} = \frac{1}{2T}\left(h^*N_l^{-1}\right)^{\frac{2}{l}}.\end{aligned}$$

It only remains to mention the inequality

$$\mathbf{P}_{\theta_0}^{(\varepsilon)}\{\mathcal{A}^C\} \le \sum_{j=1}^{3}\mathbf{P}_{\theta_0}^{(\varepsilon)}\{\mathcal{A}^C\}.$$

We now consider the expansion concerning the first terms of (7.57). For the normed difference

$$u_\varepsilon^* = \varepsilon^{-1}\sigma^{-1}(\theta_\varepsilon^* - \theta)$$

by direct calculation (on the set $\mathcal{A}$) we have

$$\begin{aligned}u_\varepsilon^* &= \sigma^{-1}J^{-1}\langle x^{(1)}, \dot{x}\rangle + \frac{1}{2}\varepsilon\sigma^{-1}J^{-3}\left\{2J\langle x^{(1)}, \dot{x}\rangle\langle x^{(1)}, \ddot{x}\rangle\right. \\ &- \left. 3\langle \dot{x}, \ddot{x}\rangle\langle x^{(1)}, \dot{x}\rangle^2 + J^2\langle x^{(2)}, \dot{x}\rangle\right\} + \eta_1\varepsilon^{3/2}.\end{aligned}$$

We can modify the set $\mathcal{A}$ in such a way that $|\eta_1| < 1$.

For simplicity of exposition we suppose that the observed process is

$$dX_t = S(\theta, X_t)\,dt + \varepsilon\,dW_t, \quad X_0 = x_0,\ 0 \le t \le T,$$

and $S(\theta, x) > 0$ for all $t \in [0, T]$. Then the processes $x_t^{(1)}$ and $x_t^{(2)}$ (see (3.43)) have the representations

$$\begin{aligned} x_t^{(1)} &= S(\theta, x_t) \int_0^t \frac{dW_s}{S(\theta, x_s)}, \\ x_t^{(2)} &= S(\theta, x_t) \int_0^t \frac{S(\theta, x_s)''}{S(\theta, x_s)} (x_s^{(1)})^2\,ds \end{aligned}$$

We shall follow the demonstration presented in Section 3.3. Introduce notations

$$\begin{aligned} \xi_1 &= \sigma^{-1} J^{-1} \langle x^{(1)}, \dot{x} \rangle &\sim \mathcal{N}(0,1), \\ \xi_2 &= J^{-1} \langle x^{(1)}, \ddot{x} \rangle &\sim \mathcal{N}(0, \sigma_2^2), \\ \xi_3 &= \sigma^{-1} J^{-1} \langle x^{(2)}, \dot{x} \rangle, \\ \rho_{12} &= \sigma J^{-1} \langle \dot{x}, \ddot{x} \rangle \end{aligned}$$

The Fubini theorem allows us to rewrite these random variables as follow.

$$\begin{aligned} \xi_1 &= \sigma^{-1} J^{-1} \int_0^T x_t^{(1)}\, \dot{x}_t\, d\mu_t = \\ &= \sigma^{-1} J^{-1} \int_0^T S(\theta, x_t) \dot{x}_t(\theta) \int_0^t \frac{dW_s}{S(\theta, x_s)}\, d\mu_t = \\ &= \int_0^T \int_s^T \sigma^{-1} J^{-1} S(\theta, x_t) \dot{x}(\theta) d\mu_t\, S(\theta, x_s)\, dW_s = \\ &= \int_0^T f_s(\theta)\, dW_s, \qquad \int_0^T f_s(\theta)^2 ds = 1 \end{aligned}$$

and by a similar way

$$\xi_2 = \int_0^T g_s(\theta)\, dW_s, \qquad \int_0^T g_s(\theta)^2\, dt = \sigma_2^2.$$

We have another representation for ξ_3.

$$\begin{aligned} \xi_3 &= \sigma^{-1} J^{-1} \int_0^T \dot{x}_t(\theta) S(\theta, x_t) \int_0^t \frac{S''(\theta, x_s)}{S(\theta, x_s)} (x_s^{(1)})^2 ds\, d\mu_t = \\ &= \int_0^T S''(\theta, x_s) S(\theta, x_s) \left(\int_0^s \frac{dW_v}{S(\theta, x_v)} \right)^2 \int_s^T S(\theta, x_t)) \dot{x}_t(\theta) \sigma^{-1} J^{-1} d\mu_t\, ds = \\ &= \int_0^T h_s(\theta) \left(\int_0^s \frac{dW_v}{S(\theta, x_v)} \right)^2 ds. \end{aligned}$$

To obtain the expansion of the distribution function of the random variable

$$u_\varepsilon^* = \xi_1 + \varepsilon \{ \xi_1 \xi_2 - \frac{3}{2} \rho_{12} \xi_1^2 + \frac{1}{2} \xi_3 \} \chi_{\{\mathcal{A}\}} + \zeta \chi_{\{\mathcal{A}^c\}}$$

we can restrict ourselves as in Section 3.3 by the expansion of "truncated" sum:

$$G_\varepsilon(x) = \mathbf{P}_\theta^{(\varepsilon)}\{\xi_1 + \varepsilon[\xi_1\xi_2 - \frac{3}{2}\rho_{12}\,\xi_1^2 + \frac{1}{2}\xi_3] < x\}.$$

The expansion of its characteristic function is

$$\begin{aligned}\Phi_\varepsilon(\lambda) &= \mathbf{E}_\theta \exp\{i\lambda\xi_1 + i\lambda\varepsilon[\xi_1\xi_2 - \frac{3}{2}\rho_{12}\,\xi_1^2 + \frac{1}{2}\xi_3]\} = \\ &= e^{-\lambda^2/2} + i\lambda\varepsilon\{\mathbf{E}_\theta\xi_1\xi_2 e^{i\lambda\xi_1} - \frac{3}{2}\rho_{12}\mathbf{E}_\theta\xi_1^2 e^{i\lambda\xi_1} + \frac{1}{2}\mathbf{E}_\theta\xi_3 e^{i\lambda\xi_1}\} + r_\varepsilon(\lambda)\end{aligned}$$

Using the decomposition $\xi_2 = a\xi_1 + \xi'$ with $\mathbf{E}\xi_1\xi' = 0$ and

$$a = \mathbf{E}\xi_1\xi_2 = \int_0^T f_s(\theta)g_s(\theta)ds,$$

we can write

$$\begin{aligned}\mathbf{E}_\theta(\xi_1\xi_2 &- \frac{3}{2}\rho_{12}\xi_1^2)e^{i\lambda\xi_1} = (a - \frac{3}{2}\rho_{12})\mathbf{E}_\theta\xi_1^2 e^{i\lambda\xi_1} = \\ &= (a - \frac{3}{2}\rho_{12})(1-\lambda^2)e^{-\lambda^2/2}.\end{aligned}$$

To calculate the last mathematical expectation we introduce additional notations:

$$\begin{aligned}\xi(s) &= \int_0^s f_t(\theta)dW_t, \qquad \zeta(s) = \int_0^s \frac{dW_v}{S(\theta,x_v)}, \\ q(s,T) &= \int_s^T f_t(\theta)^2 dt, \qquad \zeta(s) = a(s)\xi(s) + \xi'(s), \\ a(s) &= q(0,s)^{-1}\int_0^s \frac{f_t(\theta)}{S(\theta,x_t)}\,dt \\ b(s) &= \mathbf{E}_\theta\xi'(s)^2, \qquad \sigma(s)^2 = q(0,s),\end{aligned}$$

where $\mathbf{E}_\theta\xi(s)\xi'(s) = 0$.

We have

$$\begin{aligned}\mathbf{E}_\theta\xi_3 e^{i\lambda\xi_1} &= \int_0^T h_s(\theta)\mathbf{E}_\theta\zeta(s)^2 e^{i\lambda\xi_1}\,ds = \int_0^T h_s(\theta)\exp\{-q(s,T)\}\mathbf{E}_\theta\zeta(s)^2 e^{i\lambda\xi(s)}\,ds = \\ &= \int_0^T h_s(\theta)\exp\{-q(s,T)\}[a(s)^2\mathbf{E}_\theta\xi(s)^2 e^{i\lambda\xi(s)} + \mathbf{E}_\theta\xi'(s)^2\mathbf{E}_\theta e^{i\lambda\xi(s)}]ds = \\ &= \int_0^T h_s(\theta)\exp\{-q(s,T)\}[a(s)^2\sigma(s)^2(1-\lambda^2\sigma(s)^2) + b(s)]\exp\{-\lambda^2\sigma(s)^2\}ds = \\ &= \int_0^T h_s(\theta)[a(s)^2\sigma(s)^2 + b(s)]ds\; e^{-\lambda^2/2} - \int_0^T h_s(\theta)a(s)^2\sigma(s)^4 ds\;\lambda^2 e^{-\lambda^2/2} = \\ &= 2(A - \lambda^2 B)e^{-\lambda^2/2}\end{aligned}$$

in obvious notations. Joining all these equalities, we obtain

$$\begin{aligned}\Phi_\varepsilon(\lambda) &= e^{-\lambda^2/2} + \varepsilon[a - \frac{3}{2}\rho_{12} + A]\; i\lambda e^{-\lambda^2/2} + \\ &\quad + \varepsilon[a - \frac{3}{2}\rho_{12} + B](i\lambda)^3 e^{-\lambda^2/2} + r_\varepsilon(\lambda) = \\ &= e^{-\lambda^2/2} + \varepsilon R_1 i\lambda e^{-\lambda^2/2} + \varepsilon Q_1(i\lambda)^3 e^{-\lambda^2/2} + r_\varepsilon(\lambda) = G(\lambda) + r_\varepsilon(\lambda)\end{aligned}$$

The inverse Fourier transform of the function $G(\lambda)$ is the function

$$F(x) = \int_{-\infty}^{x} n(y)dy + \varepsilon R_1 n(x) + \varepsilon Q_1(x^2-1)n(x)$$

($n(x)$ is a density function of $\mathcal{N}(0,1)$). The difference $G_\varepsilon(x) - F(x)$ could be estimated by the same way as it was done in Section 3.3.

Hence the following estimate is valid

$$\left|\mathbf{P}_\theta^{(\varepsilon)}\{u_\varepsilon^* < x\} - F(x)\right| \leq C_* \varepsilon^{3/2}$$

with some positive constant C_*.

The representation (7.57) can be also obtained for another MDE as well as being generalized for more general models, as was done in Section 3.2. Moreover, the expansions of the moments like Section 3.3 can be also realized.

Disorder problem. We consider below the problem of Section 6.1 but without "smooth parameter" θ. So the observed diffusion process has the following structure:

$$\begin{aligned} dX_t &= h_t(X_t)dt + \varepsilon dW_t, \quad X_0 = x_0, \quad 0 \leq t \leq \tau; \\ dX_t &= g_t(X_t)dt + \varepsilon dW_t, \quad X_\tau, \quad \tau \leq t \leq T. \end{aligned}$$

The limiting process $x_t = x_t(\tau)$ is a solution of switching differential equation

$$\frac{dx_t}{dt} = \begin{cases} h_t(x_t), & 0 \leq t < \tau, \quad x_0, \\ g_t(x_t), & \tau \leq t \leq T, \end{cases} \tag{7.60}$$

and we have a family (on τ) of solutions $\{x_t(\tau), 0 \leq t \leq T\}, \tau \in (0,T)$. The MDE τ_ε^* of the instant of switching we define as usual:

$$\tau_\varepsilon^* = \arg \inf_{\tau \in (0,T)} \|X - x(\tau)\|,$$

where $\|\cdot\|$ is $L_2(\mu)$ norm with some measure $\mu(\cdot)$ such that $\mu([0,T]) < \infty$.

Note, that for $t < \tau$

$$\dot{x}_t(\tau) = \frac{\partial}{\partial \tau} x_t(\tau) = 0$$

and for $t \geq \tau$

$$\dot{x}_t(\tau) = h_\tau(x_\tau) - g_\tau(x_\tau) + \int_0^t g_s'(x_s)\dot{x}_s(\tau)ds.$$

Here $g_s'(x) = \frac{\partial}{\partial x} g_s(x)$. We can solve this equation and obtain

$$\dot{x}_t(\tau) = [h_\tau(x_\tau) - g_\tau(x_\tau)] \exp\left\{\int_\tau^t g_s'(x_s)ds\right\}$$

and even simplify this expression if $g_s'(x_s) = g'(x_s) > 0$ for $s \in [\tau, T]$, because

$$\int_\tau^t g'(x_s)ds = \int_\tau^t \frac{g'(x_s)x_s'}{g(x_s)}ds = \int_{x_\tau}^{x_t} \frac{dg(x)}{g(x)} = \ln \frac{g(x_t)}{g(\tau_T)},$$

so

$$\dot{x}_t(\tau) = [h_\tau(x_\tau) - g(x_\tau)]\frac{g(x_t)}{g(x_\tau)}.$$

The Gaussian process $x_t^{(1)} = x_t^{(1)}(\tau), t \in [\tau, T]$ is taken from the equation

$$dx_t^{(1)} = g_t'(x_t)x_t^{(1)}dt + dW_t,$$

with the initial value $x_\tau^{(1)}$ defined by the value of the solution

$$dx_t^{(1)} = h_t'(x_t)x_t^{(1)}dt + dW_t, \quad x_0^{(1)} = 0, \quad 0 \le t \le \tau$$

at point τ. Introduce the Gaussian random variable

$$\zeta = \left(\int_\tau^T \dot{x}_t(\tau)^2 d\mu_t\right)^{-1} \int_\tau^T x_t^{(1)}(\tau)\dot{x}_t(\tau)d\mu_t,$$

the function

$$g(\nu) = \inf_{|\tau-\tau_0|>\nu} \|x(\tau) - x(\tau_0)\|$$

and the quantity

$$J(\tau) = \int_\tau^T \dot{x}_t(\tau)^2 d\mu_t.$$

Theorem 7.16. *Let the functions $h_t(x)$ and $g(x)$ have two continuous bounded derivatives on x, the function $g(\nu) > 0$ for $\nu > 0$ and $J(\tau_0) > 0$, then*

$$\mathbf{P}_{\tau_0} - \lim_{\varepsilon\to 0} \varepsilon^{-1}(\tau_\varepsilon^* - \tau_0) = \zeta.$$

Proof. First we note that Theorem 7.5 cannot be applied directly, because the right side of Equation (7.60) is not differentiable on τ at the point $t = \tau$. So we introduce the stochastic process

$$\begin{aligned} \eta_\varepsilon(u) &= \varepsilon^{-2}\|X - x(\tau_0 + \varepsilon u)\|^2 = \|\varepsilon^{-1}(X - x(\tau_0)) - \varepsilon^{-1}(x(\tau_0 + \varepsilon u) - x(\tau_0))\|^2, \\ & \quad u \in U_\varepsilon = \{(-\tau\varepsilon^{-1}, (T - \tau)\varepsilon^{-1})\} \end{aligned}$$

and the limit process

$$\eta(u) = \|x^{(1)}(\tau_0) - u\dot{x}(\tau_0)\|, \quad u \in \mathbf{R}$$

and, as was done in misspecified and nonconsistent estimation introducing the function $\nu_\varepsilon = \varepsilon\lambda_\varepsilon \to 0, \lambda_\varepsilon \to \infty$, we prove the uniform convergence

$$\mathbf{P}_{\tau_0} - \lim_{\varepsilon\to 0} sup_{|u|<\lambda_\varepsilon}|\eta_\varepsilon(u) - \eta(u)| = 0.$$

Hence, we have the convergence

$$\varepsilon^{-1}(\tau_\varepsilon^* - \tau_0) = \arg \inf_{|u|<\lambda_\varepsilon} \eta_\varepsilon(u) \rightarrow \arg\inf_u \eta(u) = \zeta.$$

on the set $\mathcal{A}_1 = \{\omega : |\tau_\varepsilon^* - \tau_0| < \nu_\varepsilon\}$. The probability $\mathbf{P}_\tau^{(\varepsilon)}\{\mathcal{A}_1^C\}$ is estimated as usual.

REMARK 7.8 The MDEs τ_ε^* are, of course, worse than the MLE and BE in this problem (see Section 5.1).

Partially observed systems. In the case of the two-dimensional system (as in Section 6.1)

$$\begin{aligned} dX_t &= f_t(\theta_0)Y_t dt + \varepsilon dW_t, \quad X_0 = 0, \quad 0 \le t \le T, \\ dY_t &= b_t(\theta_0)Y_t dt + \varepsilon dV_t, \quad Y_0 = y_0, \quad 0 \le t \le T, \end{aligned} \tag{7.61}$$

with $\theta_0 \in \Theta \subset \mathbf{R}^d, \Theta$ is an open bounded set, the only observed component is $\{X_t, 0 \le t \le T\}$. Recall that W_t and V_t are independent Wiener processes. We have, as usual, several possibilities of the constructing MDEs. The first one is to define

$$\theta_\varepsilon^* = \arg\inf_{\theta\in\Theta} \|X - x(\theta)\|,$$

where $x_t(\theta) = x_t$ is a solution of the system (7.61) as $\varepsilon = 0$, i.e.

$$x_t(\theta) = y_0 \int_0^t f_s(\theta) \exp\left\{\int_0^s b_v(\theta)dv\right\} ds$$

and under the conditions of smoothness of the function $f(\cdot)$ and $b(\cdot)$ w.r.t. θ, we can apply Theorem 7.5 and to prove the asymptotic normality:

$$\mathbf{P}_{\theta_0} - \lim_{\varepsilon\to 0} \varepsilon^{-1}(\theta_\varepsilon^* - \theta_0) = J_*(\theta_0)^{-1}\langle x^{(1)}(\theta_0), \dot{x}(\theta_0)\rangle,$$

where the Gaussian process $x_t^{(1)}$ satisfies the equation

$$dx_t^{(1)} = f_t(\theta_0)y_t^{(1)}dt + dW_t, \quad x_0^{(1)} = 0, \quad 0 \le t \le T$$

with another Gaussian process

$$dy_t^{(1)} = b_t(\theta_0)y_t^{(1)}dt + dV_t, \quad y_0^{(1)} = 0, \quad 0 \le t \le T.$$

The matrix

$$J_*(\theta_0) = \langle \dot{x}(\theta_0), \dot{x}(\theta_0)^T\rangle.$$

Another MDE is defined as

$$\theta_\varepsilon^{**} = \arg\inf_{\theta\in\Theta} \|X - \hat{X}(\theta)\|,$$

where

$$\hat{X}_t(\theta) = \int_0^t f_s(\theta)m_s(\theta)ds$$

and $m_t(\theta)$ is the conditional mathematical expectation satisfying the equation of optimal filtration

$$dm_t = b_t(\theta)m_t dt + \varepsilon^{-2}\gamma_t(\theta)f_t(\theta)[dX_t - f_t(\theta)m_t dt].$$

In this case, we can apply Proposition 7.1 and establish the convergence

$$\mathbf{P}_{\theta_0} - \lim_{\varepsilon\to 0}\varepsilon^{-1}(\theta^*_\varepsilon - \theta_0) = J_{**}(\theta_0)^{-1}\langle W, \hat{\dot{x}}(\theta_0)\rangle,$$

where W_t is the Wiener process from the first equation of (7.61), the vector

$$\hat{\dot{x}}_t(\theta_0) = \int_0^t [\dot{f}_s(\theta_0)y_s(\theta_0) + f_s(\theta_0)\dot{y}_s(\theta_0)]ds,$$

the matrix

$$J_{**}(\theta_0) = \langle \hat{\dot{x}}(\theta_0), \hat{\dot{x}}(\theta_0)\rangle$$

and the vector $\dot{y}_s(\theta_0)$ satisfies (6.4). The nonlinear partially observed system

$$\begin{aligned} dX_t &= f_t(\theta, Y)dt + \varepsilon dW_t, \quad X_0 = x_0, \quad 0 \le t \le T,\\ dY_t &= b_t(\theta, Y)dt + \varepsilon dV_t, \quad Y_0 = y_0, \quad 0 \le t \le T \end{aligned}$$

with the observed component $\{X_t, 0 \le t \le T\}$, can be treated in a similar way. If we denote $y_t = y_t(\theta)$ as the solution of

$$\frac{dy_t}{dt} = b_t(\theta, y)$$

and define

$$x_t(\theta) = x_0 + \int_0^t f_s(\theta, y(\theta))ds,$$

then the MDE

$$\theta^*_\varepsilon = \arg\inf_{\theta\in\Theta} \|X - x(\theta)\|$$

and, under regularity conditions, this estimator is consistent and asymptotically normal (Theorem 7.5).

In this problem, we can also take one step in the expansion of the system in the vicinity of θ, i.e. to put

$$\begin{aligned} f^*_t(\theta, Y_t) &= f_t(\theta, y_t(\theta)) + f'_t(\theta, y_t(\theta))(Y_t - y_t),\\ b^*_t(\theta, Y_t) &= b_t(\theta, y_t(\theta)) + b'_t(\theta, y_t(\theta))(Y_t - y_t) \end{aligned}$$

and to write the equations of filtration for this linear (misspecified) model and substitute there the observed process from (7.61). Then we can also define the MDE

$$\theta^{**}_\varepsilon = \arg\inf_{\theta\in\Theta} \|X - \hat{X}(\theta)\|$$

with

$$\hat{X}_t(\theta) = x_0 + \int_0^t f_s(\theta, y_s(\theta)) + f'_s(\theta, y_s(\theta))[m_s(\theta) - y_s(\theta)]ds$$

and prove its consistence and asymptotic normality.

REMARK 7.9 It is also interesting to consider the problem of the optimal choice of the measure μ in the construction of the MDE. For the estimator θ_ε^*, say, the limit variance (in one-dimensional case) is (see Theorem 7.5)

$$\sigma^2 = \frac{\int_0^T \int_0^T R_{ts}\dot{x}_t(\theta)\dot{x}_s(\theta)d\mu_t d\mu_s}{\left(\int_0^T \dot{x}_t(\theta)^2 d\mu_t\right)^2}.$$

So if we put $R(t,s) = R_{ts}\dot{x}_t(\theta)^+\dot{x}_s(\theta)^+$ ($a^+ = a^{-1}$ if $a \neq 0$ and $a^+ = 0$ if $a = 0$) and

$$d\mu_t^* = \left(\int_0^T \dot{x}_t(\theta)^2 d\mu_t\right)^{-1} \dot{x}_t(\theta)^2\, d\mu_t$$

and denote by τ_1 and τ_2 two independent random variables with the distribution function $\mu^*(\cdot)$, then this variance could be presented as the following mathematical expectation

$$\sigma^2 = \mathbf{E}R(\tau_1, \tau_2).$$

The optimal μ corresponds to μ_0^* solving the problem

$$\inf_{\mu^*} \mathbf{E}R(\tau_1, \tau_2).$$

Of course, μ_0^* depends on θ and the optimal θ_ε^* should be adaptive.

Goodness of fit tests. The minimum distance approach allows us also to construct some "natural" goodness of fit tests to check the hypothesis that the observed process $\{X_t, 0 \leq t \leq T\}$ corresponds to the small white noise perturbation

$$dX_t = S_t(\theta, X)dt + \varepsilon dW_t, \quad X_0 = x_0, \quad 0 \leq t \leq T$$

of the deterministic system

$$\frac{dx_t}{dt} = S_t(\theta, x), \quad x_0, \quad 0 \leq t \leq T, \tag{7.62}$$

where $\theta \in \Theta \subset \mathbf{R}^d$. As a testing statistic, we can choose one of the following :

$$\begin{aligned}
\Delta_*(X) &= \inf_{\theta\in\Theta} \|X - x(\theta)\|, \\
\Delta_{**}(X) &= \inf_{\theta\in\Theta} \|X - \hat{X}(\theta)\|, \\
\Delta_{***}(X) &= \inf_{\theta\in\Theta} \|\hat{S} - S(\theta, x(\theta))\|, \\
\check{\Delta}(X) &= \inf_{\theta\in\Theta} \sup_{0\leq t\leq T} |X_t - x_t(\theta)|, \\
\Delta^o(X) &= \inf_{\theta\in\Theta} \int_0^T |X_t - x_t(\theta)| d\mu_t.
\end{aligned}$$

All of them define the consistent test in the following sense, which we illustrate in a test based on $\Delta_*(X)$. Denote $S(\Theta)$ as the family of solutions of the system (7.62) and suppose that $S(\Theta)$ and μ are such that $S(\Theta)$ is complete in the following sense: if

$$\inf_{\theta\in\Theta} \|x - x(\theta)\| = 0$$

for the function $x_t, 0 \le t \le T$, then there exists $\theta^* \in \Theta$ such that $x_t = x_t(\theta^*), 0 \le t \le T$.

The goodness of fit test is defined by its critical region

$$\{X: \Delta_*(X) > C_{\alpha,\varepsilon}\}.$$

The hypothesis should be rejected for these trajectories X . The constant $C_{\alpha,\varepsilon}$ we present as εC_α and the value C_α we choose from the asymptotic distribution of $\varepsilon^{-1}\Delta_*(X)$. Namely, under the regularity conditions, if the observed process corresponds to $\theta = \theta_0$, then

$$\varepsilon^{-2}\Delta_*(X)^2 \to \|x^{(1)}(\theta_0) - (\xi, \dot{x}(\theta_0))\|^2 = \|x^{(1)}(\theta_0)\|^2 - (J_*(\theta_0)\xi, \xi). \tag{7.63}$$

So under the hypothesis

$$\lim_{\varepsilon\to 0} \mathbf{P}_{\theta_0}^{(\varepsilon)}\{\varepsilon^{-1}\Delta_*(X) > C_\alpha\} =$$
$$= \mathbf{P}\{\|x^{(1)}(\theta_0)\|^2 - (J_*(\theta_0)\xi, \xi) > C_\alpha\}$$

and C_α can be chosen as $1-\alpha$ quantile of the distribution function of the random variable (7.63).

If the observed process does not belong to the prescribed parametric family and its limit function $x_t, 0 \le t \le T$ is such that

$$g = \inf_{\theta\in\Theta} \|x - x(\theta)\| > 0,$$

then

$$\mathbf{P}\left\{\inf_{\theta\in\Theta} \|X - x(\theta)\| < \varepsilon C_\alpha\right\} \le$$
$$\le \mathbf{P}\left\{\inf_{\theta\in\Theta}(\|x - x(\theta)\| - \|X - x\|) < \varepsilon C_\alpha\right\} \le$$
$$\le \mathbf{P}\left\{g - C\varepsilon \sup_{0\le t\le T} |W_t| < \varepsilon C_\alpha\right\} =$$
$$= \mathbf{P}\left\{\sup_{0\le t\le T} |W_t| > \frac{g - \varepsilon C_\alpha}{C^2\varepsilon}\right\} \le$$
$$\le 2\exp\left\{-\frac{(g - \varepsilon C_\alpha)^2}{2TC^2\varepsilon^2}\right\}.$$

So the probability of rejecting the hypothesis is

$$\mathbf{P}\left\{\inf_{\theta\in\Theta} \|X - x(\theta)\| > \varepsilon C_\alpha\right\} \le 1 - 2\exp\left\{-\frac{g^2}{3C^2T\varepsilon^2}\right\}$$

for $\varepsilon < \varepsilon_0$ with $\varepsilon_0 = (1 \wedge g(1 - \sqrt{\tfrac{2}{3}})C_\alpha^{-1})$.

7.5 Asymptotic efficiency of the MDE

Below, we study the asymptotic properties of the MDE θ_ε^* in the situation when the observed process does not belong to a prescribed parametric family of diffusion processes but is in a nonparametric vicinity of it. First, we derive an asymptotic minimax low bound on the risks of all estimators in this situation and then prove that, under certain conditions, this bound is achieved by the MDE.

In Chapter 2, we have considered the problem of parameter estimation by the observation of the diffusion-type process

$$dX_t = S_t(\theta, X)dt + \varepsilon dW_t, \quad X_0 = x_0, \quad 0 \le t \le T \tag{7.64}$$

and in the regular case, when the corresponding family of measures $\{\mathbf{P}_\theta^{(\varepsilon)}, \theta \in \Theta\}$ is LAN, we have the Hajek-Le Cam's bound (Theorem 2.1)

$$\begin{aligned}\liminf_{\varepsilon\to 0} \inf_{\theta_\varepsilon} \sup_{|\theta-\theta_0|<\varepsilon\lambda_\varepsilon} \int l(\varepsilon^{-1}(\theta_\varepsilon - \theta))d\mathbf{P}_\theta^{(\varepsilon)} \ge \\ \ge \frac{1}{\sqrt{2\pi}} \int l(xI(\theta_0)^{-\frac{1}{2}})e^{-\frac{x^2}{2}}dx,\end{aligned} \tag{7.65}$$

where $l(\cdot) \in \mathbf{W}, d = 1$ and $I(\theta_0)$ is a Fisher information

$$I(\theta_0) = \int_0^T \dot{S}_t(\theta_0, x)^2 dt.$$

Here $x = (x_t, 0 \le t \le T)$ is a solution of the limit "true" equation

$$\frac{dx_t}{dt} = S_t(\theta_0, x), \quad x_0. \tag{7.66}$$

Note, that *sup* in (7.65) is taken over the members of the parametric family $\mathbf{P}_\theta^{(\varepsilon)}, \theta \in \Theta$ close to $\mathbf{P}_{\theta_0}^{(\varepsilon)}$.

Under a strengthening of the conditions, we prove the LAM of MLE and BE, i.e. for the risks of these estimators we have equality in (7.65). So we say that these estimators are asymptotically efficient.

In the misspecified case of Section 2.6, when the observed process was

$$dX_t = b_t(X)dt + \varepsilon dW_t, \quad X_0 = x_0, \quad 0 \le t \le T \tag{7.67}$$

and $b_t(X) \notin \{S_t(\theta, X), \theta \in \Theta\}$, the measure $\mathbf{P}_b^{(\varepsilon)}$ induced by this process does not belong to the family $\{\mathbf{P}_\theta^{(\varepsilon)}, \theta \in \Theta\}$. It was shown in Section 2.6 that the MLE and BE converge to the value

$$\theta^* = \arg\inf_{\theta\in\Theta} \int_0^T [S_t(\theta, x^b) - b_t(x^b)]^2 dt, \tag{7.68}$$

where $x^b = (x^b, 0 \le t \le T)$ is a solution of the limit true value equation

$$\frac{dx_t^b}{dt} = b_t(x^b), \quad x_0^b = x_0 \tag{7.69}$$

(see Theorems 2.13 and 2.17), but the question of optimality was left open.

In this section, we consider an intermediate situation when the observed process (7.67) has the trend coefficient $b_t(\cdot)$ near some function $(S_t(\theta_0, X), 0 \leq t \leq T)$.

We suppose that the basic parametric family of diffusion processes (7.64) is given by the equation

$$dX_t = S_t(\theta_0, X_t)dt + \varepsilon dW_t, \quad X_0 = x_0, \quad 0 \leq t \leq T, \tag{7.70}$$

where $\theta_0 \in \Theta = (\alpha, \beta), |\alpha| + |\beta| < \infty$. The observed process (7.67) has the differential

$$dX_t^H = [S_t(\theta_0, X_t^H) + H_t(X_t^H)]dt + \varepsilon dW_t, \quad X_0^H = x_0, \quad 0 \leq t \leq T. \tag{7.71}$$

The corresponding limit equations (7.66) and (7.69) are

$$\frac{dx_t}{dt} = S_t(\theta_0, x_t), \quad x_0, \tag{7.72}$$

$$\frac{x_t^H}{dt} = S_t(\theta_0, x_t^H) + H_t(x_t^H), \quad x_0^H = x_0. \tag{7.73}$$

Remember, that under

Condition I.

$$|S_t(\theta_0, x) - S_t(\theta_0, y)| + |H_t(x) - H_t(y)| \leq L|x - y|$$
$$|S_t(\theta_0, x)| + |H_t(x)| \leq L(1 + |x|)$$

Equations (7.70)– (7.73) have unique solutions and, with probability 1,

$$\sup_{0 \leq t \leq T} |X_t^H - x_t^H| \leq C(L, T)\varepsilon \sup_{0 \leq t \leq T} |W_t|, \tag{7.74}$$

where the constant $C(L, T)$ does not depend on $H_t(\cdot)$.

Fix some measures μ and introduce the MDE

$$\theta_\varepsilon^* = \arg \inf_{\theta \in \Theta} \|X^H - x(\theta)\|,$$

where, as before, $\|\cdot\|$ is $L_2(\mu)$ norm and $x_t(\theta), 0 \leq t \leq T, \theta \in \Theta$, the family of solutions of Equation (7.72) with $\theta_0 = \theta$.

The role of "true value" of θ (with observations (7.71)) will play

$$\theta_H = \arg \inf_{\theta \in \Theta} \|x^H - x(\theta)\|.$$

Denote $\mathcal{K}(L)$ as the class of functions $H_t(x)$ satisfying condition I and introduce the set

$$U_\delta(\theta_0) = \{H_t(x) : \|H(x^H)\|_{L_2} + \|H'(x^H)\| < \delta, \quad H_t(x) \in \mathcal{K}(L)\},$$

where $\|\cdot\|_{L_2}$ is $L_2[0, T]$ norm.

Denote also

$$q(s) = J^{-1}(\theta_0) \int_s^T \dot{x}_t(\theta_0) \exp\left\{\int_s^t S_v'(\theta_0, x_v(\theta_0))dv\right\} d\mu_t \tag{7.75}$$

and put

$$I_* = \|q\|_{L_2}^{-2}.$$

Condition II. *The functions $S_t(\theta,x)$ and $H_t(x)$ have the following uniformly continuous and bounded derivatives*

$$|S_t'(\theta,x)| + |\dot S_t(\theta,x)| + |\ddot S_t(\theta,x)| + |\dot S_t'(\theta,x)| + |H_t'(x)| \le L.$$

Condition III. *The function*

$$g(\nu) = \inf_{|\theta-\theta_0|>\nu} \|x(\theta) - x(\theta_0)\| > 0$$

for $\nu > 0$.

Below, $l(\cdot) \in \mathbf{W}_{e,2}$.

Theorem 7.17. *Let condition II be fulfilled and $J(\theta_0) > 0$, then for any function $\lambda_\varepsilon \to \infty$ such that $\delta_\varepsilon = \varepsilon\lambda_\varepsilon \to 0$, we have*

$$\liminf_{\varepsilon\to 0} \inf_{\theta_\varepsilon} \sup_{H\in U_{\delta_\varepsilon}(\theta_0)} \int l(\varepsilon^{-1}(\theta_\varepsilon - \theta_H))d\mathbf{P}_H^{(\varepsilon)} \ge \frac{1}{\sqrt{2\pi}} \int l(xI_*^{-\frac{1}{2}})e^{-\frac{x^2}{2}}dx. \tag{7.76}$$

Proof. Fix the function $\psi_t(x) \in \mathcal{K}(L)$ such that

$$I_\psi = \|\psi(x)\|_{L_2}^2 > 0$$

and introduce a parametric family of the diffusion processes

$$dX_t^h = [S_t(\theta_0, X_t^h) + (h-\theta_0)\psi_t(X_t^h)]dt + \varepsilon dW_t, \quad X_0^h = x_0, \quad 0 \le t \le T. \tag{7.77}$$

Denote also x_t^h as the solution of

$$\frac{dx_t^h}{dt} = S_t(\theta_0, x_t^h) + (h-\theta_0)\psi_t(x_t^h), \quad x_0^h = x_0 \tag{7.78}$$

and suppose that the real parameter h is from the interval

$$|h - \theta_0| < \frac{\delta_\varepsilon}{2\|\psi(x)\|} = \tilde\delta_\varepsilon.$$

Then, for small ε, we have

$$(h-\theta_0)\psi_t(x) \in U_{\delta_\varepsilon}(\theta_0). \tag{7.79}$$

The family of measures $\{\mathbf{Q}_h^{(\varepsilon)}, h \in (\theta_0 - \tilde\delta_\varepsilon, \theta_0 + \tilde\delta_\varepsilon)\}$ induced by the processes (7.77) admits the representation

$$\frac{d\mathbf{Q}_{\theta_0+\varepsilon u}^{(\varepsilon)}}{d\mathbf{Q}_{\theta_0}^{(\varepsilon)}}(X) = \exp\left\{u\int_0^T \psi_t(x_t)dW_t - \frac{1}{2}u^2\int_0^T \psi_t(x_t)^2dt + r_\varepsilon(u)\right\}, \quad |u| < \lambda_\varepsilon,$$

where $x_t = x_t(\theta_0)$ and for any $u \in \mathbf{R}^1$

$$\mathbf{Q}_{\theta_0} - \lim_{\varepsilon \to 0} r_\varepsilon(u) = 0.$$

So this family is LAN and, by Theorem 2.1, we can write

$$\liminf_{\varepsilon \to 0} \inf_{\theta_\varepsilon} \sup_{|h-\theta_0|<\tilde{\delta}_\varepsilon} \int l(\varepsilon^{-1}(\theta_\varepsilon - h)) d\mathbf{Q}_h^{(\varepsilon)} \geq$$
$$\geq \frac{1}{\sqrt{2\pi}} \int l(x I_\psi^{-\frac{1}{2}}) e^{-\frac{1}{2}x^2} dx. \tag{7.80}$$

Using Equation (7.78) we can expand the function

$$\theta_h = \arg \inf_{\theta \in \Theta} \| x^h - x(\theta) \|$$

in the vicinity of the point $h = \theta_0$.

Denote $\overset{\circ}{x}_t(\theta_0)$ as the derivative of x_t^h on h at the point $h = \theta_0$, then from (7.78), for $\overset{\circ}{x}_t(\theta_0)$ we have the equation

$$\frac{d\, \overset{\circ}{x}_t(\theta_0)}{dt} = S_t'(\theta_0, x_t)\, \overset{\circ}{x}_t(\theta_0) + \psi_t(x_t), \quad \overset{\circ}{x}_0(\theta_0) = 0.$$

Its solution is the function

$$\overset{\circ}{x}_t(\theta_0) = \int_0^t \psi_s(x_s) \exp\left\{ \int_s^t S_v'(\theta_0, x_v) dv \right\} ds.$$

So we can write

$$\begin{aligned} \theta_h &= \arg \inf_{\theta \in \Theta} \| x(\theta_0) + (h - \theta_0)\, \overset{\circ}{x}(\theta_0) - x(\theta) + o(|h - \theta_0|) \| = \\ &= \theta_0 + (h - \theta_0) J(\theta_0)^{-1} < \overset{\circ}{x}(\theta_0), \dot{x}(\theta_0) > + o(|h - \theta_0|). \end{aligned}$$

Introduce now the condition

$$J^{-1}(\theta_0) \langle \overset{\circ}{x}(\theta_0), \dot{x}(\theta_0) \rangle = 1 \tag{7.81}$$

and denote $\mathcal{K}_1(L)$ as the class of functions $\psi_t(x)$ satisfying this condition. Then for $\psi_t(x) \in \mathcal{K}_1(L)$, we have

$$\theta_h = h + o(|h - \theta|). \tag{7.82}$$

Recall that the inequality (7.80) also holds if we replace $l(\varepsilon^{-1}(\theta_\varepsilon - h))$ by $l(\varepsilon^{-1}(\theta_\varepsilon - h - o(1))$.

Consider now the problem of minimization of I_ψ in the class $\mathcal{K}_1(L)$. Below, we change the order of integration

$$\begin{aligned} 1 &= J^{-1}(\theta_0) \int_0^T \overset{\circ}{x}_t(\theta_0) \dot{x}_t(\theta_0) d\mu_t = \\ &= \int_0^T \int_0^t \psi_s(x_s) \exp\left\{ \int_s^t S_v'(\theta_0, x_v) dv \right\} ds J^{-1}(\theta_0) \dot{x}_t(\theta_0) d\mu_t = \\ &= \int_0^T \psi_s(x_s) \int_s^T J^{-1}(\theta_0) \exp\left\{ \int_s^t S_v'(\theta_0, x_v) dv \right\} \dot{x}_t(\theta_0) d\mu_t ds = \\ &= \int_0^T \psi_s(x_s) q(s) ds. \end{aligned}$$

So by Schwarz inequality, we have

$$I_\psi = \int_0^T \psi_s(x_s)^2 ds \geq \left(\int_0^T q(s)^2 ds \right)^{-1}. \tag{7.83}$$

The function $\psi_s(x_s), 0 \leq t \leq T$ is bounded, so we have equality here if

$$\psi_s^*(x_s) = q(s)\|q\|^{-2}, \quad 0 \leq s \leq T.$$

Note that the right-hand side of (7.83) does not depend on $\psi(\cdot)$, so

$$\inf_{\psi \in \mathcal{K}_1(L)} I_\psi = I_{\psi^*} \equiv I_* \tag{7.84}$$

in our notations.

Using (7.79), (7.80), (7.82) and (7.84), we can write

$$\liminf_{\varepsilon \to 0} \inf_{\theta_\varepsilon} \sup_{H \in U_{\delta_\varepsilon}(\theta_0)} \int l(\varepsilon^{-1}(\theta_\varepsilon - \theta_H)) d\mathbf{P}_H^{(\varepsilon)} \geq$$
$$\geq \liminf_{\varepsilon \to 0} \inf_{\theta_\varepsilon} \sup_{|h-\theta_0|<\tilde{\delta}_\varepsilon} \int l(\varepsilon^{-1}(\theta_\varepsilon - \theta_h)) d\mathbf{Q}_h^{(\varepsilon)} \geq$$
$$\geq \frac{1}{\sqrt{2\pi}} \int l(x I_\psi^{-\frac{1}{2}}) e^{-\frac{1}{2}x^2} dx \geq$$
$$\geq \frac{1}{\sqrt{2\pi}} \int l(x I_*^{-\frac{1}{2}}) e^{-\frac{1}{2}x^2} dx = \mathbf{E}l(\xi), \qquad \mathcal{L}\{\xi\} = \mathcal{N}(0, I_*^{-1}).$$

REMARK 7.10 Note that the worst (minimum Fisher information) function $\psi_s^*(x_s) = \psi_s^*$ is *deterministic* and we cannot replace, say, x_v in $S_v'(\theta_0, x_v)$ in Definition (7.75) of $q(s)$ by X_v, because in this case the trend $S_t(\theta_0, X_t) + (h - \theta_0)\psi_t(X)$ will depend on the future.

We say that the estimator $\bar{\theta}_\varepsilon$ is *locally asymptotically minimax* (LAM) in the problem with the observations (7.71) if the conditions of Theorem 7.17 are fulfilled and for some $l(\cdot) \in \mathbf{W}$ and $\lambda_\varepsilon \to \infty$, we have

$$\lim_{\varepsilon \to 0} \sup_{H \in U_{\delta_\varepsilon}(\theta_0)} \int l(\varepsilon^{-1}(\bar{\theta}_\varepsilon - \theta_H)) d\mathbf{P}_H^{(\varepsilon)} = \mathbf{E}l(\xi) \tag{7.85}$$

for all $\theta_0 \in \Theta$.

To prove the LAM of MDE θ_ε^*, we begin with its uniform on $U_{\delta_1}(\theta_0)$ consistence.

Theorem 7.18. *Let condition I be fulfilled, $g(\nu) > 0$ for $\nu > 0$, then for any $\nu > 0$, there exists $\delta_1 > 0$ such that*

$$\sup_{H \in U_{\delta_1}(\theta_0)} \mathbf{P}_H^{(\varepsilon)}\{|\theta_\varepsilon^* - \theta_H| > \nu\} \leq 2\exp\left\{-\frac{c^2 g^2(\frac{\nu}{2})}{2\varepsilon^2 T}\right\} \tag{7.86}$$

with some constant $C = C(L, T)$.

Proof. First, we note that the inequality

$$|\theta_H - \theta_0| \geq \frac{\nu}{2}$$

is equivalent to

$$\inf_{|\theta-\theta_0|<\frac{\nu}{2}} \|x^H - x(\theta)\| > \inf_{|\theta-\theta_0|\geq\frac{\nu}{2}} \|x^H - x(\theta)\|$$

which allows us to write

$$\inf_{|\theta-\theta_0|<\frac{\nu}{2}} (\|x^H - x(\theta_0)\| + \|x(\theta) - x(\theta_0)\|) >$$
$$> \inf_{|\theta-\theta_0|\geq\frac{\nu}{2}} (\|x(\theta) - x(\theta_0)\| - \|x^H - x(\theta_0)\|).$$

So we have

$$\|x^H - x(\theta_0)\| > \frac{1}{2} g\left(\frac{\nu}{2}\right).$$

On the other hand, $H \in U_{\delta_1}(\theta_0)$ and this inequality provides

$$\delta_1 > \frac{1}{2} g\left(\frac{\nu}{2}\right).$$

Hence, if we choose $\delta_1 = \frac{1}{3} g\left(\frac{\nu}{2}\right)$, then $|\theta_H - \theta_0| < \frac{\nu}{2}$.

For the probability (7.86), we can write

$$\begin{aligned}
&\mathbf{P}_H^{(\varepsilon)}\{|\theta_\varepsilon^* - \theta_H| > \nu\} = \\
&= \mathbf{P}_H^{(\varepsilon)}\left\{\inf_{|\theta-\theta_H|<\nu} \|X^H - x(\theta)\| > \inf_{|\theta-\theta_H|<\nu} \|X^H - x(\theta)\|\right\} \leq \\
&\leq \mathbf{P}_H^{(\varepsilon)}\left\{\|X^H - x^H\| + \|x^H - x(\theta_0)\| + \inf_{|\theta-\theta_H|<\nu} \|x(\theta) - x(\theta_0)\| > \right. \\
&\quad > \left. \inf_{|\theta-\theta_H|>\nu} \|x(\theta) - x(\theta_0)\| - \|X^H - x^H\| - \|x^H - x(\theta_0)\| \right\} \leq \\
&\leq \mathbf{P}_H^{(\varepsilon)}\left\{2\|X^H - x^H\| + 2\delta_1 > \inf_{|\theta-\theta_0|>\frac{\nu}{2}} \|x(\theta) - x(\theta_0)\|\right\} = \\
&= \mathbf{P}_H^{(\varepsilon)}\left\{\|X^H - x^H\| > \frac{1}{2} g\left(\frac{\nu}{2}\right) - \delta_1\right\} = \\
&= \mathbf{P}_H^{(\varepsilon)}\left\{\|X^H - x^H\| > \frac{1}{6} g\left(\frac{\nu}{2}\right)\right\} \leq \\
&\leq \mathbf{P}\left\{\sup_{0\leq t\leq T} |W_t| > \frac{cg\left(\frac{\nu}{2}\right)}{\varepsilon}\right\} \leq 2\exp\left\{-\frac{c^2 g^2\left(\frac{\nu}{2}\right)}{2\varepsilon^2 T}\right\},
\end{aligned}$$

because, we have the estimate (7.74) for the difference $X_t^H - x_t^H$, where the constant $C(L,T)$ could be chosen as not depending on $H \in U_{\delta_1}(\theta_0)$.

So the estimator θ_ε^* of θ_H is consistent uniformly on $H \in U_{\delta_1}(\theta_0)$ as $\varepsilon \to 0$.

Theorem 7.19. *Let the functions $S_t(\theta, x), H_t(x)$ have two continuous bounded derivatives on θ and x, $g(\nu) > 0$ and $J(\theta_0) > 0$, then the MDE θ_ε^* is LAM for the loss functions $l(\cdot) \in \mathbf{W}_{e,2}$, i.e.*

$$\lim_{\varepsilon\to 0} \sup_{H\in U_{\delta_\varepsilon}(\theta_0)} \mathbf{E}_H l(\varepsilon^{-1}(\theta_\varepsilon^* - \theta_H)) = \mathbf{E} l(\xi),$$

where $\mathcal{L}\{\xi\} \simeq \mathcal{N}(0, I_*^{-1})$.

Proof. First, we introduce the "good set"

$$\mathcal{A}_1 = \{\omega : \ |\theta_\varepsilon^* - \theta_H| < \delta_\varepsilon\},$$

where $\delta_\varepsilon = \varepsilon\lambda_\varepsilon \to 0$ and $\lambda_\varepsilon \to \infty$ as $\varepsilon \to 0$. Then, for $\omega \in \mathcal{A}_1$ the MDE $\theta_\varepsilon^* = \theta_H + \varepsilon u_\varepsilon^*$, where u_ε^* is one of the solutions of the minimum distance equation

$$\langle X^H - x(\theta_H + \varepsilon u), \dot{x}(\theta_H + \varepsilon u)\rangle = 0, \quad |u| < \delta_\varepsilon.$$

Rewrite this equation as

$$\begin{aligned}&\langle x(\theta_H + \varepsilon u) - x(\theta_H), \dot{x}(\theta_H + \varepsilon u)\rangle = \\ &\qquad \langle X^H - x^H, \dot{x}(\theta_H + \varepsilon u)\rangle + \langle x^H - x(\theta_H), \dot{x}(\theta_H + \varepsilon u)\rangle\end{aligned}$$

and, by the Taylor formula, we have

$$u_\varepsilon^*\langle \dot{x}(\tilde{\theta}_H), \dot{x}(\theta_\varepsilon^*)\rangle = \langle \varepsilon^{-1}(X^H - x^H), \dot{x}(\theta_\varepsilon^*)\rangle + \varepsilon^{-1}\langle x^H - x(\theta_H), \dot{x}(\theta_\varepsilon^*)\rangle.$$

Remember that

$$\langle x^H - x(\theta_H), \dot{x}(\theta_H)\rangle = 0$$

and we have the estimate (see (7.28) with $a = 1$)

$$\sup_{0\le t\le T} |\varepsilon^{-1}(X_t^H - x_t^H) - y_t^{(1)}| \le C\varepsilon \sup_{0\le t\le T} |W_t|^2,$$

where y_t is a Gaussian process:

$$dY_t^{(1)} = [S_t'(\theta_0, x_t^H) + H_t'(x_t^H)]y_t^{(1)}dt + dW_t, \quad y_0^{(1)} = 0, \quad 0 \le t \le T$$

and the constant C could be chosen as not depending on $H \in \mathcal{K}(L)$. We also have $|\theta_\varepsilon^* - \theta_H| < \delta_\varepsilon$ on $\mathcal{A}_1$, so using standard arguments we prove the convergence

$$\lim_{\varepsilon\to 0} \sup_{H\in U_{\delta_\varepsilon}(\theta_0)} \mathbf{P}_H^{(\varepsilon)}\{|u_\varepsilon^* - \xi_H| > \delta\} = 0.$$

The uniform integrability of the risk $\mathbf{E}_H l(u_\varepsilon^*)$ follows from Theorem 7.17 and the estimate $g(\nu) \ge c\nu$ because if we denote $F_{\varepsilon,H}(x) = \mathbf{P}_H^{(\varepsilon)}\{|u_\varepsilon| < x\}$, then

$$\begin{aligned}\int l(u_\varepsilon^*)d\mathbf{P}_H^{(\varepsilon)} &< \int e^{\gamma u_\varepsilon^2}d\mathbf{P}_H^{(\varepsilon)} = -\int_0^\infty e^{\gamma x^2}d[1 - F_{\varepsilon,H}(x)] = \int_0^\infty e^{\gamma x^2}dF_{\varepsilon,H}(x) = \\ &= 2\gamma\int_0^\infty xe^{\gamma x^2}[1 - F_{\varepsilon,H}(x)]dx \le 2\gamma\int_0^\infty xe^{-(C-\gamma)x^2}dx < N.\end{aligned}$$

So for $\gamma < C$ this integral is bounded uniformly on $H \in U_{\delta_\varepsilon}(\theta_0)$.

REMARK 7.11 Similar bounds and LAM could be established for other MDEs. For example, if we consider

$$\theta_\varepsilon^{**} = \arg\inf_{\theta\in\Theta} \|X^H - \hat{X}(\theta)\|,$$

where observations $\{X_t^H, 0 \le t \le T\}$ are taken from (7.71) and the process

$$\hat{X}_t(\theta) = x_0 + \int_0^t S_v(\theta, X_v^H)dv, \quad \theta \in \Theta,$$

then we choose θ_H (the "true value") of θ) as

$$\theta_H = \arg \inf_{\theta \in \Theta} \|x^H - \hat{x}(\theta)\|$$

with the limit for the $\hat{X}_t(\theta)$ function

$$\hat{x}_t(\theta) = x_0 + \int_0^t S_t(\theta, x_v^H)dv.$$

The parametric family is the same as (7.77), so $\overset{\circ}{x}_t\ (\theta_0)$ has the same form and the difference lies in the definition of $\hat{\dot{x}}_t(\theta)$. We can put

$$\psi_s^* = q^*(s)\|q^*\|^{-2},$$

where $q^*(s)$ is as in (7.75) if we replace $\dot{x}_t(\theta_0)$ by $\hat{\dot{x}}_t(\theta_0)$ and so on.

REMARK 7.12 Suppose now that the "true value" of θ with observations (7.71) is understood as θ^* in (7.68). Then a similar bound is defined as follows. The expansion of

$$\theta_h = \arg \inf_{\theta \in \Theta} \int_0^T [S_t(\theta_0, x_t^h) + (h - \theta_0)\psi_t(x_t^h) - S_t(\theta, x_t^h)]^2 dt$$

in the vicinity of the point θ_0 has the form

$$\theta_h = \theta_0 + (h - \theta_0)I(\theta_0)^{-1} \int_0^T \psi_t(x_t)\dot{S}_t(\theta_0, x_t)dt + o(|h - \theta_0|)$$

and the condition (7.81) is now

$$I(\theta_0)^{-1} \int_0^T \psi_t(x_t)\dot{S}_t(\theta_0, x_t)dt = 1.$$

Hence

$$\int_0^T \psi_t(x_t)^2 dt \ge I(\theta_0)$$

and

$$\min_{\psi \in \mathcal{K}_1(L)} I_\psi = I(\theta_0).$$

So we can choose

$$\psi_t(x_t) = \dot{S}_t(\theta_0, x_t)$$

and the worst parametric family could be written as

$$dX_t^h = [S_t(\theta_0, X_t^h) + (h - \theta_0)\dot{S}_t(\theta_0, X_t^h)]dt + \varepsilon dW_t.$$

This coincides with the expansion of the trend $X_0 = x_0, 0 \le t \le T$ of original model.

The corresponding minimax bound is

$$\liminf_{\varepsilon\to 0} \inf_{\theta_\varepsilon} \sup_{H\in U_{\delta_\varepsilon}(\theta_0)} \int l(\varepsilon^{-1}(\theta_\varepsilon - \theta_H))d\mathbf{P}_H^{(\varepsilon)} \geq \mathbf{E}l(\xi), \tag{7.87}$$

where $\mathcal{L}\{\xi\} = \mathcal{N}(0, I(\theta_0)^{-1})$.

In Theorem 2.13 we proved that the MLE $\hat{\theta}_\varepsilon$ is consistent uniformly on the parametric vicinity ($\mu \in \mathbf{K}_\mu$) and asymptotically normal, i.e. for the observations (7.71) with fixed $H \in \mathcal{K}(L)$, in our notations the corresponding limit is

$$\varepsilon^{-1}(\hat{\theta}_\varepsilon - \theta^*) \rightarrow \xi_H = \ddot{G}^{-1}(\theta^*)\left\{\int_0^T \dot{S}_t(\theta^*, x_t^H)dW_t - \right.$$
$$\left. - \int_0^T [\dot{S}_t(\theta^*, x_t^H)\Delta_t'(\theta^*, x_t^H) + \dot{S}_t'(\theta^*, x_t^H)\Delta_t(\theta^*, x_t^H)]y_t^{(1)}dt\right\}, \tag{7.88}$$

where $\Delta_t(x) = S_t(\theta^*, x) - H_t(x) - S_t(\theta_0, x)$,

$$\ddot{G}(\theta^*) = -\int_0^T \dot{S}_t(\theta^*, x_t^H)^2 dt - \int_0^T \ddot{S}_t(\theta^*, x_t^H)\Delta_t(\theta^*, x_t^H)dt.$$

If we now assume , that $H \in U_{\delta_\varepsilon}(\theta_0)$ with $\delta_\varepsilon \rightarrow 0$, then

$$\lim_{\varepsilon\to 0} \mathbf{E}_H|\xi_H - \xi|^p = 0$$

for any $p > 1$. So it only remains to prove the uniform on $H \in U_{\delta_\varepsilon}(\theta_0)$ asymptotic normality (7.88).

Remarks

Firstly, I would like to mention several problems of estimation closely related to the observations of diffusion processes, but not considered in this work. First of all, this is a theory of parametric and nonparametric estimation for the model of the so called *signal in white Gaussian noise* (SWGN) :

$$dX_t = S_t(\theta)\, dt \;+\; \varepsilon\, dW_t, \qquad X_0 = 0,\ 0 \le t \le T$$

This theory is now well developed through the works of I. Ibragimov, R. Khas'minskii and their successors (see [51] - [53] and references therein). Their results for this model are usually obtained by applications of certain powerful methods worked out by the authors for general models. The methods are rather universal and, particularly in our work, we often use their theorems 1.4 - 1.7 to solve certain problems for diffusion-type observations. Another also well-developed theory of estimation for diffusion observations, can be found in the book by R. Liptser and A. Shiryayev [97], where the problem of the calculation of conditional mathematical expectation of nonobserved component of partially observed system is considered in detail. Some problems of parametric estimation are also presented there.

There are several works on parametric estimation for the models of diffusion processes in situations which do not coincide with the one considered in our work. These are the works of A. Tsitovich [111] with observations of diffusion process with small diffusion coefficient observed up to first exit time from certain domain, D. Florens-Zmirou [37] with observations of zero-crossing times only, V. Genon-Catalot and C. Laredo [41] with observations of the maximal value of the process, D. Dacunha-Castelle and D. Florens-Zmirou [24] with the observations in discrete times, and many others. All these works as well as our own are devoted to parameter estimation of the trend coefficients of diffusion processes. If the diffusion coefficient is a function of an unknown parameter, then the measures corresponding to the different values of this parameter are singular. That's why in these problems, the observations are taken in discrete times . We mention the following works on this subject : G. Donhal [28], D. Dacunha-Castelle, D. Florens-Zmirou [24], V. Genon-Catalot, J. Jacod [43].

Chapter 1

Besides the usual agreement concerning the notations of probability theory (Section 1.1), for the convenience of references, we give a brief extraction of the Itô calculus (Section 1.2) and asymptotic estimation theory (Section 1.3). The properties (1.2) -

(1.10) of the Itô integra, as well as Lemmas 1.10, 1.12 and Theorems 1.1 and 1.2, are taken from the book of R. Liptser and A. Shiryayev [97] (it can be found in any other book on stochastic analysis). Lemma 1.1 is from ([97], Lemma 8.8) , Lemmas 1.3 and 1.6 are from M. Burnashev [13], and Lemma 1.5 is a strengthening of Lemma 7 in [68]. Results like Lemmas 1.13 and 1.14 are known. See, for example Yu. Blagoveshenskii [11] or the book of M. Freidlin and A. Wentsell [39], where the asymptotics of such a type of processes are studied in detail. The theorems of Section 1.3 are borrowed from the book of I. Ibragimov and R. Khas'minskii [53]. Note only that the notion LAN was introduced by L.Le Cam [91] and the minimax bound (Theorem 1.3) is due to J. Hajek [48] and L. Le Cam [93].

Chapter 2

Theorem 2.1 is a version of Proposition 2 in [64] and Theorem 2.2 is a strengthening of Theorem 2 in [66] (see also I. Genis [40], Yu. Kutoyants [69] and C. Laredo [89] on similar problems). Theorem 2.3 is a version of Theorem 3.4 in [69].

Parameter estimation problems for the linear processes were considered in detail by M. Arato [3]. The case $x_0 = 0$ in Example 2.1 corresponds to "curios singularity" of P. Feigin [36]. Theorem 2.4 is due to the author. The case $\theta > 0,\ x_0 = 0$ in Theorem 2.4 is analogous to the case $\theta > 0$ and $T \to \infty$ (see [6], where such type of result was announced). " Large signal asymptotics" was also studied by S. Veres [113]. Theorems 2.5-2.7 were published in [73]. See also the work of G. Apoyan [1] for some generalizations of this result. Theorems 2.9 - 2.12 are due to the author, but note that the inequality (2.23) for locally asymptotically mixing normal families was derived by P. Jeganatan [57] (see also [67], Theorem 3.4.4, where this inequality was proved for the quadratic loss functions).

The results of Section 2.6 are due to the author. The first version of Theorems 2.13 was published in [74]. I. McKeague [99] considered this problem for ergodic diffusion with $T \to \infty$.

The results of Section 2.7 are due to the author [83]. These results were extended to more general models in [87]. See the works of R. Redner [109] and B. Clark, et al. [22] , where similar problems were considered. The convergence of the MLE of the parameter of some nonidentifiable models to sets like ours $\{\theta_1, \ldots, \theta_k\}$, was established by several authors, see , for example, H. White [117], V. Borkar and A. Bagchi [12].

The results of Section 2.8 are due to the author. The behavior of the MLE and BE in certain vicinities of the boundary (including the boundary) for the models of SWGN were studied by M. Burnashev [14].

Chapter 3

Theorem 3.1 is an extension to the diffusion-type observations of the asymptotic expansion, obtained earlier for the diffusion processes [68]. For the model of SWGN, such expansions for the MLE and BE were obtained by M. Burnashev [14]. Later, he also presented the expansions of the distribution functions and of the moments [15] , [16]. My interest in this problem was inspired by the work [14]. Particularly, the idea of using "good sets" was borrowed from there. M. Burnashev derived the expansion in two steps. Firs, he expanded the likelihood ratio by the powers of small parameters and then expanded the estimators. A direct extension of his results to the diffusion observations was announced by K. Bauer [7].

The results of Sections 3.2 and 3.3 are due to the author. Recently, N. Yoshida successfully applied the calculus of Malliavin-Watanabe to this problem and also obtained an asymptotic expansion of the MLE and BE by the powers of ε, as well as their distribution functions [118]-[1120]. A similar approach was used in the work A. Dermoune and Yu. Kutoyants [25], where first the expansion of the MLE is obtained as was done in the present work , but for misspecified diffusion-type process and then the expansion of the distribution function of the sum $\sum_{j=1}^{k} \Psi_j\ \varepsilon^j$ was realized via Malliavin calculus.

The results presented here are certain natural extensions of the classical expansions for the model of i.i.d. observations [21]. See also the article of J. Pfanzagl [106], where the role of asymptotic expansions in statistics is discussed.

Chapter 4

The results of Sections 4.1 and 4.2 were correspondingly published in [71] and [72]. The estimator $\hat{S}_t$ defined by (4.7) was introduced by I. Ibragimov and R. Khas'minskii [53] in the problem of signal estimation for the SWGN model. Theorem 4.3 is a version of the theorem of R. Farrel [35] applied to diffusion observations. The proof of it is borrowed from [53], Theorem 7.5.1. Theorems 4.8 and 4.9 are a revised version of the results published in [2]. The proof of Theorem 4.8 is based on the mentioned there general result of P. Millar [101].

Chapter 5

The results of this chapter are due to the author. Estimation of the moment of switching for the SWGN model was considered by I. Ibragimov and R. Khas'minskii [52] (see also [53], $ 7.2) and our work was inspired by these results. For the piecewise constant trend and multi-dimensional switching, this problem was treated by L.Vostrikova [114] (see also [115] and [116], where several general theorems on the convergence of likelihood ratio processes in such problems are suggested). The possibility of consistent hypothesis testing on the basis of a misspecified switching model was realized in [19].

Chapter 6

The results of this chapter are due to the author. The asymptotic expansion of the MLE of the parameter of a partially observed linear system was obtained in [75]. The consistency of the MLE of a parameter of a partially observed nonlinear system was established by M. James and F. Le Gland [56]. Mention can be made of the work of F. Campillo and F. Le Gland [18], where numerical methods of the calculation of MLE for nonlinear models are proposed. A disorder problem for a partially observed linear system with piecewise constant switching coefficients in observations or state equations was considered by F. Campillo [18]. The case when switching is present in two equations was treated in [20].

Chapter 7

The class of minimum distance estimators considered here was introduced in [76] and [77]. Theorems 7.1-7.6 were announced in [76], Proposition 7.2 and Theorems 7.7 and 7.8 are taken from [77], [81] and [84]. See also Y. Dodge [30], where the role of the L_1 approach in statistics is clarified. The LAM of the MDE for the i.i.d. and stationary sequences were established by P. Millar [101], [102]. In the proof of Theorem 7.16, we have used B. Levit's approach [95] to the estimation of certain functionals, following its exposition in [53], Chapter 4. We consider the "true value" θ_H as such a functional. The LAM of the MLE mentioned in Remark 7.9 was discussed (for the i.i.d. model) by R. Beran [8].

The minimum distance (or least-squares) estimators for diffusion processes with $T \to \infty$ asymptotic were considered by several authors (see, for example, A. Dorogovtsev [29], B. Prakasa Rao, H. Rubin [108], R.A. Kasonga [59], etc). The asymptotics ($T \to \infty$) for some of the estimators proposed in Chapter 7 were studied in the works of H. Dietz and Yu. Kutoyants [26], [27], E. Fournie and Yu. Kutoyants [38], P. Bertrand and Yu. Kutoyants [9], Yu. Kutoyants , F. Le Gland and R. Rakotozafy [85]. For Markov step and inhomogeneous Poisson processes these estimators were studied in R. Höpfner and Yu. Kutoyants [50] and Yu. Kutoyants and F. Liese [80].

References

1. Apoyan, G.T. *Parameter estimation of a nondifferentiable trend coefficient*, Uchen. Zap. Erevan Univ.1 (1986), 33-62 (in Russian).

2. Apoyan, G. and Kutoyants, Yu. *On the state estimation of a dynamical system perturbed by small noise*, Proc. 5-th Pannonian Symp. Math. Statist., Visegrád, 1985, 247-253.

3. Arato, M. *Linear Stochastic Systems with Constant Coefficients. A Statistical Approach*, Lecture Notes in Control and Inform. Sciences, **45**, Springer, New York, 1983.

4. Azencott, R. *Formule de Taylor stochastique et développement asymptotique d'intégrales de Feymann*, Séminaire de Prob. XVI, Lecture Notes in Math. **921**, Springer, Berlin, 1982, 237-285.

5. Banon, G. *Nonparametric identification for diffusion processes*, SIAM J. Control. Optim. **16**, 3 (1980), 380-395.

6. Basawa, I.V. and Scott, D.J. *Asymptotic Optimal Inference for Non-ergodic models*, Lecture Notes in Statist. **17**, Springer, Berlin, 1983.

7. Bauer, K. *On asymptotic properties of parameter estimates of trend coefficient of diffusion process*, Theory Probab. Appl., **25**, 2 (1980), 437-439.

8. Beran, R. *Efficient robust estimates in parametric models* Z. Wahrscheinlichkeitstheorie verw. Gebiete, **55** (1981), 91-108.

9. Bertrand, P. and Kutoyants, Yu.A. *A class of minimum distance estimators for partially observed linear stochastic system* (submitted for publication).

10. Billingsley, P. *Convergence of Probability Measures* , J.Wiley, New York, 1968.

11. Blagoveshenski, Yu.N. *Diffusion processes depending on small parameters*, Theory Probab. Appl. **7**, 2 (1962), 135-152.

12. Borkar, V., Bagchi, A. *Parameter estimation in continuous time stochastic processes*, Stochastics, **8** (1982), 193-212.

13. Burnashev, M.V. *On maximum likelihood estimate of signal in white Gaussian noise*, Probl. Inform. Transm. **11**, 4 (1975), 55-69 (in Russian).

14. Burnashev, M.V. *Asymptotic expansions for estimates of a signal parameter in Gaussian white noise*, Math. USSR Sbornik, **33** (1977), 150-184.

15. Burnashev, M.V. *On minimax detection of a nonexact known signal*, Theory Probab. Appl. **24**, 1 (1979), ,106-108 (in Russian).

16. Burnashev, M.V. *Study of second-order properties of parameter estimators of a signal in white gaussian noise*, Probl. Inform. Transm., **17**, 1 (1981), 17-28 (in Russian).

17. Campillo, F. *Filtrage et detection de ruptures de rrocessus partiellement observés*, Thèse, Université de Provence, Marseiile, 1984.

18. Campillo, F. and Le Gland, F. *MLE for partially observed diffusions: direct maximization vs. the EM algorithm*, Stochast. Process. Appl. **33**, 2 (1989), 245-274.

19. Campillo, F., Le Gland, F. and Kutoyants, Yu. *Asymptotics of the GLRT for the disorder problem in diffusion process*, INRIA, Rapports de Recherche N 1735, 1992.

20. Campillo, F., Le Gland, F. and Kutoyants, Yu. *Double disorder for Kalman filter with small noise*, (submitted for publication).

21. Chibisov, D.M. *An asymptotic expansion for a class of estimators including maximum likelihood estimators*, Theory Probab. Appl. **18** (1973), 295-303.

22. Clarke, B.R., Yeo, G.F. and Milne, R.K. *Local asymptotic theory for multiple solutions of likelihood equations, with application to a single ion channel model* Scand. J. Statist. **20**, 2 (1993), 133-146.

23. Dacunha-Castelle, D. and Duflo, M. *Probabilités et Statistiques 2. Problemes à temps mobile.* , Hassen, Paris, 1983.

24. Dacunha-Castelle, D. and Florens-Zmirou, D. *Estimation of of the coefficients of a diffusion from discrete observations*, Stochastics, **19** (1986), 263-284.

25. Dermoune, A. and Kutoyants, Yu. *Expansion of the distribution of maximum likelihood estimate for misspecified diffusion type observations* (submitted for publication).

26. Dietz, H.M. *Asymptotic properties of maximum likelihood estimators in diffusion type models.* Part 1 (General statements), Preprint 228,1989, Part 2 (Local asymptotics of exponential memory Ornstein-Uhlenbeck type model), Preprint 258,1990, Humboldt Univer., Berlin.

27. Dietz, H.M. and Kutoyants, Yu.A. *On minimum distance estimation by diffusion process with ergodic properties* (submitted for publication).

28. Dohnal, G. *On estimating the diffusion coefficient* , J. Appl. Probab. **24** (1987), 105-114.

29. Dorogovtsev, A.Ya. *Theory of Parameter Estimation of Random Processes*, Visha Shkola, Kiev, 1982.

30. Dodge, Y.(ed.) L_1 - *Statistical Analysis and Related Methods*, North-Holland, Amsterdam, 1992.

31. Dzaparidze K. and Valkeila E. *On large deviation probabilities for the maximum likelihood estimator*, Probab. Theory Math. Statist., vol. 1, Grigelionis B. et al (eds), VSP/Mosklas, 1990, 285-292.

32. Ebady, A.M. *On MDE in the Problems of Identification of Diffusion Type Processes*,Dissertation, Erevan State University, Erevan, 1991 (in Russian).

33. Ebady, A.M. and Kutoyants, Yu.A. *On consistency of a MDE for diffusion observations*, Uchen. Zap. Erevan. Univ. , 1 (1991), 17-21 (in Russian).

34. Essen, C.G. *Fourier analysis of distribution functions*, Acta Mathematica, **77** (1987), 1-125.

35. Farrel, R. *On the best obtainable asymptotic rates of convergence in estimation of a density function at a point*, Ann. Math. Statist., **43**,1 (1972), 170-180.

36. Feigin, P. *Some comments concerning a curios singularity*, J. Appl. Prob., **16** (1979), 440-444.

37. Florens-Zmirou, D. *Estimation de paramétres d'une diffusion á partir des temps de passages en zero*, C.R. Acad. Paris,**303**, 13 (1986), 655-658.

38. Fournie, E. and Kutoyants, Yu.A. *Estimateur de la distance minimale pour processus de diffusion ergodiques* , INRIA, Rapports de Recherche 1952, juillet, 1993, 1-33.

39. Freidlin, M.I., Wentsell, A.D. *Random Perturbations of Dynamical Systems*, Springer, New York, 1984.

40. Genis, I. *Estimation of multidimensional parameter of trend coefficient of diffusion type process*, Litovskii Math. Sbornik, **20**, 2 (1980), 3-13.

41. Genon-Catalot, V. and Laredo, C. *Limit theorems for the first hitting times processes of a diffusion and statistical applications*, Scand. J. Statist., **14** (1987), 143-160.

42. Genon-Catalot, V. and Laredo, C. *Non-parametric estimation for partially observed transient diffusion processes*, Stochastics **18** (1986), 169-196.

43. Genon-Catalot, V. and Jacod, J. *On the estimation of the coefficient for multidimensional diffusion process*, Ann. Inst. H.Poincaré, Probabilité et Statistique, 1 (1993), 119-151.

44. Greenwood, P.E. and Wefelmeyer, W. *Efficiency of estimators for partially specified filtered models*, Stochast. Proc. Appl., **36** (1990), 353-370.

45. Greenwood, P.E. and Wefelmeyer, W. *Maximum likelihood estimator and Kullback-Leibler information in misspecified Markov chain models*, University of Cologne , Preprints in Statistics 141, 1993.

46. Gikhman, I.I., Skorohod, A.V. *Stochastic Differential Equations*, Springer, Berlin, 1972.

47. Gikhman, I.I., Skorohod, A.V. *Introduction to Theory of Random Processes*, W.B. Saunders, Philadelphia, 1969.

48. Hajek, J. *Local asymptotic minimax and admissibility in estimation* Proc.6-th Berkeley Symp. on Math. Statist. Probab., 1972, 1, 175-194.

49. Khas'minskii R.Z. *Stability of Stochastic Differential Equations.* Sijthoff and Noordhoff, Amsterdam, 1980.

50. Höpfner, R. and Kutoyants, Yu. *On minimum distance estimation in recurrent Markov step process 1* (submitted for publication).

51. Ibragimov, I.A. and Khas'minskii R.Z. *An estimator of the parameter of a signal in Gaussian white noise*, Probl. Inform. Transm., 10 (1975), 31-46.

52. Ibragimov, I.A. and Khas'minskii, R.Z. *Parameter estimation for a discontinues signal in white Gaussian noise*, Probl.Inform.Transm., 11 (1975), 203-212.

53. Ibragimov, I.A. and Khas'minskii, R.Z. *Statistical Estimation: Asymptotic Theory* , Springer, Berlin, 1981.

54. Ibragimov, I.A. and Khas'minskii, R.Z. *Asymptotically normal families of distributions and efficient estimation*, Ann. Statist., **19**, 4 (1991), 1681-1724.

55. Ito, K. and McKean, H.P., Jr. *Diffusion Processes and their Sample Paths*, Springer, Berlin, 1965.

56. James, M. and Le Gland, F. *Consistent parameter estimation for partially observed diffusions with small noise*, INRIA, Rapport de Recherche 1223, 1990.

57. Jeganatan, P. *Some properties of risk functions in estimation when the limit of experiment is mixed normal*, Sankhya, ser.A, **45** (1983), 66-86.

58. Kallianpur, G. and Selukar, R.S. *Parameter estimation in linear filtering* , J. Multivariate. Anal., **39** (1991), 284-304.

59. Kasonga R.A. *Parameter estimation by deterministic approximation of a stochastic differential equation*, Commun. Statist. Stochastic Models, **6**,1 (1990), 59-67.

60. Konecny, F.*Maximum likelihood estimation of a drift parameter from a partially observed diffusion in a case of small measurement noise*, Statistics and Decisions, **8** (1990), 115-130.

61. Koshevnik, Yu.A. and Levit, B.Ya. *On a nonparametric analog of the information matrix*, Theory Probab. Appl., **21**, 4 (1976), 738-753.

62. Kutoyants, Yu.A. *Local Asymptotic Normality for Stochastic Processes and its Applications in Statistics*, Dissertation, Moscow Phys-Techn. Inst., Moscow, 1975.

63. Kutoyants, Yu.A. *Local asymptotic normality for processes of diffusion type*, Izv. Akad. Nauk Arm. SSR, Mathematika, **10** (1975), 103-112 (in Russian).

64. Kutoyants, Yu.A. *On a problem of testing hypotheses and asymptotic normality of stochastic integrals*, Theory Probab. Appl.,**20** (1975), 376-384.

65. Kutoyants, Yu.A. *On a property of parameter estimator of the trend coefficient of diffusion process* , Izv. Akad. Nauk Arm. SSR, Mathematika, **12**, 4 (1977), 245-251 (in Russian).

66. Kutoyants, Yu.A. *Estimation of parameter of a diffusion type process*, Theory Prob. Appl. **23** (1978), 641-649.

67. Kutoyants, Yu.A. *Parameter Estimation for Stochastic Processes*, Armenian Academy of Sciences, Erevan, 1980 (in Russian).

68. Kutoyants, Yu.A. *Expansion of a maximum likelihood estimate by diffusion powers*, Theory Probab. Appl., **29** (1984), 465-477.

69. Kutoyants, Yu.A. *Parameter Estimation for Stochastic Processes*, Heldermann, Berlin, 1984.

70. Kutoyants, Yu.A. *Parameter estimation for diffusion type processes of observations*, Math. Operationsforsch. u. Statist., ser. Statistics, **15**, 4 (1984), 541-551.

71. Kutoyants, Yu.A. *On nonparametric estimation of the trend coefficient of the diffusion process*, in Statistics and Control of Stochastic Processes (Steklov Seminar 1984), Optimization Software, New York, 1985, 230-250.

72. Kutoyants, Yu.A. *On efficient nonparametric estimation of trend coefficient*, Statistics and Control of Stochastic Processes (Steklov Seminar 1985), Optimization Software, New York, 1988, 196-207.

73. Kutoyants, Yu.A. *An example of estimating a parameter of a nondifferentiable drift*, Theory Probab. Appl.,**33**, 1 (1988), 175-179.

74. Kutoyants, Yu.A. *On identification problem of dynamical system with small noise*, Soviet J. Contemp. Math. Ana., **23**, 3 (1988), 79-95.

75. Kutoyants, Yu.A. and Pohlmann, H. *Parameter estimation for Kalman-Bucy filter with small noise*, Math. Operationsforsch. u. Statist., ser. Statist., **25** (1994).

76. Kutoyants, Yu.A. *Minimum distance parameter estimation for diffusion type observations*, C.R. Acad. Paris, 312,sér I (1991), 637-642.

77. Kutoyants, Yu.A. and Pilibossian, P. *On minimum uniform metric estimate of parameters of diffusion type processes*, Univ. Paris VI, Rapport Techhnique 142, 1991, (Stochastic Processes and Their Applications , 1994)

78. Kutoyants, Yu.A., Mourid, T. and Bosq, D. *Estimation paramétrique d'un processus de diffusion avec retards*, Ann. Inst. Henri Poincaré, 1992, 28, 1, 95-106.

79. Kutoyants, Yu.A. and Mourid, T. *Estimation par la distance minimale pour un processus de type diffusion avec retards*, Publ. Inst. Stat. Univ. Paris, XXXVIII, 2 (1994), 3-18.

80. Kutoyants, Yu.A., Liese, F. *On minimum distance estimation for spatial Poisson processes*, Ann. Academiae Scient. Fennicae, ser.A,I, **17** (1992), 65-71.

81. Kutoyants, Yu.A. and Pilibossian, P. *On minimum L_1-norm estimates of the parameter of Ornstein-Uhlenbeck process*, Statistics Probab. Lett., **20** (1994).

82. Kutoyants, Yu.A. *Minimum distance approach in parametric estimation for diffusion processes*, in Applications of Time Series Analysis in Astronomy and Meteorology, Padova, 1993, 37-46.

83. Kutoyants, Yu.A.*Nonconsistent estimation by diffusion type observations*. Statistics Probab. Lett., **20** (1994).

84. Kutoyants, Yu.A., Nercessian, A. and Pilibossian, P. *On limit distribution of the minimum sup-norm estimate of the parameter of the Ornstein-Uhlenbeck process*, Rev. Roumaine Math. Pures Appl., **39** (1994).

85. Kutoyants, Yu.A., Le Gland, F. and Rakotozafy R. *Identification d'un système non-lineaire partiellement observé non-linear par la méthode de la distance minimale*, INRIA, Rapports de Recherche 2125, 1993.

86. Kutoyants, Yu.A. *Parameter estimation for misspecified Poisson process*, Math. Meth. Statistics, 1994.

87. Kutoyants, Yu.A. and Vostrikova, L.Yu. *On non-consistency of estimatores*, (submitted for publication).

88. Lanska, V. *Minimum contrast estimation in diffusion processes*, J. Appl. Probab.,**16** (1979), 65-75.

89. Laredo, C. *A sufficient condition for asymptotic sufficiency of incomplete observations of a diffusion process*, Ann. Statist., **18**, 3 (1990), 1158-1171.

90. Le Breton, A. *Sur l'Estimation de Paramétres dans les Modeles Differentiels Stochastiques Multidimensionels*, Thèse, Grenoble, 1976.

91. Le Cam, L. *Locally asymptotically normal families of distributions*, Univ. Calif. Publ. Statist. 1960, 3, 27-98.

92. Le Cam, L. *Limits of experiments*, Proc. 6-th Berkeley Symp. Math. Statist., 1972, 1, 245-261.

93. Le Cam, L. *Asymptotic Methods in Statistical Decision Theory*, Springer, New York, 1986.

94. Le Cam, L. and Yang, G.L. *Asymptotics in Statistics* Springer, New York, 1990.

95. Levit, B. *On optimality of some statistical estimates,* Proc. Prague Symp. Asymptotic. Statistics, v. II, 215-238, 1973.

96. Lin'kov, Yu.N. *On parameters estimates of the processes of diffusion type*, Theory of Stochastic Processes, Kiev, 1981, 9, 71-82(in Russian)

97. Liptser, R.S and Shiryayev, A.N. *Statistics of Random Processes*, I, II, Springer, New York, 1977, 1978.

98. Liptser, R.S and Shiryayev, A.N. *Theory of Martingales*, Kluwer, Dordrecht, 1989.

99. McKeague, I. *Estimation for diffusion processes under misspecified models*, J. Appl. Probab., **21** (1984), 511-520.

100. McKean, H.P. *Stochastic Integrals*, Academic Press, New York, 1969.

101. Millar, P.W. *The minimax principle in asymptotic statistical theory*, Lecture Notes in Math. **976**, 1983, 76-265.

102. Millar, P.W. *A general approach to the optimality of minimum distance estimators*, Trans. Amer. Math. Soc., **286** (1984), 377-418.

103. Nguyen, H.T. and Pham, D.T. *Identification of nonstationary diffusion model by the method of sieves*, SIAM J. Control Optim., **20**, 5 (1982), 603-611.

104. Novikov, A.A. *On stopping times for a Wiener process*, Theory Probab. Appl., **16** (1971), 449-456.

105. Parr, C. *Minimum distance estimation : bibliography*, Commun. Statist.-theor. Meth., A10(12) (1981), 1205-1224.

106. Pfanzagl, J. *Asymptotic expansions in parametric statistical theory*, in Developments in Statistics (P.R. Krishnaiah, ed), v. 3, Academic Press, New York, 1980, 1-97.

107. Pollard, D. *The minimum distance method of testing* , Metrika, **27** (1980), 43-70.

108. Prakasa Rao, B.L.S. and Rubin, H. *Asymptotic theory of estimation in nonlinear stochastic differential equations* Sankhya, **43** , Ser. A, Pt. 2 (1981), 170-189.

109. Redner, R. *Note on the consistency of the maximum likelihood estimate for nonidentifiable distributions*, Ann. Statistics, **9**, 1 (1981), 225-228.

110. Terent'yev, A.S. *Probability distribution of a time location of an absolute maximum at the output of a synchronized filter*, Radioengineering and Electronics, 1968, 13, 4, 652- 657 (in Russian)..

111. Tsitovich, I.I. *On estimating the drift of a diffusion process from observations in a domain*, Theory Probab. Appl., **22** 1977, 851-858.

112. Strasser, H. *Mathematical Theory of Statistics*, De Gruyter, Amsterdam, 1985.

113. Veres, S. *Parameter dependent asymptotic normality for vector elementary Gaussian processes*, Proc.5-th Pannonian Symp. Math. Statist., Visegrád, 1985, 371-380.

114. Vostrikova, L.Yu. *Detection of a "disorder" in a Wiener process*, Theory Probab. Appl. **26**, 2 (1981), 356-362.

115. Vostrikova, L.Yu. *Functional limit theorems for the disorder problem*, Stochastics, **9** (1983), 103-124.

116. Vostrikova, L.Yu. *On weak convergence of likelihood ratio process of general statistical parametric models*, Stochastics, **23** (1988), 277-298.

117. White, H. *Maximum likelihood estimation of misspecified models*, Econometrica, **50** (1982), 1-25.

118. Yoshida, N. *Asymptotic expansions of maximum likelihood estimators for small diffusions via the theory of Malliavin-Watanabe*, Prob. Theory Related Fields, 1992, 92, 275-311.

119. Yoshida, N. *Asymptotic expansion of Bayes estimators for small diffusion*, Probab. Theory Related Fields, **95** (1993), 429-450.

120. Yoshida, N. *Asymptotic expansion for statistics related to small diffusion*, Research Memorandum, v. 406, 1991. (to appear in J. Japan Statistical Society).

Index

Other *Mathematics and Its Applications* titles of interest:

P.M. Alberti and A. Uhlmann: *Stochasticity and Partial Order. Doubly Stochastic Maps and Unitary Mixing.* 1982, 128 pp. ISBN 90-277-1350-2

A.V. Skorohod: *Random Linear Operators.* 1983, 216 pp. ISBN 90-277-1669-2

I.M. Stancu-Minasian: *Stochastic Programming with Multiple Objective Functions.* 1985, 352 pp. ISBN 90-277-1714-1

L. Arnold and P. Kotelenez (eds.): *Stochastic Space-Time Models and Limit Theorems.* 1985, 280 pp. ISBN 90-277-2038-X

Y. Ben-Haim: *The Assay of Spatially Random Material.* 1985, 336 pp. ISBN 90-277-2066-5

A. Pazman: *Foundations of Optimum Experimental Design.* 1986, 248 pp. ISBN 90-277-1865-2

P. Kree and C. Soize: *Mathematics of Random Phenomena. Random Vibrations of Mechanical Structures.* 1986, 456 pp. ISBN 90-277-2355-9

Y. Sakamoto, M. Ishiguro and G. Kitagawa: *Akaike Information Criterion Statistics.* 1986, 312 pp. ISBN 90-277-2253-6

G.J. Szekely: *Paradoxes in Probability Theory and Mathematical Statistics.* 1987, 264 pp. ISBN 90-277-1899-7

O.I. Aven, E.G. Coffman (Jr.) and Y.A. Kogan: *Stochastic Analysis of Computer Storage.* 1987, 264 pp. ISBN 90-277-2515-2

N.N. Vakhania, V.I. Tarieladze and S.A. Chobanyan: *Probability Distributions on Banach Spaces.* 1987, 512 pp. ISBN 90-277-2496-2

A.V. Skorohod: *Stochastic Equations for Complex Systems.* 1987, 196 pp. ISBN 90-277-2408-3

S. Albeverio, Ph. Blanchard, M. Hazewinkel and L. Streit (eds.): *Stochastic Processes in Physics and Engineering.* 1988, 430 pp. ISBN 90-277-2659-0

A. Liemant, K. Matthes and A. Wakolbinger: *Equilibrium Distributions of Branching Processes.* 1988, 240 pp. ISBN 90-277-2774-0

G. Adomian: *Nonlinear Stochastic Systems Theory and Applications to Physics.* 1988, 244 pp. ISBN 90-277-2525-X

J. Stoyanov, O. Mirazchiiski, Z. Ignatov and M. Tanushev: *Exercise Manual in Probability Theory.* 1988, 368 pp. ISBN 90-277-2687-6

E.A. Nadaraya: *Nonparametric Estimation of Probability Densities and Regression Curves.* 1988, 224 pp. ISBN 90-277-2757-0

H. Akaike and T. Nakagawa: *Statistical Analysis and Control of Dynamic Systems.* 1998, 224 pp. ISBN 90-277-2786-4

Other *Mathematics and Its Applications* titles of interest:

A.V. Ivanov and N.N. Leonenko: *Statistical Analysis of Random Fields.* 1989, 256 pp. ISBN 90-277-2800-3

V. Paulauskas and A. Rackauskas: *Approximation Theory in the Central Limit Theorem. Exact Results in Banach Spaces.* 1989, 176 pp. ISBN 90-277-2825-9

R.Sh. Liptser and A.N. Shiryayev: *Theory of Martingales.* 1989, 808 pp. ISBN 0-7923-0395-4

S.M. Ermakov, V.V. Nekrutkin and A.S. Sipin: *Random Processes for Classical Equations of Mathematical Physics.* 1989, 304 pp. ISBN 0-7923-0036-X

G. Constantin and I. Istratescu: *Elements of Probabilistic Analysis and Applications.* 1989, 488 pp. ISBN 90-277-2838-0

S. Albeverio, Ph. Blanchard and D. Testard (eds.): *Stochastics, Algebra and Analysis in Classical and Quantum Dynamics.* 1990, 264 pp. ISBN 0-7923-0637-6

Ya.I. Belopolskaya and Yu.L. Dalecky: *Stochastic Equations and Differential Geometry.* 1990, 288 pp. ISBN 90-277-2807-0

A.V. Gheorghe: *Decision Processes in Dynamic Probabilistic Systems.* 1990, 372 pp. ISBN 0-7923-0544-2

V.L. Girko: *Theory of Random Determinants.* 1990, 702 pp. ISBN 0-7923-0233-8

S. Albeverio, PH. Blanchard and L. Streit: *Stochastic Processes and their Applications in Mathematics and Physics.* 1990, 416 pp. ISBN 0-9023-0894-8

B.L. Rozovskii: *Stochastic Evolution Systems. Linear Theory and Applications to Non-linear Filtering.* 1990, 330 pp. ISBN 0-7923-0037-8

A.D. Wentzell: *Limit Theorems on Large Deviations for Markov Stochastic Process.* 1990, 192 pp. ISBN 0-7923-0143-9

K. Sobczyk: *Stochastic Differential Equations. Applications in Physics, Engineering and Mechanics.* 1991, 410 pp. ISBN 0-7923-0339-3

G. Dallaglio, S. Kotz and G. Salinetti: *Distributions with Given Marginals.* 1991, 300 pp. ISBN 0-7923-1156-6

A.V. Skorohod: *Random Processes with Independent Increments.* 1991, 280 pp. ISBN 0-7923-0340-7

L. Saulis and V.A. Statulevicius: *Limit Theorems for Large Deviations.* 1991, 232 pp. ISBN 0-7923-1475-1

A.N. Shiryaev (ed.): *Selected Works of A.N. Kolmogorov, Vol. 2: Probability Theory and Mathematical Statistics.* 1992, 598 pp. ISBN 90-277-2795-X

Yu.I. Neimark and P.S. Landa: *Stochastic and Chaotic Oscillations.* 1992, 502 pp. ISBN 0-7923-1530-8

Other *Mathematics and Its Applications* titles of interest:

Y. Sakamoto: *Categorical Data Analysis by AIC.* 1992, 260 pp.
ISBN 0-7923-1429-8

Lin Zhengyan and Lu Zhuarong: *Strong Limit Theorems.* 1992, 200 pp.
ISBN 0-7923-1798-0

J. Galambos and I. Katai (eds.): *Probability Theory and Applications.* 1992, 350 pp. ISBN 0-7923-1922-2

N. Bellomo, Z. Brzezniak and L.M. de Socio: *Nonlinear Stochastic Evolution Problems in Applied Sciences.* 1992, 220 pp. ISBN 0-7923-2042-5

A.K. Gupta and T. Varga: *Elliptically Contoured Models in Statistics.* 1993, 328 pp. ISBN 0-7923-2115-4

B.E. Brodsky and B.S. Darkhovsky: *Nonparametric Methods in Change-Point Problems.* 1993, 210 pp. ISBN 0-7923-2122-7

V.G. Voinov and M.S. Nikulin: *Unbiased Estimators and Their Applications. Volume 1: Univariate Case.* 1993, 522 pp. ISBN 0-7923-2382-3

V.S. Koroljuk and Yu.V. Borovskich: *Theory of* U-*Statistics.* 1993, 552 pp.
ISBN 0-7923-2608-3

A.P. Godbole and S.G. Papastavridis (eds.): *Runs and Patterns in Probability: Selected Papers.* 1994, 358 pp. ISBN 0-7923-2834-5

Yu. Kutoyants: *Identification of Dynamical Systems with Small Noise.* 1994, 298 pp. ISBN 0-7923-3053-6

MIX
Papier aus verantwortungsvollen Quellen
Paper from responsible sources
FSC® C105338

If you have any concerns about our products,
you can contact us on
ProductSafety@springernature.com

In case Publisher is established outside the EU,
the EU authorized representative is:
Springer Nature Customer Service Center GmbH
Europaplatz 3, 69115 Heidelberg, Germany

Printed by Libri Plureos GmbH
in Hamburg, Germany